Silicon

Silicon

Electrochemistry, Production, Purification and Applications

Eimutis Juzeliūnas

WILEY-VCH

Author

Prof. Eimutis Juzeliūnas
Center for Physical Sciences and Technology
Department of Electrochemical Materials
Sauletekio Str 3
10257 Vilnius
Lithuania

Cover Image: © Georgy Shafeev/Shutterstock

Library of Congress Card No.: applied for

British Library Cataloguing-in-Publication Data
A catalogue record for this book is available from the British Library.

Bibliographic information published by the Deutsche Nationalbibliothek
The Deutsche Nationalbibliothek lists this publication in the Deutsche Nationalbibliografie; detailed bibliographic data are available on the Internet at <http://dnb.d-nb.de>.

Print ISBN: 978-3-527-34897-8
ePDF ISBN: 978-3-527-83190-6
ePub ISBN: 978-3-527-83192-0
oBook ISBN: 978-3-527-83191-3

Typesetting Straive, Chennai, India
Printing and Binding CPI Group (UK) Ltd, Croydon, CR0 4YY

C9783527348978_171023

To the memory of my parents – mother Zofija and father Enrikas
and
To my beloved family – wife Katerina and sons Laimis and Povilas,
who inspired and supported.

Contents

Theodor von Grotthuss medal. Author: Petras Repšys. Produced by Lietuvos monetų kalykla (Lithuanian Mint)

"Though light illuminates darkness, nothing is darker than the light."

Theodor von Grotthuss (1785–1822)

Preface

Silicon lies at the heart of modern technology. Silicon can be used in various fields, such as optoelectronics, sensors, batteries, optical fibers, photoelectrochemical water splitting, terahertz emitters, and numerous other applications. As an abundant, non-toxic, efficient, and robust material, silicon will dominate the solar energy market at least for the next few decades.

Electrochemistry deals with the chemical transformations, which are induced by an electric current, or *vice versa* – with the transformations, which generate an electric current. These processes provide an opportunity to store or produce electricity with a minimum carbon footprint. Electrochemistry can, therefore, significantly contribute to low-carbon economy; it offers an advancement in sustainable energy solutions and environment-friendly technologies.

In the early 2000s, V. Lehman (2002) and X. G. Zhang (2004) published several books on silicon electrochemistry. Since then, various breakthrough directions in silicon electrochemistry have emerged. For instance, luminescent porous silicon nanoparticles were electrochemically produced and applied as the carriers of the drug payload *in vivo*. Electrochemical silicon surface modifications increased the efficiency of photovoltaic devices used for solar energy harvest or for the production of solar fuel. Silicon photoelectrodes have been successfully developed for hydrogen and oxygen production by water splitting as well as CO_2 reduction. Electrochemically produced Si surface nano-architectures showed an intrinsic quantum confinement effect. Environment-friendly and secure solutions offered silicon electrochemistry in high-temperature molten salts. Electrochemical silicon reduction from silica in high-temperature molten salts has been discovered. Electrochemical deposition of doped silicon as well as formation of p-–n junction have also been demonstrated.

This book aims to summarize the experimental and technological work done in recent decades on silicon electrochemistry, production, and purification, highlighting subjects of technological significance and future perspectives. The book aims to be highly beneficial to the communities of chemists and material scientists working in academia and industrial sectors, especially in the field of sustainable energy development: photovoltaics, light harvesting efficiency, solar-to-chemical conversion, production of solar-grade silicon as well as production of batteries, photoelectrodes, or silicon-based semiconductors. The secondary market of this book includes the education and socio-economic sectors with focal points on such topicalities as the reduction in global climate change, replacement of fossil fuels by renewable energy, and strategies of low-carbon economy.

Vilnius, Lithuania *Eimutis Juzeliūnas*
June, 2022

References

Lehman, V. (2002). *Electrochemistry of Silicon. Instrumentation, Science, Materials and Applications*. Wiley-VCH.

Zhang, X.G. (2004). *Electrochemistry of Silicon and Its Oxide*. Kluwer Academic Publishers.

List of Abbreviations

3PI	three-phase interface (interlines)
AFM	atomic force microscopy
AL	acetonitrile
Al-BSF	aluminum backside field technology
ALD	atomic layer deposition
BCE	before the common (or current) era
BMIm	1-butyl-3-methylimidazolium
BMPy	1-butyl-3-methylpyridinium
BMPyrr	*N*-butyl-*N*-methylpyrrolidinium
b-Si	black silicon
CE	contacting electrode
CNT	carbon nanotubes
COP 21	Paris Climate Conference
CV	cyclic voltammetry
CVD	chemical vapor deposition
CZ	Czochralski process
DMAE	2-dimethylaminoethanethiol
DMS	dimethyl sulfide
DRC	Democratic Republic of the Congo
DRE	damage removal etching
EC	The European Commission
EDX (EDS)	energy-dispersive X-ray spectroscopy
EIS	electrochemical impedance spectroscopy
EMIm	1-ethyl-3-methylimidazolium
EMPyrr	N-ethyl-*N*-methylpyrrolidinium
EQCM	electrochemical quartz crystal microbalance
EU	the European Union
FAP	tris(pentafluoroethyl)-trifluorophosphate
FBR	fluidized bed reactor
fs	femtosecond
FTIR	Fourier transform infrared spectroscopy
FTO	fluorine-doped tin oxide
GDMS	glow discharge mass spectrometry

GDP	Gross domestic product
GI-XRD	grazing incidence X-ray diffractometry
ICE	initial Coulombic efficiency
ICP	inductive coupled plasma
IL	ionic liquid
IPA	isopropyl alcohol
IRENA	International Renewable Energy Agency
ISFET	ion-sensitive field effect transistor
ITO	indium tin oxide
LCD	liquid-crystal display
LIB	lithium-ion battery
M, Me	metal, metallic
MACE	metal-assisted chemical etching
MG-Si	metallurgical-grade silicon
MOE	molten oxide electrolysis
MS	magnetron sputtering
MT	metric ton
MWT	metal wrap through
NHE	normal hydrogen electrode
NMR	nuclear magnetic resonance
NPs	nanoparticles
ns	nanosecond
NTD	neutron transmutation doping
NWs	nanowires
P-Si	porous silicon
PEC	photoelectrochemical cells, photoelectrochemistry
PEDOT:PSS	poly(3,4-ethylenedioxythiophene):poly(styrenesulfonate)
PERC	passivated emitter and rear contact
PIII	plasma immersion and ion implantation
PV	photovoltaic(s)
PVD	physical vapor deposition
QCN	quartz crystal nanobalance
R&D	research and development
RE	reference electrode
RFMS	radio frequency magnetron sputtering
RIE	reactive ion etching
Si-H	hydrogen-terminated silicon surface
Si-F	fluoride-terminated silicon surface
Si-OH	hydroxide terminated silicon surface
SCR	space-charge region
SEM	scanning electron microscopy
SERS	surface-enhanced Raman spectroscopy
SoG-Si	solar-grade silicon
SP	solubility product
STC	silicon tetrachloride

SWIR	short-wavelength infrared
TBAB	tetrabutylammonium bromide
TBAC	tetrabutylammonium chloride
TBMA	tributyl(methyl)ammonium
TCS	trichlorsilane
TEAC	tetraethylammonium chloride
TEOS	tetraethylorthosilicate
TFO	trifluoromethylsulfonate
TFSA	bis(trifluoromethylsulfonyl)amide
TFSI	bis(trifluoromethylsulfonyl)imide
TG-DTA	thermogravimetry and differential thermal analysis
THF	tetrahydrofuran
TMAH	tetramethylammonium hydroxide
TMHA	*N*-trimethyl-*N*-hexylammonium
TPAC	tetrapropylammonium chloride
TRL	technology readiness level
UPD	underpotential deposition
UTE	ultra-high temperature electrochemistry
VLS	vapor–liquid–solid deposition
XRD	X-ray diffraction
XPS	X-ray photoelectron spectroscopy

About the Author

Eimutis Juzeliūnas, photo by J. Stacevičius/ Lithuanian National Radio and Television, LRT.

Professor Dr. Eimutis Juzeliūnas is a principal research associate and a head of the Department of Electrochemical Materials Science at the Centre for Physical Sciences and Technology in Vilnius, Lithuania. He worked previously as a rector of Klaipėda university (2014–2018) and a director of the Institute of Chemistry in Vilnius (2001–2009). He was a Marie Curie International Fellow of the European Commission at the University of Cambridge, UK (2009–2011, 2013–2014) where he carried out research on silicon electrochemistry in high-temperature molten salts. He also was a Fulbright fellow at the Vanderbilt University (USA), a fellow of the American Chemical Society at the Pennsylvania State University (USA), and an Alexander von Humboldt fellow at the company DECHEMA e.V. (Germany). His main research area is electrochemical materials science; current research interests are silicon electrochemistry for energy applications, environmental and microbiological degradation of metals (corrosion), physical vapor deposition of resistant alloys, and nanogravimetry of electrochemical processes.

1 Introduction

World production of silicon (Si) reached (2010–2020) about eight millions of metric tonnes in the last decade. This quantity was produced mainly by the carbothermic silica (SiO_2) reduction. The process requires a large supply of energy and emits carbon oxides (CO_x). A fundamental challenge is the electrochemical silicon extraction from silica or other solids using electricity instead of harmful chemistries. Zero carbon footprint could be attained when using electrons as absolutely clean reduction agents generated by renewable sources. Electrochemical methods can be used on a wide scale of applications: extraction, purification, surface engineering, or thin-film technologies. Thus, silicon electrochemistry has the potential to significantly contribute to low-carbon economy; this field offers an advancement in environmentally friendly and secure technologies of energy generation and storage.

Breakthrough research topics have emerged in silicon electrochemistry in recent decades. The electrochemical formation of porous silicon (P-Si) was discovered in 1956 (Uhlir 1956). Canham reported in 1990 that a visible room-temperature photoluminescence from P-Si layer formed electrochemically on Si wafer (Canham 1990). The discovery inspired wide studies of P-Si for applications in optoelectronics, lasers, and sensors. Luminescent porous silicon nanoparticles were applied as the carriers of the drug payload, whose infrared luminescence enabled monitoring of the particles *in vivo* (Park et al. 2009). Electrochemical nano-micro-structuring of silicon has been widely investigated. The surface modifications increased the efficiency of photovoltaic (PV) devices used for solar energy harvest or for production of solar fuel. Silicon photoelectrodes have been successfully used for hydrogen and oxygen production by water splitting as well as CO_2 reduction (Sun et al. 2014). Electrochemically produced Si surface nano-architectures showed intrinsic quantum confinement effect. Electrochemical reduction of silicon dioxide to silicon in a molten salt electrolyte has been reported, which formed the basis for new processes in silicon semiconductor technology and high-purity silicon production (Nohira et al. 2003). Environmentally friendly and secure solutions offered silicon electrochemistry in high temperature molten salts (Juzeliūnas and Fray 2020). Electrochemical deposition of doped silicon as well as formation of p–n junction has been demonstrated (Zou et al. 2017, 2019; Peng et al. 2018). The approach has

Silicon: Electrochemistry, Production, Purification and Applications, First Edition. Eimutis Juzeliūnas.

the potential of reducing capital cost and energy consumption for fabrication of solar cells when compared with the conventional manufacturing process.

This book features recent achievements in silicon electrochemistry, particularly, in electrochemical silicon extraction, purification, and processing in high-temperature molten salts. The introductory part of the book (Chapters 2–4) is devoted to general aspects of silicon application. A historical overview of silicon production is provided, and its importance in a low-carbon economy is considered. Chapter 4 addresses the physical and chemical properties of silicon, which are most relevant for electrochemical materials science. The subsequent material is more specific. Chapter 5 describes the major technologies used for silicon purification such as Siemens, Union Carbide, or Ethyl Corporation processes. This chapter also provides the principles of electrorefining in high-temperature molten salts, highlighting the advantages and disadvantages when compared with conventional industrial processes.

Chapter 6 addresses electrodeposition of thin layers and discusses the possibility of replacing multiple processes of Si wafer fabrication with one-step electrochemical deposition. Traditional manufacturing entails an energy-intensive and environmentally unfriendly production of metallurgical grade silicon (MG-Si), as well as its upgrade to solar grade silicon (SoG-Si), ingot casting, and slicing. Electrodeposition from molten fluoride, chloride, and oxide electrolytes on various substrates is discussed. A recently proposed strategy for electrodeposition of photoactive silicon and p–n junction is highlighted in detail. Silicon deposition from ionic liquids – the room-temperature molten salts – is also discussed in this chapter. Significant attention is given to the purity level of silicon electrodeposits, which are essential for photo-electrochemical applications.

Chapter 7 discloses photoelectrochemical (PEC) properties of silicon-oxide electrodes coated with ultrathin films of silica (SiO_2), hafnia (HfO_2), and alumina (Al_2O_3). The pivotal concept of PEC methodology is to obtain information, which correlates with that of the solid-state cells so that there is no prior need to design a solar cell that characterizes Si surface photo-responsiveness. Significant attention is given to studies of Si-oxide interfacial stability by the quartz crystal nanobalance (QCN) – a sensitive mass detector, which provides information about the electrode mass change with nanogram resolution *in situ* and in real time.

Deoxidation of metal oxides in a molten salt electrolyte was discovered in the year 2000 (Chen et al. 2000). The process was named the FFC Cambridge process. Simplicity and rapidity of the process have attracted global interest. Over 30 metals or semimetals were extracted from solid compounds by this energy-efficient and environment-friendly route. Chapter 8 addresses the FFC principle and its application in silicon reduction from silica. The electrochemical extraction provides a green alternative to conventional carbo-thermic silicon production. Chapters 9–12 provide further details on Si–SiO_2 conversion in molten salts. Voltammetry, basic reactions, and *in situ* studies by synchrotron X-ray diffraction are discussed, and experimental conditions used by many authors are summarized.

Technological opportunities carry out the operation at ultra-high temperatures and at liquid state of silica feedstock. Such processes are referred to as molten oxide electrolysis (MOE). Chapter 13 discusses the MOE principles of silicon extraction in a liquid state.

This study focused majorly on electrochemical surface engineering. Chapter 14 discusses the chemical–physical methods of silicon surface structuring, such as laser engineering and various etchings: chemical, photoelectrochemical, reactive ion, plasma immersion ion implantation, and metal-assisted chemical. The vapor–liquid–solid method is also discussed.

Chapter 15 features a comprehensive material obtained on electrochemical Si structuring at high-temperature molten salts including formation of black silicon (B-Si). B-Si is a nano-micro-porous material, which effectively absorbs the light on a wide range of wavelengths. Electrochemical Si structuring in molten salts is attractive due to its environmental friendliness, technical simplicity, and cost-effectiveness.

The book also outlines the perspectives of electrochemical synthesis of semiconductors (Chapter 16), the basic principles and materials for photo-electrodes, and the preservation of solar-fuel generators (Chapter 17).

In conclusion, while silicon electrochemistry offers a range of technological opportunities, most of the developments are still on the conceptual or bench-scale level. As a result, viable technological developments are still pending.

References

Canham, L.T. (1990). Quantum wire array fabrication by electrochemical and chemical dissolution. *Appl. Phys. Lett.* 57: 1046–1048. https://doi.org/10.1063/1.103561.

Chen, G.Z., Fray, D.J., and Farthing, T.W. (2000). Direct electrochemical reduction of titanium dioxide to titanium in molten calcium chloride. *Nature* 407: 361–364. https://doi.org/10.1038/35030069.

Juzeliūnas, E. and Fray, D. (2020). Silicon electrochemistry in molten salts. *Chem. Rev.* 120: 1690–1709. https://doi.org/10.1021/acs.chemrev.9b00428.

Nohira, T., Yasuda, K., and Ito, Y. (2003). Pinpoint and bulk electrochemical reduction of insulating silicon dioxide to silicon. *Nat. Mater.* 2: 397–401. https://doi.org/10.1038/nmat900.

Park, J.-H., Gu, L., Maltzahn, G. et al. (2009). Biodegradable luminescent porous silicon nanoparticles for *in vivo* applications. *Nat. Mater.* 8: 331–336. https://doi.org/10.1038/NMAT2398.

Peng, J.J., Yin, H.Y., Zhao, J. et al. (2018). Liquid-tin-assisted molten salt electrodeposition of photoresponsive n-type silicon films. *Adv. Funct. Mater.* 28: 1703551. https://doi.org/10.1002/adfm.201870194.

Sun, K., Shen, S., Liang, Y. et al. (2014). Enabling silicon for solar-fuel production. *Chem. Rev.* 114: 8662–8719. https://doi.org/10.1021/cr300459g.

Uhlir, A. (1956). Electrolytic shaping of germanium and silicon. *Bell System Tech. J.* 35: 333. https://doi.org/10.1002/j.1538-7305.1956.tb02385.x.

Zou, X., Ji, L., Yang, X. et al. (2017). Electrochemical formation of a p–n junction of thin film silicon deposited in molten salt. *J. Am. Chem. Soc.* 139: 16060–16063. https://doi.org/10.1021/jacs.7b09090.

Zou, X., Ji, L., Ge, J. et al. (2019). Electrodeposition of crystalline silicon films from silicon dioxide for low-cost photovoltaic applications. *Nat. Commun.* 10: 5772. https://doi.org/10.1038/s41467-019-13065-w.

2

Silicon Electrochemistry – Toward Low-Carbon Economy

Climate change is one the greatest challenges the world faces today. The renewable energy (solar, wind, water, biomass, geothermal) is considered as a climate change imperative today. Excessive exploitation of fossil energy sources had a negative impact on climate, economy, and everyday lives. The growing threat from climate change is apparent, such as natural disasters, halved mass of inland glaciers, rise of sea level, and extinction of numerous terrestrial species. Volatile oil, gas, and coal prices in recent decades and concerns about their supply from politically unstable countries are amongst the drivers of the need for alternatives, such as renewables.

The European Green Deal plan was unveiled in 2019 – a three-decade effort to make the climate neutral by 2050 (https://ec.europa.eu/info/strategy/priorities-2019-2024/european-green-deal; Tamma et al. 2019; Simon 2019). The Commission's President Ursula von der Leyen characterized the initiative as "Europe's man on the moon moment" adding that "the growth model based on fossil fuels and pollution is out of date and out of touch with our planet" (Simon 2019). The overarching objective in the plan is "Climate neutral Europe," which aims to reach net-zero greenhouse gas emissions by 2050. The ambition for 2030 is cutting-off the emission by 50–55%. The plan addresses major actions such as circular economy, building renovation, pollution-free environment, strategy of ecosystems and biodiversity, green and healthier agriculture, electric vehicles, and sustainable fuels, such as biofuels and hydrogen.

Another global initiative is the Paris Agreement – an international treaty on mitigation of climate change (https://unfccc.int/process-and-meetings/the-paris-agreement/the-paris-agreement). In 2015, 196 parties in the Paris climate conference (COP 21) adopted an agreement to limit global warming below 2 °C compared to pre-industrial levels.

Such highly ambitious plans led to fundamental changes in the way we generate and use the energy. Electrochemical technologies have shown great potential in advancing the economy's transition toward climate neutrality. Electrochemistry provides sustainable solutions in such fields as green-energy storage and solar-to-chemical or solar-to-electricity conversion.

Materials used in solar devices play a major role in the cost breakdown of the overall utilization process of solar energy. Creation and synthesis of new effective materials for solar-energy applications, therefore, is very high on the agenda of materials

Silicon: Electrochemistry, Production, Purification and Applications, First Edition. Eimutis Juzeliūnas.

scientists and engineers. Vast majority of the solar cells are produced from silicon where the total wafer cost dominates in the overall cell cost balance. It is assumed that silicon, being a nontoxic, efficient, and robust material, will play a key role in the solar energy market for the next few decades (Schmalensee et al. 2015; Green 2016; Polman et al. 2016). Mitigation of climate change, as a global task, could be achieved by using technologies based on the Earth-abundant materials. The abundance of silicon in the Earth's crust (27%) makes it possible to expand the application of this material to the terawatt scale by 2050 (Schmalensee et al. 2015).

2.1 Silicon for Energy Storage – Electrochemical Batteries

Electrochemistry deals with relationship between electricity and chemical change. Typical example of an electrochemical device used in everyday life is a battery – the device, which generates electricity by chemical reactions. Batteries are widely used in portable electronics, medical devices, or e-mobility including electric cars. A vice versa process is when electricity generates chemical reactions, for instance, electroplating of metals. The process is widely applied for the production of coatings in order to protect metals and alloys from corrosion or to improve their aesthetic appearance, as well as to decorate them.

Batteries represent a key technology for low-carbon economy to reduce CO_2 emissions from transport, power, and industrial sectors. Batteries are essential devices used to store stationary energy from sustainable sources such as solar or wind. To reach the sustainability targets, batteries must exhibit ultra-high energy and power performance close to theoretical limits. Other requirements include outstanding lifetime, reliability, safety, and recyclability. Important requirement is also scalability to electricity grid level, as well as cost-effectiveness and sustainable battery production. The cost target set by the European Commission (EC) for the next-generation batteries of stationary energy storage is below 0.05 €/kWh/cycle by 2030. The growth in global battery demand is anticipated to multiplied by a factor 14× from 2018 to 2030. The greatest part of this demand goes to the electric mobility sector (Edström et al. 2020).

Lithium-ion (Li-ion) based technology dominates the current battery market. Lithium-ion batteries (LIBs) are state-of-the-art technology for portable electronics and electric vehicles. The Nobel Prize in chemistry has been awarded to J. B. Goodenough, M. S. Whittingham, and A. Yoshito in 2019 for the development of LIBs. These batteries, however, have several shortcomings, especially for stationary energy storage applications.

Apart from Li, cobalt (Co) is a key electrode material in Li-ion batteries. For cathode production, various co-containing materials have recently been investigated: $LiCoO_2$, $LiNi_xCo_yAl_{1-x-y}O_2$, $LiNi_xMn_yCo_{1-x-y}O_2$, etc. (Huang et al. 2021; Chu et al. 2020; Liu et al. 2019a,b). Presently, about 60% of mined cobalt is used to produce the LIBs electrodes. Congo (The Democratic Republic of the Congo, the DRC) along with Zambia, Madagascar, Zimbabwe and the Republic of South Africa are mining about 70% of world cobalt. The DRC is a main global producer, which mines about 60% of the world's cobalt feedstock. The Amnesty International organization reported in 2016 on the violation of human rights in this country – child labor in health-endangering mines (https://www.amnesty.org/en/documents/afr62/3183/2016/en).

The EC issues periodically the communications with a list of critical raw materials, which specifies the materials that are most important for the EC economically, alongside having high supply risk. As of 2020, the list also includes the materials, from which key components of LIBs are produced: lithium, cobalt, and graphite (Communication 2020). The list indicates 100% reliance of the European Union (EU) import on lithium, 86% on cobalt, and 98% on natural graphite. The communication states that the EU demand for lithium for electric vehicle batteries and energy storage will increase by 18 times in 2030 and almost up to 60 times in 2050. The corresponding figures for cobalt are 5 and 15 times, respectively. The limited availability of lithium makes it doubtful whether LIBs manufacturing can scale up to significantly larger production volumes.

The World Bank projected that the scenario of global warming below 2° (COP 21) will increase the demand in metals for battery applications at the level of 1000% by 2050 (World Bank 2017). The list of relevant materials for battery manufacturing includes metals such as Al, Co, Fe, Pb, Li, Mn, and Ni.

An important constraint of the stationary energy storage using LIBs is the continuous consumption of Li-ion electrolyte, which limits both cycle and calendar life. Another restriction lies in the limitations of operating temperature window. The operation requires complex thermal management, which is impractical for stationary storage applications, particularly in "hot" countries. These batteries also contain hardly recyclable materials, such as lithium. LIBs are still facing safety issues. Also, LIBs are relatively expensive.

Thus, there is great demand and a great challenge to create effective post-Li electrochemical energy storage systems. An alternative suggests sodium with an electrochemical charge transfer reaction $Na \leftrightarrow Na^+ + e^-$, which could be performed in a solid compound, for instance, in a chloride. Oceans provide unlimited sodium source, which is for free.

Metal-air batteries explore air oxygen as a major reactant:

$$O_2 + 2H_2O + 4e^- \leftrightarrow 4OH^- \tag{2.1}$$

This approach enables battery weight reduction, at the same time, increasing the capacity for energy storage. Air–metal technologies are attractive because they are green, safe, and cost-efficient in terms of feedstock. The technologies propose usage of an unlimited source of oxygen from the atmosphere, which is for free. The metal–air batteries can utilize various metal electrodes, e.g. Li, Zn, Al, Fe, Mg (Tong et al. 2021). Of these metals, Li-air battery has the highest theoretical energy density and practical operation voltage (13 000 W h kg^{-1} and 2.4–2.7 V vs. standard hydrogen electrode (SHE), respectively). Lithium is followed by aluminum (8073 W h kg^{-1} and 1.2–1.6 V) and magnesium (6815 W h kg^{-1} and 1.2–1.4 V). Iron has the least performance (764 W h kg^{-1} and 0.8 V), however, the cost of iron metal anode is substantially less when compared to other metals. Tong et al. summarized that the cost of iron is about 200 times less when compared with the cost of lithium and about 15 times less when compared with magnesium or aluminum (Tong et al. 2021). The authors also estimated that Mg, Al, and Zn electrodes are nearly 10 times cheaper than the Li metal electrodes. Li–air batteries are limited due to dendrite formation, poor-cycling efficiency, and difficulties in finding a suitable highly stable electrolyte. Fe–air battery is very attractive in terms of excellent resource-efficiency. Note also that the electrodes can be prepared combining them in the form of alloys, for instance, Mg–Al, Mg–Al–Zn, Mg–Li, Mg–Zn, etc.

Silicon is an attractive material for LIBs anode production. The Li–Si binary system is characterized by exceptionally high Li insertion capacity. One Si atom can accommodate 4.4 Li atoms forming the alloy $Li_{22}Si_5$ with theoretical specific insertion capacity of 4200 mA h g^{-1}. The analogous value for $Li_{15}Si_4$ is 3576 mA h g^{-1} (Ashuri et al. 2016; Huang et al. 2021). Such capacity values are the highest among all anode materials used in LIBs. Silicon outperforms in terms of capacity the conventionally used graphite (372 mA h g^{-1} for LiC_6) (Liang et al. 2014). It is assumed conceptually that electrode capacity can be substantially increased when moving from classical intercalation reaction to alloying reaction (Ma et al. 2014). Lithium can react with Si-forming $Li_{22}Si_5$ alloy, while graphite accommodates much less lithium forming LiC_6. However, a challenge is disintegration of the electrode due to nearby 300–400% volume change during the lithiation–delithiation process, that is, the expansion–contraction cycles (Ashuri et al. 2016; Li et al. 2017; Zuo et al. 2017). This detrimental process reduces the intrinsic electrical conductivity and causes the interfacial instability, which leads to capacity fading.

The restoration of electric properties after the damage, so called self-healing, is of paramount importance in energy-storage device. This was stated in the BATTERY 2030+ Roadmap of the EC: "Establishing a new research community that includes a wide range of R&D disciplines to develop self-healing functionalities for batteries" (Edström et al. 2020). Scientists are in search of effective means of surface engineering that buffers the volume changes, for instance, by graphene films, silicon-carbon nanocomposites, Si nanowires, nanotubes, and other nanoparticles, solid core–shell structures, porous Si designs, controllable pores, or patterned silicon films on foreign metal substrates, etc. These efforts have been summarized in review articles (Liang et al. 2014; Ma et al. 2014; Ashuri et al. 2016; Li et al. 2017; Zuo et al. 2017; Huang et al. 2021).

Oxidized silicon (SiO_x) has also been proposed as an alternative to pure Si. Such electrodes show a lower volume change when compared to pure silicon (Liu et al. 2019a). However, silicon monoxide possesses some problematic features, such as low intrinsic electrical conductivity, non-negligible volume change, and low initial Coulombic efficiency (ICE) (Liu et al. 2019a). Thus, there is need to overcome these drawbacks in order to use the material for LIBs practically. The improvement of ICE could be achieved by pre-litiation methods, for instance, using stabilized lithium metal powder (Huang et al. 2021). Various strategies have been proposed to reduce the SiO volume change: (i) SiO disproportionation into a fraction of Si nanocrystals by high-energy milling (Hwa et al. 2013) or heating at 900 °C (Huang et al. 2021);

(ii) synthesis of porous SiO_x via chemical etching (Yu et al. 2014); (iii) composites of disproportionated Si–SiO_x with carbon (Si–SiO_x–C) (Yamada et al. 2011; Choi et al. 2012; Kim et al. 2013); (iv) core–shell porous silicon with nitrogen-doped carbon layer (p-Si/NC) (Xing et al. 2018); (v) silicon monoxide coated with N-doped carbon (SiO-NC) produced using N-containing ionic liquid (Lee et al. 2013) and analogous system using pitch and melanine as carbon and nitrogen sources (Huang et al. 2021).

The materials based on porous silica (SiO_2) have also been studied as an alternative to graphite anodes in LIBs. Porous silica has shown improved cycling stability; its discharge potential was similar to that of pure Si. Low Coulombic efficiency of SiO_2 was modified by preparing a carbon-coated C/SiO_2 composition (Buga et al. 2021). The thin carbon layer of the composition diminished interfacial impedance of silica, whereby the capacity increases to 714 mA h g^{-1}.

Despite intensive research, most of the proposed methods of the volume change buffering remain on a bench scale. Vital industrial applications are still pending.

2.2 Silicon for Energy Conversion – Photovoltaic Devices

2.2.1 Solar to Electricity

Solar energy could effectively act as a substitute to fossil fuels. The solar constant, which measures the quantity of solar energy to a square meter of the Earth surface, is estimated to be as high as 1367 W m^{-2}, according to the World Energy Council. The constant translates into total amount of 3 400 000 EJ in a year (EJ – Exajoule, 1 EJ = 10^{18} J) (Breeze 2019). The solar radiation, which reaches the earth, covers more than 7000 times the world's energy needs, that is, the annual global primary energy consumption. Breeze estimates that if 0.1% of the solar energy was converted into electricity with 10% efficiency, it would provide around 10 000 GW energy, which exceeds the world energy needs (Breeze 2019). Solar energy provides the opportunity for decentralized supply of energy actually at any place around the globe. It can be accumulated (as a heat) and/or converted to electricity with no polluting emissions to the environment. Other advantages include no-noise operation and unproblematic decommissioning.

Photovoltaic (PV) electricity generation is a rapidly expanding industry, which aims to accelerate the development of clean, sustainable, and efficient energy technologies with the scenarios to reach solar energy domination in the electricity market. In fact, creation of efficient and low-cost devices of solar energy harvesting plays a determining role in political-social landscape of twenty-first century. There is a public awareness in the PV technology as a provider of clean, sustainable, and secure energy from the most abundant source, which is for free. It is not surprising, therefore, that world PV production has recently experienced close to exponential growth.

It is now close to 70 years since the invention of the first reasonably efficient silicon cells in 1954 (Green 2005). The first silicon-based module for outdoor applications was produced at Bell Laboratories in 1955. The module was an assembly of 48 sub-modules and attained 2% efficiency. The reader interested in historical development of silicon-based PV devices is referred to M. Green's paper, which reviews improvement in energy conversion as well as prices of commercial modules over a 50-year period (Green 2005).

The International Renewable Energy Agency (IRENA) comprising of about 170 member countries declares its mission "supporting countries in their transition to a sustainable energy future" (www.irena.org). IRENA is an intergovernmental organization – a driver of actions, which advance the transformation of the global energy system in the pursuit of low-carbon economy. This organization promotes adoption of all forms of renewable energy. IRENA issued a global energy transformation paper, which aims to highlight the future of solar PV, including the key aspects, such as the deployment, investment, technology, grid integration, and socio-economic aspects (IRENA 2019). IRENA's analysts foresee that solar PV by 2050 would represent the second-largest power source behind wind power covering a quarter of the total electricity needs. The total growth of solar PV capacity is expected to rise from 480 GW produced in 2018 to 8519 GW projected by 2050. Asia (mostly China) has a dominating position in terms of installed capacity and will continue to dominate in the PV market reaching over 50% share in the world's production by 2050. The paper shows evolution of PV industry starting from mass production of solar cells in 1963 to 480 GW global capacity by the end of 2018. Important milestones in the provided timeline of PV development include the beginning of mass production of solar cells (1963), the first solar building (1973), the first solar plane flight around the world (2016), and the attaining of the global installed solar capacity of 480 GW (2018). Efforts have been made to beat the cost of photo-electricity down, below that of the power generated from fossil or nuclear fuels. Cost reduction of photo-electricity is permanently high on agenda of PV engineers and materials scientists. The IRENA's paper shows a three-times reduction in the auction price during an eight-year period, that is, from 241 USD/MWh in 2010 to 85 USD/MWh in 2018 (IRENA 2019).

Analysts estimated the PV capacity growth until 2050 (Figure 2.1) It is expected that installations will reach 2840 GW globally by 2030. The figure will rise up to 8519 GW by 2050 assuming utility-scale (60–80%) and distributed rooftop (40–20%). This means 18 times higher production in 2050 when compared to the amount produced in 2018.

Vast majority of the solar cells are produced from silicon. Various structural forms of silicon are used in solar cell production: mono-Si, poly-Si, and amorphous. Crystalline silicon has about 95% share of the PV solar production (IRENA 2019; Fraunhofer 2019). Thanks to technological maturity and a fall in the price of raw material, silicon will keep its dominating position in the PV market. Increased outputs of the Si-based panels also strengthen positions of silicon over other material. For instance, an average efficiency of multi-crystalline panels increased from ~13% in 2006 to ~17% recently, and this positive tendency is expected to continue (IRENA 2019; Fraunhofer 2019). A study of the Massachusetts Institute of Technology (MIT) Energy Initiative has shown that silicon, as an abundant, non-toxic, efficient and robust material, will maintain top position in the solar

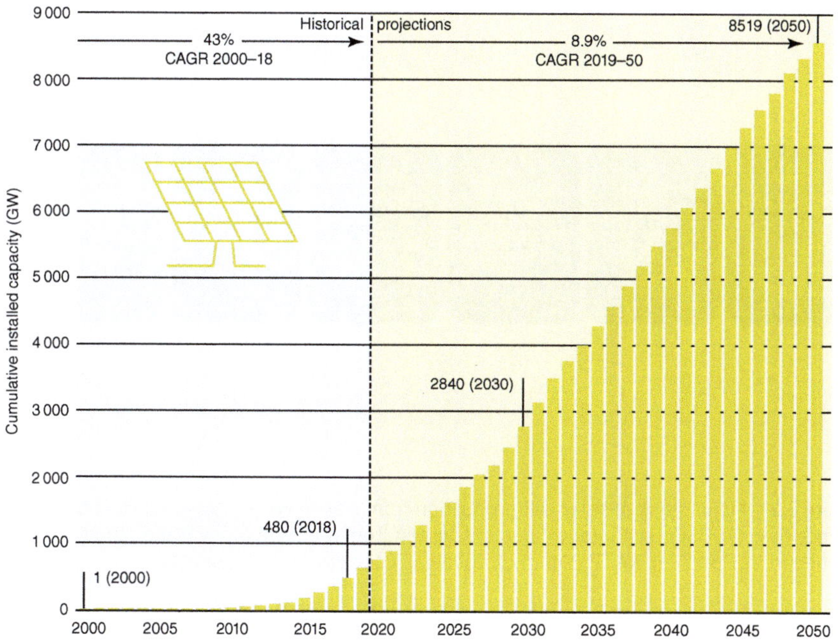

Figure 2.1 PV capacity growth expected by 2050, as reported by IRENA analysts. Source: IRENA 2019/IRENA/Public domain.

energy market at least for the next few decades (Schmalensee et al. 2015). Lowering of the cost of Si-based PV devices remains very crucial.

Production of PV module is a multistep process comprising such procedures as production of metallurgical-grade silicon (MG-Si), upgrading to solar-grade silicon (SoG-Si), ingot casting, and slicing. Traditional manufacturing includes high energy-intensive production and purification of silicon, such as carbothermic silicon reduction from silica and purification to solar grade by the Siemens process. The cost also increases due to significant material loss when sawing the silicon ingot into wafers. Subsequent production of the PV cell from Si wafer also includes numerous steps.

First-generation technologies (conventional solar cell architecture) explore crystalline silicon PV panels, whose evolution has covered a whole scope of technical maturity (Technology Readiness Levels, TRLs): basic research and technology development (R&D), demonstration, pilot lines, system launch and operations, enter into market and penetration, and market maturity. There are many other kinds of solar panel technologies. One modification becoming more common is the passivated emitter and rear contact (PERC) cells. Efficiency of the PERC cells is 6–12% higher when compared with conventional cells. The modification has an extra layer within the backside, which allows the reflection of some rays back into the cell. The advantages of the PERC cell includes the reduction in recombination, enhancement of light absorption, and higher internal reflectivity. From an economical point of view, it is important that the PERC cells, being a modification of the conventional cells,

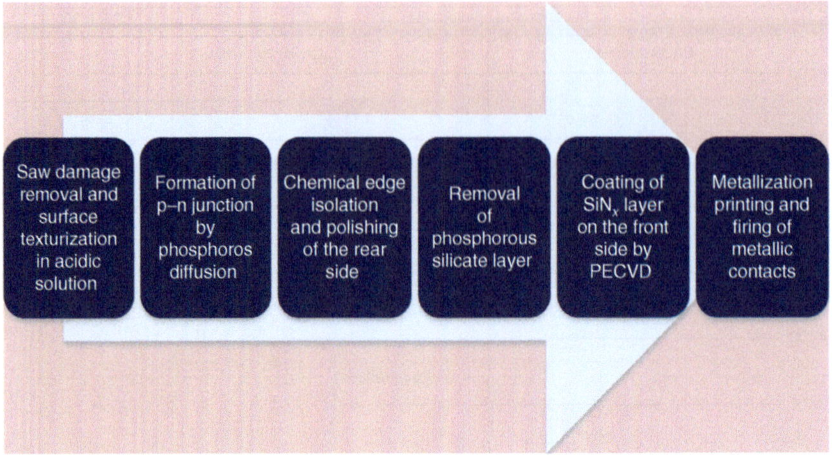

Figure 2.2 Major steps of solar cell production from silicon wafer according to Al back surface field (Al-BSF) technology. Source: Adopted from the technology route scheme applied in the "Soli Tek R&D" (Vilnius, Lithuania). Courtesy of Dr. J. Denafas.

do not require great investment in acquiring additional equipment. These cells have recently become the new industry standard.

Figure 2.2 demonstrates the major stages of the aluminum backside field (Al-BSF) technology. The process starts from thorough wafer quality check: inspection of as-cut wafer material, geometry, surface defects, and micro-cracks. Bare silicon reflects more than 30% of the incident light (Singh et al. 2010; Szlufcik et al. 2005), thus, antireflection textures are formed in acidic (typically in hydrofluoric acid, HF) solutions. Further steps include p–n junction formation (e.g. by phosphorous diffusion), chemical edge isolation, removal of phosphorous silicate layer, passivation of silicon (e.g. by SiN_x layer), and preparation of metallic contacts. The BSF formation provides a good contact between Si and Al with limited penetration of the AlSi alloy that is formed during the firing process, and it enables a significant reduction in the thickness of Al deposited in the rear contact. The technology of passivated emitter and rear cell (PERC) includes some additional steps: (i) edge isolation and polishing of the rear side, (ii) back side passivation by the AlO_x layer, and (iii) its selective removal (e.g. by laser contact opening).

An innovative approach is metal wrap through (MWT) back-contact solar cell technology. In this technology, the positive and negative electrodes are arranged on the rear side of the cell. The back contacts enable avoiding of application of busbars on the front side. Due to this, the shaded area is reduced, and the conversion efficiency can be increased by a few percent. The MWT processing is close to conventional fabrication sequences; thus, it does not require substantial additional investment. Photograph in Figure 2.3 demonstrates the examples of the solar cells produced in "Soli Tek R&D" (the limited liability company (LLC) in Vilnius, Lithuania) using MWT and Al-BSF technologies.

It is obvious that each of the technological steps discussed above contributes to the cell cost and, in turn, to the cost of photo-electricity. Thin-film technologies provide

Figure 2.3 Solar cells produced according to Al-BSF technology (MWT cell on the left side and the other two are Al-BSF samples). The photograph also shows the production unit in the "Soli Tek R&D" (Vilnius, Lithuania). Source: Courtesy of Dr. J. Denafas.

the potential for reduction of the photo-electricity production cost. Depositing thin silicon layers instead of thick individual cells, which have to be constructed, framed, and wired together, could reduce the consumption of material and energy.

Thin films provide opportunity to produce large complete modules with improved appearance for many visual applications (e.g. architectural glass, liquid crystal displays [LCD]). Mercaldo et al. presented analysis of architectural issues of thin-film PVs, in particular, those related to applications of transparent and conductive oxides and films of silicon solar cells (Mercaldo et al. 2009). Modern architectural theories look at buildings as living organisms that should be able – during their life – to generate the energy needed to be in operation. Thin-film technology is suitable to satisfy such advanced architectural theory. It is a challenging goal to form thin silicon films with effective light absorbance on technically important metallic substrates such as steel, aluminum alloys, and copper. This makes it possible to combine structural materials with silicon for facing panels on buildings, which could be used to harness solar energy. Putting the silicon on thin metallic foils electrochemically would expand the application fields to a wide range of subjects with flexible geometries as well as on the metal-coated glass. Electrodeposition of thin Si layers, including the electrochemical formation of a p–n junction, which could replace multiple processes of Si wafer treatment, will be discussed in Chapter 6.

The main disadvantage of silicon is its poor intrinsic ability to absorb the light – more than 30% of incident light is reflected if silicon surface is not specifically textured or coated by antireflection coatings. Also, high-processing complexity of silicon significantly contributes to the cost of PV devices. Light-harvesting efficiency of the Si-based photoelectrochemical (PEC) cells can be achieved by creating silicon nano-micro architectures, which enhance light absorbance. To this end, various surface engineering techniques have been applied, such as femtosecond laser engineering, annealing in vacuum, coating by atomic layer deposition (ALD) as well as various etchings: chemical, reactive ion, inductively coupled plasma, and cryogenic. These methods, however, are costly and technically sophisticated. As a result, the cost-competitive solutions for industrial-scale solar energy applications are limited. Furthermore, etchings usually involve toxic chemicals, such as hydrofluoric acid, and rather expensive catalysts, such as gold.

Electrochemical methods of silicon surface engineering are very attractive due to environmental friendliness, technical simplicity, and cost-effectiveness. Various methods of electrochemical Si surface texturing in aqueous and molten salt electrolytes will be surveyed in Chapters 14 and 15.

Non–silicon based thin-film technologies have been intensively investigated as alternatives to silicon PVs. Particularly, there have been great expectations from perovskite-based cells. This type refers to a broad class of ABX_3 structures, which originate from the mineral calcium titanium oxide ($CaTiO_3$). The component A can be either organic ions, such as methylamonium (MA, $CH_3NH_3^+$) or formamidinium (FA, $CH(NH_2)_2^+$), or inorganic ions, such as cesium (Cs^+) or rubidium (Rb^+). The component B refers to a smaller divalent cation, in most cases lead (Pb^{2+}) or sometimes tin (Sn^{2+}). The component X refers to halogen anions, typically iodide (I^-), bromide (Br^-), or chloride (Cl^-). Such ABX_3 structures show a high light absorption under visible light as well as an extremely long charge diffusion length. These features presuppose the high PV performance of the perovskite-based cells. More than a dozen of firms from the USA, South Korea, China, UK, Poland, Japan Switzerland, and Netherlands strived to commercialize the perovskite solar cells (Extance 2019). The challenge to overcome, perovskite-cells have limited lifetime due to sensitivity to air and moisture. Another issue lies in the structural transformations and performance fading when the crystals warm up and cools down. To be competitive, the operational durability of perovskite cell should be comparable with the durability of the silicon-based cells with a 25-year warranty. This is "now looking increasingly unlikely" according to Martyn Green from the

University of New South Wales in Sydney, Australia (Extance 2019). Another limitation lies in scaling up matters: efficiency of the cells does not replicate when their size increased. Increased efficiency up to theoretical limit of 45% can be achieved in tandem cells, that is, a system of perovskite and silicon layers. Such strategy is under active investigation (Extance 2019).

The rapid expansion of PV resulted in an increasing amount of end-of-life solar panels, which poses a hazard for future PV development. It is a challenging goal to find solutions for PV modules recycling. An innovative opportunity has recently proposed a group consisting of 25 researchers from research establishments in Singapore, China, and Japan (Cao et al. 2022). Their idea lies in the conversion of PV waste into feedstock for thermoelectrics (Figure 2.2). Such conversion is in line with the principles of circular economy: elimination of waste and pollution as well as circulation of products and materials.

Figure 2.4 outlines the major conversion steps of Si-based solar cells into Si suitable for thermoelectric application. Polycrystalline Si solar cells were used as a feedstock. Aluminum and silver were removed from the cell by leaching in HCl, HNO_3, and H_3PO_4. The dried cells were pulverized into fine powder by a ball milling using the balls made of tungsten carbide. This was done in an argon atmosphere to avoid silicon oxidation by air. Furthermore, dopants (P and Ge) were added and the powders were homogenized by ball milling. Finally, the powders were consolidated by spark plasma sintering at 1150 °C, and the specimens were cut into desired shapes. The strategy provided low-cost and environmentally friendly recycling of silicon for circular economy applications. The economic issues of the proposed process relate to minimizing the cost associated with the application of Ge and P dopants as well as the ball milling and spark plasma sintering. Cost minimization is expected when the technology is used on a large-scale application (economy of scale).

2.2.2 Solar-to-Chemical Conversion

Collection of the solar energy in chemical form helps to conserve energy and makes it sustainable. Electrochemical solar energy capturing in chemical bonds (e.g. hydrogen production from water) mimics the process of photosynthesis in plants. Production of hydrogen in combination with solar energy occurs with near-zero greenhouse gas emissions. When captured in chemical form, the energy can be converted to power in fuel cells with specifically high efficiency (up to 85%) or in combustion engines. It is also possible to combine hydrogen generation with a fuel cell to generate electricity and replace the water that has been consumed. Electrochemical hydrogen generation means the usage of electrons as absolutely clean agents, albeit at slightly lower efficiency but without pollutant emissions. The electrochemical process offers such advantages as high purity, flexibility, supply on-site and on-demand as well as fast start-up and shutdown opportunities. Of particular interest is solar-to-fuel process, which occurs directly (wireless) on the photo-electrodes.

Solar-to-chemical conversion offers important opportunity for solar energy accumulation and transmission. In this respect, the most attractive process is hydrogen

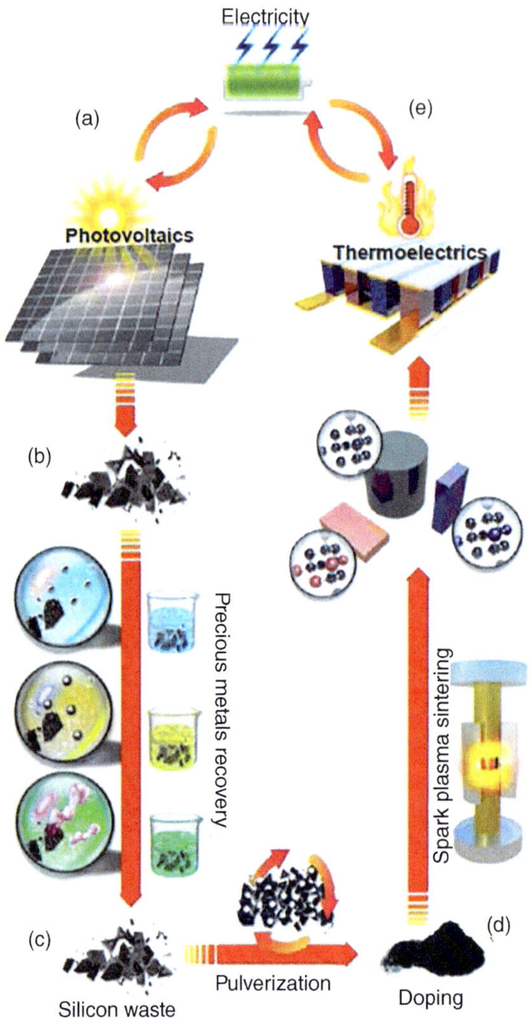

Figure 2.4 The process of conversion of the waste of solar cells into thermoelectrics: (a) – solar cells; (b) – recovering of Al and Ag; (c) – pulverizing of Si in a ball milling machine in Ar atmosphere; (d) – doping of Si powder with phosphorous and germanium; (e) – final pellets for thermoelectrics. Source: Reproduced from (Cao et al. 2022) with permission from the publisher, ©Wiley, 2021.

generation by PEC splitting of water. This process has recently attracted the attention of researchers due to hydrogen production with a very low carbon footprint. Almost zero greenhouse emission is achieved when hydrogen is generated by splitting water using renewable energy sources such as solar, wind, hydro, or ocean. However, renewable energy is irregular and sporadic in nature; thus, hydrogen generation is applicable for energy storage and load management. Technologies demonstrate the connection of the electrolyzers to electricity and gas grids (Bertucioli et al. 2014). With the power-to-gas approach, excess electricity generated from renewable energy is converted into hydrogen, stored, and when needed, it is reconverted into electricity. Certainly, the produced hydrogen can also be used as a feedstock in chemical industry applications. The hydrogen can also be converted to hydrocarbon fuels, for instance, methane. An international review was published about the power-to-gas pilot plants for stationary applications (Gahleitner 2012). The review evaluated 41 already realized and 7 planned power-to-gas large-scale systems.

Silicon, as a photosensitive, chemically inert, low-cost, and robust material, is attractive in PEC applications. Effectiveness of solar energy conversion to chemical bonds depends largely on the ability of a PEC device to absorb the light. Numerous efforts have been undertaken to increase silicon absorbance by increasing its actual surface area – the roughness at micro and nano scale. These techniques include femtosecond laser engineering with subsequent sample etching, deposition of transparent conducting oxides using ALD, reactive ion etching, chemical etching using procures metal catalysts, inductively coupled plasma and cryogenic etching, vapor–liquid–solid reactions, etc. These techniques are surveyed in detail in Section 14.2. Electrochemical methods of silicon surface engineering in aqueous and molten salt electrolytes are discussed in Chapters 14 and 15.

Of great interest is nanoporous silicon, known as black silicon, which effectively absorbs light in a wide range of wavelengths due to internal trapping by porous structures. Black silicon photocathode for hydrogen production has been reported (Oh et al. 2011) and patented (Oh and Branz 2014). The authors showed that nanostructured black silicon significantly accelerates hydrogen production with reduced need for surfactants. Techniques to fabricate black silicon, such as laser texturization, plasma treatment, metal-assisted etching, or cryogenic deep reactive ion etching are discussed in Section 14.3. These techniques, however, are costly. There is great demand for cost-effective and green technologies of black silicon mass production. More cost-effective methods based on electrochemical Si surface texturing in aqueous and molten salt electrolytes are surveyed in Chapters 14 and 15.

Black silicon is also a promising material for solar thermo-generating applications. Flexible black silicon on insulator has been formed, which is a potential candidate for use as an absorber for the solar-thermo generators such as concentrating solar power parabolic trough (Mei et al. 2011). This is an alternative to increasing the absorption efficiency of sunlight using antireflection coating and the surface anodized metal on the heat transfer fluid container. Black silicon can absorb sunlight from the ultraviolet (UV) and near-infrared (IR) regions non-sensitively to the angles of incoming light.

References

Amnesty International report (2016): "This Is What We Die For", https://www.google.com/url?sa=t&rct=j&q=&esrc=s&source=web&cd=&ved=2ahUKEwjx7tL0mKj5AhXlsYsKHeiVBAYQFnoECAgQAQ&url=https%3A%2F%2Fwww.amnesty.org%2Fen%2Fwp-content%2Fuploads%2F2021%2F05%2FAFR6231832016ENGLISH.pdf&usg=AOvVaw15we0aXd_0NG0WUN_xbbXQ.

Ashuri, M., He, Q., and Shaw, L.L. (2016). Silicon as a potential anode material for Li-ion batteries: where size, geometry and structure matter. *Nanoscale* 8: 74–103. https://doi.org/10.1039/C5NR05116A.

Bertucioli, L., Chan, A., Hart, D., et al. (2014). Study on development of water electrolysis in the EU. Final report. E4tech Sàrl, Switzerland; element energy, Cambridge, UK.

Breeze, P. (2019). Solar power, Chapter 13. In: *Power Generation Technologies*, 3e, 293–321. Elsevier Ltd https://doi.org/10.1016/C2017-0-03267-6.

Buga, M.-R., Spinu-Zaulet, A.A., Ungureanu, C.G. et al. (2021). Carbon-coated SiO_2 composites as promising anode material for Li-ion batteries. *Molecules* 26: 4531. https://doi.org/10.3390/molecules26154531.

Cao, J., Sim, Y., Tan, X.Y. et al. (2022). Upcycling silicon photovoltaic waste into thermoelectrics. *Adv. Mater.* 2110518. https://doi.org/10.1002/adma.202110518.

Choi, L., Lee, M.J., Oh, S.M., and Kim, J.J. (2012). Fading mechanisms of carbon-coated and disproportionated Si/SiO_x negative electrode ($Si/SiO_x/C$) in Li-ion secondary batteries. Dynamics and component analysis by TEM. *Electrochim. Acta* 85: 369–376. https://doi.org/10.1016/j.electacta.2012.08.098.

Chu, B., Liu, S., You, L. et al. (2020). Enhancing the cycling stability of Ni-rich $LiNi_{0.6}Co_{0.2}Mn_{0.2}O_2$ cathode at a high cutoff voltage with Ta doping. *ACS Sustainable Chem. Eng.* 8: 3082–3090. https://doi.org/10.1021/acssuschemeng.9b05560.

Communication from the Commission to the European Parliament, the Council, the European Economic and Social Committee and the Committee of the Regions (2020). Critical raw materials resilience: charting a path towards greater security and sustainability. Brussels. COM(2020) 474 final.

Edström, K; Dominko, R.; Fichtner, M., et al. Batteries 2030+ roadmap, inventing the sustainable batteries of the future. Research needs and future actions. https://battery2030.eu/research/roadmap/ (Retrieved 02.08.2022), 2020.

Extance, A. (2019). The reality behind solar power's next star material. *Nature* 570: 429–432. https://doi.org/10.1038/d41586-019-01985-y.

Fraunhofer ISE (2019). Photovoltaics report. Fraunhofer ISE and PSE Conferences & Consulting GmbH, Freiburg, 14 March 2019.

Gahleitner, G. (2012). Hydrogen from renewable electricity: an international review of power-to-gas pilot plants for stationary applications. *Int. J. Hydrogen Energy* 38: 2039–2061. https://doi.org/10.1016/j.ijhydene.2012.12.010.

Green, M.A. (2005). Silicon photovoltaic modules: a brief history of the first 50 years. *Prog. Photovoltaics Res. Appl.* 13: 447–455. https://doi.org/10.1002/pip.612.

Green, M. (2016). Commercial progress and challenges for photovoltaics. *Nat. Energy* 1: 15015. https://doi.org/10.1038/nenergy.2015.15.

Huang, B., Chu, B., Huang, T., and Yu, A. (2021). Nitrogen-doped carbon-coating disproportionated SiO materials as long cycling stable anode for lithium ion batteries. *Molecules* 26: 1536. https://doi.org/10.3390/molecules26061536.

Hwa, Y., Park, C.M., and Sohn, H.J. (2013). Modified SiO as a high performance anode for Li-ion batteries. *J. Power Sources* 222: 129–134. https://doi.org/10.1016/j.jpowsour.2012.08.060.

IRENA (2019). Future of Solar Photovoltaic: Deployment, Investment, Technology, Grid Integration and Socio-Economic Aspects (A Global Energy Transformation Paper). International Renewable Energy Agency, Abu Dhabi. ISBN 978-92-9260-156-0. http://www.Irena.org/publications

Kim, K.W., Park, H., Lee, J.G. et al. (2013). Capacity variation of carbon-coated silicon monoxide negative electrode for lithium-ion batteries. *Electrochim. Acta* 103: 226–230. https://doi.org/10.1016/j.electacta.2013.04.040.

Lee, D.J., Ryou, M.-H., Lee, J.-N. et al. (2013). Nitrogen-doped carbon coating for a high-performance SiO anode in lithium-ion batteries. *Electrochem. Commun.* 34: 98–101. https://doi.org/10.1016/j.elecom.2013.05.029.

Li, J.-Y., Xu, Q., Li, G. et al. (2017). Research progress regarding Si-based anode materials towards practical application in high energy density Li-ion batteries. *Mater. Chem. Front.* 1: 1691–1708. https://doi.org/10.1039/C6QM00302H.

Liang, B., Liu, Y., and Xu, Y. (2014). Silicon-based materials as high capacity anodes for next generation lithium ion batteries. *J. Power Sources* 267: 469–490. https://doi.org/10.1016/j.powsour.2014.05.096.

Liu, D., Liu, S., Zhang, C. et al. (2019a). Revealing the effect of Ti doping on significantly enhancing cyclic performance at a high cutoff voltage for Ni-rich $LiNi_{0.8}Co_{0.15}Al_{0.05}O_2$ cathode. *ACS Sustainable Chem. Eng.* 7: 10661–10669. https://doi.org/10.1021/acssuschemeng.9b01312.

Liu, S., Wang, L., Zhang, C. et al. (2019b). Dynamic evolution of cathode-electrolyte interface of $LiNi_{0.6}Co_{0.2}Mn_{0.2}O_2$ during the initial charge-discharge process. *J. Power Sources* 438: 226979. https://doi.org/10.1016/j.powsour.2019.226979.

Ma, D., Cao, Z., and Hu, A. (2014). Si-based anode materials for Li-ion batteries: a mini review. *Nano-Micro Lett.* 6: 347–358. https://doi.org/10.1007/s40820-014-0008-2.

Mei, H., Wang, C., Yao, J. et al. (2011). Development of novel flexible black silicon. *Opt. Commun.* 284: 1072–1075. https://doi.org/10.1016/j.optcom.2010.10.024.

Mercaldo, L.V., Addonizio, M.L., Noce, M.D. et al. (2009). Thin film silicon photovoltaics: architectural perspectives and technologies. *Appl. Energy* 86: 1836–1844. https://doi.org/10.1016/j.apenergy.2008.11.034.

Oh, J. and Branz, H.M. (2014). Anti-reflective nanoporous silicon for efficient hydrogen production. Patent US8729798.

Oh, J., Deutsch, T.G., Yuan, H.-C., and Branz, H.B. (2011). Nanoporous black silicon photocathode for H_2 production by photoelectrochemical water splitting. *Energy Environ. Sci.* 4: 1690–1694. https://doi.org/10.1039/C1EE01124C.

Polman, A., Knight, M., Garnett, E.C. et al. (2016). Photovoltaic materials: present efficiencies and future challenges. *Science* 352 (6283): https://doi.org/10.1126/science.aad4424.

Schmalensee, R., Bulovič, V., Armstrong, R. et al. (2015). *The Future of Solar Energy: An Interdisciplinary MIT Study*. Massachusetts Institute of Technology, MIT Energy Initiative https://energy.mit.edu/research/future-solar-energy/ (Retrieved 02.08.2022).

Simon, F. (2019). EU Commission unveils "European Green Deal": The key points. EURACTIV. https://www.euractiv.com/section/energy-environment/news/eu-commission-unveils-european-green-deal-the-key-points (accessed 28 June 2022).

Singh, P., Sharma, S.N., and Ravindra, N.M. (2010). Applications of porous silicon thin films in solar cells and biosensors. *JOM* 62: 15–24.

Szlufcik, J., Agostinelli, G., Duerinckx, F. et al. (2005). Low cost industrial technologies of crystalline silicon solar cells. In: *Solar Cells: Materials, Manufacture and Operation* (ed. T. Markvart and L. Castaner), 90–110. Elsevier Ltd.

Tamma, P., Schaart, E., and Gurzu, A. (2019). Europe's Green Deal plan unveiled. Politico. www.politico.eu (accessed 28 June 2022).

Tong, F., Wei, S., Che, X., and Gao, W. (2021). Magnesium alloys as anodes for neutral aqueous magnesium-air batteries. *J. Magnes. Alloy.* 9: 1861–1883. https://doi.org/10.1016/j.jma.2021.04.011.

World Bank Group (2017). The Growing Role of Minerals and Metals for a Low Carbon Future. World Bank, Washington, DC. © World Bank. https://openknowledge.worldbank.org/handle/10986/28312 License: CC BY 3.0 IGO.

Xing, Y., Zhang, I., Mao, S. et al. (2018). Core-shell structure of porous silicon with nitrogen-doped carbon layer for lithium-ion batteries. *Mater. Res. Bull.* 108: 170–175. https://doi.org/10.1016/j.materresbull.2018.09.004.

Yamada, M., Ueda, A., Matsumoto, K., and Ohzuku, T. (2011). Silicon-based negative electrode for high-capacity lithium-ion batteries "SiO"-carbon composite. *J. Electrochem. Soc.* 158: A417–A421. https://doi.org/10.1149/1.3551539.

Yu, B.-C., Hwa, Y., Park, C.-M., and Sohn, H.-J. (2014). A new approach to synthesis of porous SiO_x anode for Li-ion batteries via chemical etching of Si crystallites. *Electrochim. Acta* 117: 426–430.

Zuo, X., Zhu, J., Müller-Buschbaum, P., and Cheng, Y.-J. (2017). Silicon based lithium-ion battery anodes: a chronicle perspective review. *Nano Energy* 31: 113–143. https://doi.org/10.1016/j.nanoen.2016.11.013.

3

Brief Historical Overview of Silicon Production. Metallurgical-Grade Silicon

Silicon is the second most abundant element after oxygen in the Earth's crust. This element very rarely occurs in its pure form. Consequently, it has been used since the existence of mankind in the form of naturally occurring compounds. Flint or quartz, for instance, has been used to prepare tools and weapons. The use of natural glass, especially of volcanic origin, dates back to thousands of years over the Stone Age. The earliest archeological discovery of glassware dates back to the middle of the 3rd-millennium BCE.

There are uncertainties as to where and when the man-made glass precisely originates. It is believed that the first man-made glass was produced as a by-product of metalworking or faience making. The term faience basically refers to a tin-glazed pottery. Egyptian faience originates from the Ancient Egypt; the material is produced from quartz by its vitrification at high temperatures. The first man-made glass may have appeared in the Middle East regions, e.g. Egypt and Mesopotamia 3000–2000 BCE (Macfarlane and Martin 2004). Glass is thought to have been invented by accident. Phoenician traders used the nitrate blocks for cooking, and during the process, they noticed the formation of a clear liquid when the blocks melted and mixed with the sand of the beach. A method for producing glass vessels invented by Egyptian craftsmen dates back to about 1500 BCE (Macfarlane and Martin 2004). The first manual of glassmaking has been "published" on Assyrian stone tablets at about 650 BCE. Invention of glassblowing is credited to Syrian craftsmen, who launched this technology about 2000 years ago. Furthermore, the Romans adopted this "know-how" and transferred their skills and technology to Western Europe. Venice in Italy became a glassmaking center in the thirteenth century where the guild of crystal workers was established in 1284.

It is difficult to overestimate the significance of the glass to the development of mankind. The importance of this material to everyday life and technological progress is enormous. The first uses of glass do not appear very important: glass was used in the manufacture of toys, vessels and vases, or in jewelry as a substitute for precious stones. The importance of glass in everyday life has essentially been on the increase when it was applied in the making of windows. Another key

Silicon: Electrochemistry, Production, Purification and Applications, First Edition. Eimutis Juzeliūnas.

area of application was making mirrors by coating glass with silver. Justus von Liebig[1] discovered the chemical process of glass silvering in 1835. Production of lenses marks another key milestone in glass application – people were given the opportunity to improve eyesight. The mass production of eyeglasses had begun in Venice by the end of the fourteenth century. The products were exported to all over Europe; the demand increased, and the possession of eyeglasses became a sign of intelligence, prestige, and status.

Glass has contributed immensely to the acquisition of scientific knowledge. This material led to the development of new scientific instruments, such as thermometers, microscopes, or telescopes. Glass was a key material in the development of powerful and precise laboratory instruments. Glass instrumentation revolutionized studies of living things. Important glass-based instruments have been developed for life sciences, such as micro- or molecular biology, cell biology, histology, and bacteriology. With regards to glass, advanced methods of diagnosis and medical treatment have been developed. Astronomers are grateful to glass discoverers and manufacturers for developing the sky observation techniques; chemists – for the rich laboratory glassware, electrochemists – for selective glass electrodes enabling measurement of fundamental solution characteristics, such as pH or metal ion concentration. Macfarlane and Martin say: "glass literally opened people's eyes and minds to new possibilities and turned western civilization from an aural to a visual mode of interpreting experience"; glass is "a molten liquid that has shaped our world" (Macfarlane and Martin 2004).

Start of the age of glass science is associated with the names of Otto Schott (1851–1935) and Ernst Abbe (1840–1905). These glass technologists and scientists created in Jena (Germany) numerous glass compositions that possessed novel optical properties. Both scientists started collaboration in 1879. They studied systematically the optical performance of glasses depending on their chemical composition. Working together with Calr Zeiss (1816–1888), they developed the glasses for optical instruments, such as microscopes and telescopes. Calr Zeiss was a German instrument-maker and businessman who established, in 1846, a workshop in Jena known as Carl Zeiss AG. Now, the company branded as ZEISS is active in business worldwide. A catalog appeared in 1886, which referred to 44 optical glass compositions. The work was named (in German) as "Glastechnisches Laboratorium, Schott und Genossen, Jena" (see the photo in (Kurkjian and Prindle 1998)). In the twentieth century, studies were conducted to understand glass as a material and to predict its properties. To this end, numerous advanced methods have been used: X-ray techniques, optical spectroscopy, or physical investigations. Kurkjian

1 Justus Freiherr von Liebig (May 12, 1803–April 20, 1873) was a German chemist who contributed significantly to organic chemistry, particularly, to agricultural and bioorganic chemistry. He studied pharmacy at Bonn and Erlangen Universities and worked at the Sorbonne with L. J. Gay-Lussac. Liebig was a professor at the Universities of Giessen, Heidelberg, and Munich. He demonstrated the existence of free radicals, clarified basics of acids, and analyzed body fluids. He also proved that plants use carbon dioxide, water, and ammonia. Liebig discovered glass silvering for making mirrors. The mineral liebigite $Ca_2(UO_2)(CO_3)_3$ $11H_2O$ was named after him. (Retrieved from Visuotinė Lietuvių Enciklopedija, and the Encyclopedia Britannica Online, March 2022).

and Prindle surveyed the development of glass compositions; they provided a brief review of the early history of glass, the start of glass science, and the development of more recent glass compositions, such as special glasses and optical fibers (Kurkjian and Prindle 1998).

While natural silicon materials have been used for thousands of years, application of silicon in its pure form began much latter. Silicon isolation for chemists was not a trivial task because of high affinity of this element to oxygen. Jöns Jakob Berzelius[2] (1779–1848) reduced potassium fluorosilicate (K_2SiF_6) in 1823 with molten potassium metal and obtained amorphous silicon. Berzelius is usually given credit for silicon discovery. Other scientists are also thought to have significantly contributed to silicon discovery. In 1811, Joseph Louis Gay-Lussac[3] (1778–1850) and Louis Jacques Thénard[4] (1777–1857) heated potassium metal with silicon tetrafluoride (SiF_4) and obtained amorphous silicon. However, the authors did not identify the product as a new element. Berzelius also was the first who prepared silicon tetrachloride ($SiCl_4$). Silicon was produced when passing silicon tetrachloride over heated potassium. Silicon tetrachloride (or tetrachlorsilane) can be obtained by silica chlorinating. Silicon tetrachloride is a very important compound used in silicon purification technologies. Due to low boiling point of $SiCl_4$ (57.6 °C), silicon purification can be achieved by $SiCl_4$ rectification and its thermal decomposition. Silicon purification principles will be discussed in Chapter 5.

At present, silicon is mainly produced by deoxidation of silica (SiO_2). Deoxidation is a process in which metal is converted from an oxide (Me_xO_y) to a metallic form (Me)

$$Me_xO_y + 2ye^- \leftrightarrow x\mathrm{Me} + yO^{2-} \tag{3.1}$$

Basically, this reaction means electron (e^-) transfer from a reducing agent to a metal. This rather simple by-form reaction was and remains of paramount importance. Historically, advancement in utilization of this reaction marks the periods of development of mankind known as Bronze and Iron ages. The Bronze Age relates to the advancement in smelting of copper-tin bronze, which replaced stone in

2 Jöns Jakob Berzelius (August 20, 1779–August 7, 1848) was a Swedish chemist, one of the founders of modern chemistry. Berzelius studied medicine at Uppsala University and was a professor of medicine and pharmacy at the Karolinska Institute (1807–1832). His important contributions to chemistry include determination of atomic weights, chemical symbols, electrochemical theory, and discovery of several elements including the silicon. He also studied isomerism and catalysis and developed analytical techniques. (Retrieved from the Encyclopedia Britannica Online, March 2022).

3 Joseph-Louis Gay-Lussac (December 6, 1778–May 9, 1850) was a French chemist and physicist who made important advances in behavior of gases and created new analytical techniques. He is best known for the law of the combining volumes of gases (1808). He also determined with Alexander von Humboldt the hydrogen/oxygen ratio in water (Retrieved from the Encyclopedia Britannica Online, March 2022).

4 Louis Jacques Thénard (May 4, 1777–June 21, 1857) was a French chemist. He taught at the École Polytechnique, the Sorbonne and was a chancellor at the University of Paris. He isolated potassium and sodium metals, studied esters and organophosphorous compounds, discovered hydrogen peroxide. A lot of his research was done together with Gay-Lussac. The mineral thenardite is named after L. J. Thénard. Thenardite is non-marine evaporate – an anhydrous sodium sulfate mineral, Na_2SO_4. (Retrieved from Visuotinė Lietuvių Enciklopedija, March 2022).

production of tools and weapons (3000–1000 BCE). In the Iron Age, the technology succeeded to extract iron, which became much more used as the bronze. The transition from Bronze to Iron Age took hundreds of years (1000–500 BCE). Reversal of the reaction (3.1) means the release of electrons by a metal. Such process is of vital societal importance as well, however, usually with a negative sign. The process is known as corrosion. According to the World Corrosion Organization, direct corrosion damage to the society is estimated to be 3–4% of the gross domestic product (GDP) of industrialized countries. This percentage translates into trillions of $US (Schmitt et al. 2009). Thus, from the society's point of view, the reaction (3.1) could be either globally beneficial or detrimental depending on the reaction direction.

Many industrial processes of chemical deoxidation of metal ores are performed by the reaction of carbothermic reduction. The most common example is the process of industrial production of iron, in which iron ore is reduced with coke in a blast furnace. Another example is the Kroll processes for extraction of titanium from rutile (TiO_2). Titanium forms alloys with other metals (Al, V, Mo), which due to lightness, toughness, and high corrosion resistance are widely used in automotive, aerospace, military, and *in-vivo* applications. In the Kroll process, rutile is reduced with coke at 1000 °C and converted by chlorine gas to volatile titanium tetrachloride ($TiCl_4$). The latter is separated by fractional distillation and titanium is reduced by liquid magnesium:

$$TiO_2 + C + 2Cl_2 \rightarrow TiCl_4 + CO_2 \qquad (1000°C) \tag{3.2}$$

$$2Mg + TiCl_4 \rightarrow MgCl_2 + Ti \qquad (800 - 850°C) \tag{3.3}$$

Silicon is produced industrially by carbothermic reduction as well. Quartzite or sand is used as a feedstock and various types of coals (charcoal, coal, coke) are used as a redactor. The reduction process is performed in electric arc furnace at c. 1800 °C (Zulehner et al. 1995):

$$SiO_2 + C \rightarrow Si + CO_2 \tag{3.4}$$

Silicon carbide may form as a byproduct at such temperatures (Greenwood and Earnshaw 1997)

$$SiO_2 + C \rightarrow Si + 2CO; \qquad Si + C \rightarrow SiC \tag{3.5}$$

The reduction is performed with excess of silica, which prevents the carbide production

$$2SiC + SiO_2 \rightarrow 3Si + 2CO \tag{3.6}$$

Apart from carbon, other redactors can also be applied. For instance, silicon is reduced by aluminum in an aluminothermal process (Zulehner et al. 2000)

$$3SiO_2 + 4Al \rightarrow 3Si + 2Al_2O_3 \tag{3.7}$$

World production of silicon over two decades (2001–2020) is presented in Figure 3.1. The amount of produced silicon over the first decade increased more than two times (2001–2011) and thereafter reached some saturation at approximately 7–8 millions of metric tons (MT). According to (3.4), such amount of Si

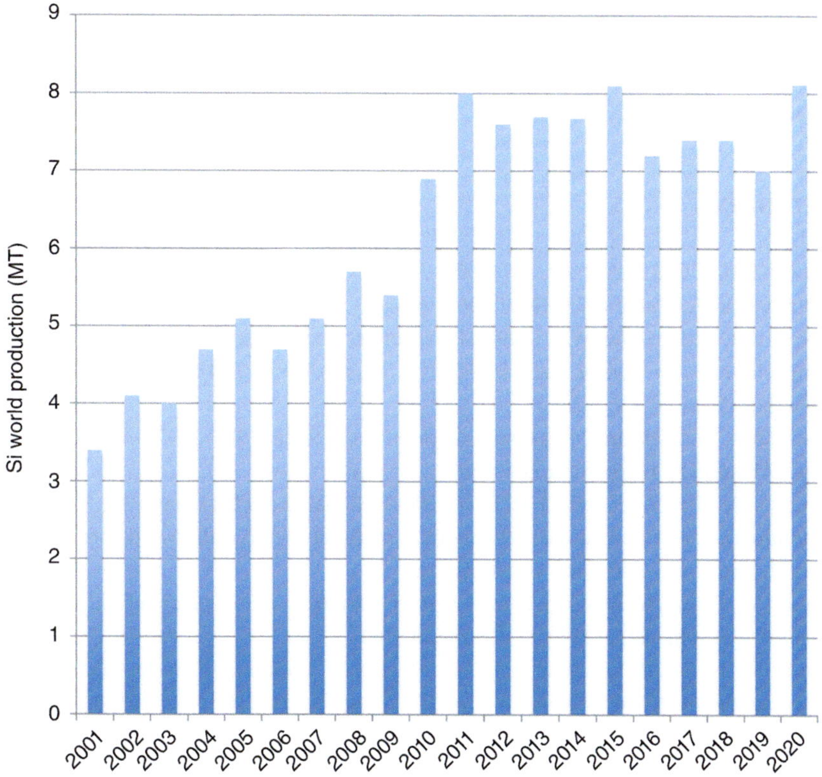

Figure 3.1 Silicon world production over two recent decades (2001–2020) in millions of metric tons (MT). The data of the Mineral Commodity Summaries 2022 – Silicon. US Department of the Interior, US Geological Survey.

production means direct emission up to 11.8 MT of CO_2 per year. Certainly, there are non-direct emissions as well, for instance, those related to heat generation. The carbothermic reduction is a very endothermic process with energy consumption of about 10.5 MWh per one ton of produced silicon.

The carbothermic reduction gives the metal-grade silicon (MG-Si), which contains a rather high content of impurities (Si – 96–99%, the rest – Fe, Al, B, P, Ca, Mg, Mn, Ti, C, etc.). To obtain pure silicon (e.g. solar grade silicon, SoG-Si), MG-Si is converted using hydrochloric acid to volatile trichlorsilane ($SiHCl_3$, the boiling point – 31.8 °C), which makes it possible to separate silicon from impurities by distillation in fractionating columns. To decompose the silane, it is passed through pure silicon bars heated to 1100 °C. The decomposition reactions will be discussed in Chapter 5.

The carbothermic deoxidation and purification requires large amount of energy and emits greenhouse gases. A fundamental challenge is to extract metals using electrons as absolutely clean reducing agents generated by sustainable renewable sources, in particular, by solar energy. Such task orientates electrochemical research toward new technologies of metals reduction in electrolytes as well as efficient and cost-competitive energy conversion and storage, e.g. solar cells, batteries, supercapacitors, and hydrogen production from water. Opportunity for electrochemical silicon reduction from oxide in high temperature molten salt is discussed in Chapter 8.

References

Greenwood, N.N. and Earnshaw, A. (1997). *Chemistry of the Elements*, 2e, 330. Elsevier Butterworth-Heinemann. ISBN: 0 7506 3365 4.

Kurkjian, C.R. and Prindle, W.R. (1998). Perspectives on the history of glass composition. *J. Am. Ceram. Soc.* 81: 795–813. https://doi.org/10.1111/j.1151-2916.1998.tb02415.x.

Macfarlane, A. and Martin, G. (2004). A world of glass. *Science* 305: 1407–1408. https://doi.org/10.1126/science.1093597.

Schmitt, G., Schütze, M., Hays, G. F., et al. (2009). Global needs for knowledge dissemination, research, and development in materials deterioration and corrosion control. The World Corrosion Organization.

Zulehner, W., Elvers, B., Hawkin, S. et al. (1995). *Ullmann's Encyclopedia of Industrial Chemistry*, 5e, vol. A23, 721–748. Weinheim: VCH.

Zulehner, W., Neuer, B., and Rau, G. (2000). Silicon. In: *Ullmann's Encyclopedia of Industrial Chemistry*. Weinheim: Wiley-VCH https://doi.org/10.1002/14356007.a23_721.

4

Physical and Chemical Properties of Silicon

We will briefly cover the physical and chemical properties of silicon, which are most relevant for electrochemical material science. We look at silicon as an electrode material to be applied in the studies of electrochemical processes.

Silicon (Si, atomic number 14) at room temperature is a crystalline brittle solid semiconductor with a blue–gray luster. Its melting point is 1414 °C (1687 K, 2577 °F) and the boiling point is 3265 °C (3538 K, 5909 °F). Density of solid silicon at room temperature is 2.329 g cm^{-3}; in a liquid state (at melting point), its density is 2.57 g cm^{-3}. Silicon band gap is 1.12 eV at 300 K, and the atomic radius is 1.18 Å. Silicon has a face-centered diamond-cubic crystal structure with the lattice constant $a = 5.43$ Å. The distance between the nearest two atom neighbors is $\sqrt{3a/4} = 2.35$ Å (Zhang 2004, p. 45). The list of other physical constants includes: the number of the atoms per unit cell – 8; the dielectric constant – 11.9; the diffusion constant for electrons – 34.6 cm^2 s^{-1}, and that for hole – 12.3 cm^2 s^{-1}; the intrinsic carrier density – 1.45×10^{10}.

Electric conductivity of silicon depends on temperature and impurities. Typically, an intrinsic conductivity of 4.3×10^{-6} Ω^{-1} cm^{-1} and resistivity of 2.3×10^{-5} Ω cm at 300 K are quoted (Zhang 2004, p. 45). Conductivity of silicon increases with temperature, and this property is at the heart of silicon electrochemistry in molten salts. While pure silicon is not active at room temperature, it transforms into active electrode at higher temperatures, such as those in high-temperature molten salts.

To increase conductivity, silicon materials are doped. Typically, phosphorous (P), antimony (Sb), or arsenic (As) is used as the donors of n-type conductivity. Boron (B), aluminum (Al), and gallium (Ga) are used as the acceptors of p-conductivity. For instance, resistivity of n-Si doped with phosphorous reduces from 3.8×10^3 Ω cm at the doping level 10^{12} cm^{-3} to 1.2×10^{-4} Ω cm at the doping level 10^{21} cm^{-3} and temperature 23 °C. Analogous values for p-Si with the boron dopant are 1.3×10^4 Ω cm and 1.3×10^4 Ω cm, respectively (Zhang 2004, p. 45).

Silicon is a tetravalent metalloid, which is a type of chemical element with the properties between metals and nonmetals. For instance, metalloids are brittle and less electrically conductive when compared with metals.

Silicon is the second most abundant element in the Earth's crust and eighth in the universe by mass. This element is mostly available as silica (SiO_2) or silicates – the

Silicon: Electrochemistry, Production, Purification and Applications, First Edition. Eimutis Juzeliūnas.

anions composed of silicon and oxygen with the general formula $\left[SiO_{4-x}^{(4-2x)-}\right]$ with x being 0, 1, or 2. It very rarely occurs in its pure form. The most common natural forms of silicon minerals are silica sand, clays, and stone.

The standard electrode potential for the Si/SiO_2 electrode is defined by the reaction

$$Si + 2H_2O \rightarrow SiO_2 + 4H^+ + 4e^- \tag{4.1}$$

The potential is $E° = -0.91$ V at temperature 25 °C and effective proton concentrations of 1 mol l^{-1}. This rather negative value implies that pure silicon is a rather active element with respect to oxidation. Thus, silicon readily oxidizes in water or humid atmosphere. However, the oxidation leads to the formation of a thin oxide film, which prevents oxidation and makes the material inert. This is true in acidic and neutral aqueous solutions. In alkaline solutions, the oxidation reaction is expressed as

$$Si + 6OH^- \rightarrow SiO_3^{2-} + 3H_2O + 4e^- \tag{4.2}$$

The standard potential of this reaction is as high as $E° = -1.697$ V. The electronegativity and the silicide formation indicate that silicon actively dissolves in alkaline medium.

Reaction (4.2) provides the basis for technologically important field – chemical silicon texturing. Chemical etching in alkaline solutions is a state-of-the-art process applied industrially. Micro-pyramid textures are produced by Si etching in KOH solution with addition of isopropyl alcohol and potassium silicate (K_2SiO_3) as well as solution of tetramethylammonium hydroxide (TMAH) (Basu et al. 2013; Khandelwal et al. 2013; Otto et al. 2015). Such etching yields the pyramidal textures, which show around 90% of light absorbance in visual and near IR spectral ranges due to light-trapping effect.

Silicon, when being in a passive state in acid solutions, dissolves in the presence of fluoride ions:

$$Si + 6F^- \rightarrow SiF_6^{2-} + 4e^- \tag{4.3}$$

Protective silica layer on silicon substrate is not stable in the presence of fluoride ions. Stability and protective ability of SiO_2 layer are determined by the following reactions:

$$SiO_2 + 6HF \longrightarrow SiF_6^{2-} + 2H^+ + 2H_2O \tag{4.4}$$

$$SiO_2 + 6F^- + 4H^+ \longrightarrow SiF_6^{2-} + 2H_2O \tag{4.5}$$

According to Pourbaix diagram, formation of the soluble hexafluoride SiF_6^{2-} is restricted by two pH boundaries at the SiO_2/SiF_6^{2-} interface. For instance, the hexafluoride solubility region is restricted by the region approximately pH 2 to 3.8 at the electrode potential $E > -1$ V vs. saturated hydrogen electrode and fluoride concentration 10^{-2} M (Osseo-Asare et al. 1996; Zhang 2004). The reaction (4.4) determines SiO_2 stability boundary at the low pH end, and the reaction (4.5) determines the stability at the high pH end. The stability boundaries also depend on silica crystallinity and crystallographic structure. Apart from SiF_6^{2-}, the presence of other silicon fluoride species, such as SiF and SiF_5^{2-}, can also be considered.

Silica instability in fluoride-containing solutions (4.4) and (4.5) forms the basis for silicon etchings technologies. For instance, a stain etching method is known, which is based on the action of hydrofluoric and nitric acid mixture (HF-HNO_3). The hydrofluoric acid removes from the surface silicon oxide and prevents its further formation, while the nitric acid oxidizes the Si and forms surface texture. Polished, textured, or porous surface structures can be obtained by this method. The light absorbance over 90% as well as luminescent nanostructures were achieved by the stain etching procedures (Cullis et al. 1997; Liu et al. 2014). Another example is the metal-assisted chemical etching (MACE). Such etching process is based on Si interaction with hydrofluoric acid in the presence of a strong oxidizer (e.g. hydrogen peroxide, H_2O_2) and a noble metal catalyst (e.g. Au, Ag) catalyst:

$$Si + H_2O_2 + 6HF \rightarrow 2H_2O + H_2SiF_6 + H_2 \quad (4.6)$$

The reaction yields the Si nanostructures of high-aspect ratio. The MACE process is discussed in Chapter 14.

Pure silicon surface contains unsaturated bonds, so-called dangling bonds, or surface states. The term surface states relates to the energy levels arising from the orbitals in Si surface plane where atoms cannot be surrounded by the tetrahedral symmetry, as that observed in the silicon bulk structure (Bard and Faulkner 2001, p. 751). The surface states are characterized by specific energies (e.g. in the band gap); they substantially influence the electronic properties of the surface junction. These surface atoms actively react with the chemical species, for instance, with environmental oxygen. Thus, silicon surface is newer and ideally pure; some layer of foreign species is an inevitable component under ambient conditions. In aqueous solutions, silicon atoms can react with hydrogen, hydroxyl, and oxygen. Typically, in water, silicon surface is covered by thin oxide film. Thickness of this film is up to several nanometers, depending on conditions. The chemical surface passivation when Si atoms are bonded to other atoms is referred to as termination. In HF solutions, hydrogen-terminated surface (Si–H) is formed. Some oxygen- and fluorine-terminated site could be present as well. One to three hydrogen atoms (mono-, di-, and tri-hydride) can terminate each Si atom depending on its position in the lattice. As assumed in classical electrochemistry, the idealized solid surface of an electrode is composed of various crystallographic singularities, such as adsorbed atoms (adatoms), kink-sites (semi-crystal positions), crystal steps, planes, and vacancies. (Vetter 1967). These positions have different activities (surface energy) and can accommodate different number of passivation species. The hydrogen-terminated surfaces are hydrophobic and luminescent. These surfaces are also amenable to further chemical modifications by various functional groups, e.g. addition of alkenes. Hydrogen termination occurs over whole pH range, including alkaline solutions under high etching rates. It is assumed that dipping in HF solution leads to termination by mono-, di-, and tri-hydride, whereas, in basic fluoride (pH 9–10), the surface is terminated by mono-hydride (Zhang 2004, p. 57). It is important to note that surface termination is modified by water rinsing after HF treatment. Water removes di- and tri-hydrides leaving mono-hydride termination behind. This effect is referred to as an atomic scale surface flattening. Fluorine termination in HF solution is typically very low when compared with hydrogen termination. However, content of the attached fluoride increased with fluoride concentration: over half surface was fluorine terminated when HF concentration in the electrolyte was increased by 50% or more (Takahagi et al. 1988).

Apart from hydrogen- or oxygen-termination, silicon surface can also adsorb electrolyte impurities. Metal cations can be reduced on Si surface, as its electrochemical potential is rather negative, that is, $E° = -0.91$ V (Eq. (4.1)). Si surfaces with chemical oxides were immersed into dilute nitric acid solutions of ten metals: Al, Ba, Ca, Cr, Cu, Fe, K, Ni, Sr, and Zn (Loewenstein and Mertens 1998). These metals were deposited on Si surface, and its content increased when pH was increased from 3 to 5.6. Some of them can occur due to physical or chemical adsorption.

Silicon surface can be hydrophobic or hydrophilic. Hydrophilicity arises from Si–OH termination, whereas hydrophobicity is caused by Si–H, Si–CH_x, or Si–F termination. Silicon dipped into HF solution becomes hydrophobic due to Si–H termination. Silicon electrode in a passive state (Si–SiO_2) is hydrophilic; however, it becomes hydrophobic when polarized negatively below the breakdown potential of passivity zone. Hydrophilic and hydrophobic properties can change due to aging: hydrophilic Si–OH surface changes to a hydrophobic one when it losses the hydroxyl groups. On the contrary, appearance of hydroxide groups on an aging hydrophobic surface leads to surface hydrophilicity.

Comprehensive materials on silicon surface chemistry, silicon-electrolyte interface, and interdisciplinary aspects are available in literature (Lehman 2002; Zhang 2004).

References

Bard, A.J. and Faulkner, L.R. (2001). *Electrochemical Methods: Fundamentals and Applications*, 2e. Wiley.

Basu, P.K., Sarangi, D., Shetty, K.D., and Boreland, M.B. (2013). Liquid silicate additive for alkaline texturing of mono-Si wafers to improve process bath lifetime and reduce IPA consumption. *Sol. Energy Mater. Sol. Cells* 113: 37–43. https://doi.org/10.1016/j.solmat.2013.01.037.

Cullis, A.G., Canham, L.T., and Calcott, P.D.J. (1997). The structural and luminescence properties of porous silicon. *J. Appl. Phys.* 82 (3): 909–965. https://doi.org/10.1063/1.366536.

Khandelwal, R., Plachetka, U., Min, B. et al. (2013). A comparative study based on optical and electrical performance of micro- and nano-textured surfaces for silicon solar cells. *Microelectron. Eng.* 111: 220–223. https://doi.org/10.1016/j.mee.2013.03.162.

Lehman, V. (2002). *Electrochemistry of Silicon. Instrumentation, Science, Materials and Applications.* Wiley-VCH Verlag GmbH & Co https://doi.org/10.1002/3527600272.

Liu, X., Coxon, P.R., Peters, M. et al. (2014). Black silicon: fabrication methods, properties and solar energy applications. *Energy Environ. Sci.* 7: 3223–3263. https://doi.org/10.1039/C4EE01152.

Loewenstein, L.M. and Mertens, P.W. (1998). Adsorption of metal ions onto hydrophilic silicon surfaces from aqueous solution: effect of pH. *J. Electrochem. Soc.* 145: 2841. https://doi.org/10.1149/1.1838723.

Osseo-Asare, K., Wei, D., and Mishra, K.K. (1996). Dissolution windows for wet chemical processing of silicon and silicon dioxide: potential–pH diagrams for the Si–F–H_2O system. *J. Electrochem. Soc.* 143: 749. https://doi.org/10.1149/1.1836512.

Otto, M., Algasinger, M., Branz, H. et al. (2015). Black silicon photovoltaics. *Adv. Optical Mater.* 3 (2): 147–164. https://doi.org/10.1002/adom.201400395.

Takahagi, T., Nagai, I., Ishitani, A., and Kuroda, H. (1988). The formation of hydrogen passivated silicon single crystal surfaces using ultraviolet cleaning and HF etching. *J. Appl. Phys.* 64: 3516. https://doi.org/10.1063/1.341489.

Vetter, K.J. (1967). *Electrochemical Kinetics: Theoretical Aspects*. New York: Academic Press Inc.

Zhang, X.G. (2004). *Electrochemistry of Silicon and its Oxide*. Kluwer Academic Publishers.

5

Silicon Refining: From Metallurgical-Grade to Electronic-Grade

Silicon is used in many industries without purification. A typical example is the production of ferrosilicon – the alloy of iron and silicon with a high silicon mass fraction, typically in the range of 15–90%. In photovoltaics (PV), a much higher purity is required. Solar-grade silicon (SoG-Si) is used in PV devices, which is as much as 99.9999 wt% (six nines) pure. In electronics, e.g. integrated circuit chips, semiconductor purity grade is necessary with silicon content of at least 99.9999999 wt% (nine nines). Satisfactory diffusion length of the carriers is attained at such purity. The requirement for the content of boron and phosphorous in the semiconductor grade silicon is less than 0.1 ppb (parts-per-billion). The ultra-pure material is commercially called polysilicon.

Silicon is extracted industrially by a carbothermic reduction. This is a conventional method of deoxidation of silicon dioxide (such as quartz, SiO_2) by coal in electric arc furnaces ($SiO_2 + 2C \rightarrow Si + 2CO$). Such process yields a metallurgical-grade silicon (MG-Si), which contains rather a high percentage of impurities (Si – 98–99%, the rest – Fe, Al, B, P, Ca, Mg, Mn, Ti, C, etc.). There are different standard grades of commercially used silicon for impurity contents (c, wt%); for instance, grade 553 predicts $c < 0.5$ of Fe and Al, and $c < 0.3$ of Ca (Habashi 1997). Typical content of boron and phosphorous in MG-Si is in the range of 20–40 ppm (Habashi 1997).

Ciftja, Engh, and Tangstad from the Norwegian University of Science and Technology (NTNU) as well as Fabry and Hesse from Wacker Chemie AG company reviewed broadly the methods of silicon refining and recycling (Ciftja et al. 2008; Fabry and Hesse 2012). Comprehensive research results on developments in silicon process technologies, resources, energy, and environmental issues could also be found in the book series "Silicon for the chemical and solar industry" (NTNU, Norway), which is published periodically as the Si-conference proceedings over three recent decades. Production of semiconductor-grade silicon as well as the routes from metal-grade to solar-grade silicon have been discussed in many publications, for example, by Ceccaroli and Lohne (2003).

While over one hundred technological processes have been proposed and tested, only a few of them have matured to a technology readiness level for industrial application. Major processes, which are viable at industrial and commercial scales, are the Siemens process, the Union Carbide process, and the Ethyl Corporation process. Other purification methods fall into the following categories: zone refining,

Silicon: Electrochemistry, Production, Purification and Applications, First Edition. Eimutis Juzeliūnas.

acid leaching, alloying process, directional solidification, refining of liquid MG-Si by evacuation, formation of volatile species, oxidation of impurities, slagging, and gas blowing. Methods based on inductive thermal plasma and treatment of waste from optical fibers wastes as well as recycling of sludge containing silicon were also well-reviewed (Ciftja et al. 2008).

Vast majority (~95%) of the SoG-Si is produced using MG-Si as a feedstock (Fabry and Hesse 2012). In this chapter, major methods of MG-Si upgrading to SoG-Si are briefly addressed. Emphasis is placed on the electrochemical Si purification in molten salt electrolytes, which is promising due to energy efficiency and environmental friendliness.

5.1 Purification Based on Direct Si Chlorination

Ultra-pure silicon is industrially obtained through the production of volatile silanes ($HSiCl_3$, $SiCl_4$), distilling them, and decomposing at high temperature on silicon rods. The method of precursor preparation for pure Si production is most widely used, which is based on direct chlorination of MG-Si by hydrochloric acid. The reaction yields gaseous Si compounds (Breneman 1987; Fabry and Hesse 2012; Dropka et al. 2006; Noda et al. 2004)

$$Si\,(s) + 3HCl\,(g) \Leftrightarrow SiHCl_3\,(g) + SiH_2Cl_2\,(g) + SiCl_4\,(g) + H_2\,(g) \tag{5.1}$$

The above reaction is a summary, which comprises several consecutive steps

$$Si + 3HCl \Leftrightarrow SiHCl_3 + H_2 \tag{5.2}$$

$$SiHCl_3 + HCl \Leftrightarrow SiCl_4 + H_2 \tag{5.3}$$

$$2SiHCl_3 \Leftrightarrow SiCl_4 + SiH_2Cl_2 \tag{5.4}$$

The concentration ratio of the yielded compounds depends on the reaction conditions. Thus, formation of the main products silicon tetrachloride (STC, $SiCl_4$) and trichlorsilane (TCS, $SiHCl_3$) is sensitive to impurities in the MG-Si: the impurities can catalyze or inhibit these reactions. For instance, manganese (Mn) decreases, whereas aluminum (Al) increases selectivity of $SiCl_4$ synthesis. Fabry and Hesse summarized the impurity effects on the selectivity of the TCS and STC synthesis (Fabry and Hesse 2012): (i) $FeCl_3$ and $NiCl_2$ promote the formation of $SiCl_4$; (ii) $CuCl_2$ catalyzes formation of dichlorsilane (SiH_2Cl_2); (iii) $MgCl_2$, $CaCl_2$, and $CrCl_3$ reduce the yield of $SiHCl_3$ but do not catalyze the formation of $SiCl_4$; (iv) $AlCl_3$, $TiCl_2$, and $ZnCl_2$ does not affect the reaction selectivity at starting temperatures. The reaction selectivity also depends on temperature: the yield of $SiCl_4$ decreases above 260 °C and selectivity of $SiHCl_3$ nearly to 100% could be achieved at temperatures around 400 °C when using pure silicon as a feedstock. Metal silicides (Me_nSi_m) are formed from the interaction between metal halides and silicon. The silicide phases reduce the yield of $SiCl_4$ and this decreases in the following sequence (Acker et al. 2000; Fabry and Hesse 2012): $Si > NiSi_2 > FeSi_2 > NiSi > FeSi > Fe_2Si > Ni_2Si$.

Thus, when chlorinating the silicon, two major competing reactions occur: formation of TCS (5.2) and STC (5.3). Both products are important during Si purification processes. Furthermore, the Siemens process will be discussed, which is based on TCS conversion, and the Union Carbide process, which utilizes STC as a feed gas.

5.2 The Siemens Process

The Siemens technology was developed in the late 1950s; this technology gained a dominant market spot for polysilicon production. Polysilicon market research specialists are of the opinion that nearly 90% of polysilicon production emanated from Siemens technology. Thus, the estimations by the company Wacker Chemie AG indicated that the Siemens process had a shear of 88%, as of 2012 (Fabry and Hesse 2012). Jäger-Waldau reported the analogous figure to be about 80%, as of 2009 (Jäger-Waldau 2012). Bernreuter stated that with respect to emerging Chinese polysilicon giants, the Siemens process will continue to take the lead in the market (Bernreuter 2020). The contribution of these low-cost giant plants to the global polysilicon production balance is more than 100 000 MT per year. As a result, the production costs will reduce to "unprecedented levels" (Bernreuter 2020).

The Siemens process is based on chlorination of MG-Si feedstock converting it to volatile TCS, its rectification and decomposition on silicon substrates used in the form of rods. The decomposition process is known as the chemical vapor deposition (CVD). The low boiling point and high volatility of TCS (T_b = 31.8 °C) make it possible to separate the silicon precursor from impurities by distillation in fractionating columns. In particular, boron and phosphorous can be removed from the silicon feedstock down to the ppb level using multiple rectification. High-purity TCS is vaporized and diluted with hydrogen (H_2) and the mixture is passed into deposition reactor where TCS is decomposed to Si on pure Si rods heated to around 1100 °C.

$$4HSiCl_3 \rightarrow Si + 3SiCl_4 + 2H_2 \tag{5.5}$$

The mechanism of the above summary reaction is rather complex; various reaction pathways with different intermediate products are possible depending on the reaction conditions such as temperature, pressure, flow rate, and the concentration ratio of the reactants $[H_2]/[HSiCl_3]$ (Dropka et al. 2006; Ciftja et al. 2008; Sirtl and Reuschel 1964; Fabry and Hesse 2012)

$$SiHCl_3 \rightarrow SiCl_2 + HCl \tag{5.6}$$

$$2SiHCl_3 \rightarrow SiH_2Cl_2 + SiCl_4 \tag{5.7}$$

$$3SiH_2Cl_2 \rightarrow SiH_4 + 2SiHCl_3 \tag{5.8}$$

$$SiH_2Cl_2 \rightarrow \mathbf{Si} + 2HCl \tag{5.9}$$

$$SiHCl_3 + H_2 \rightarrow \mathbf{Si} + 3HCl \tag{5.10}$$

$$SiHCl_3 + HCl \rightarrow SiCl_4 + H_2 \tag{5.11}$$

$$SiCl_4 + H_2 \rightarrow SiCl_2 + 2HCl \quad (5.12)$$

$$SiCl_2 + H_2 \rightarrow \mathbf{Si} + 2HCl \quad (5.13)$$

$$SiCl_2 + SiCl_2 \rightarrow \mathbf{Si} + SiCl_4 \quad (5.14)$$

The Si conversion is a CVD process; the deposition reactions (5.9), (5.10), and (5.13) are thermally induced on Si rods. The gas flow rate and temperature affect the desorption of HCl from Si surface and, thus, controls the Si deposition rate. Reaction (5.14) is a disproportionation reaction that is the reaction, in which Si^{2+} undergoes simultaneously the oxidation to Si^{4+} and the reduction to Si^0 steps. The TCS decomposition (5.6) gives STC as a byproduct. The reverse reaction (5.3), which yields TCS, is of importance in terms of economy and ecology. This reaction makes it possible TCS recycling and reusing the compound in the Si deposition reactor.

Figure 5.1 schematically outlines the basic principle of the Siemens reactor. The reactor is being composed of the silicon rods, on which silicon is deposited, e.g. (5.10), the inlet for reactants ($SiHCl_3/H_2$), and the outlet for reaction products ($SiCl_4$, HCl, etc.). The silicon rods are resistively heated to ~1100 °C.

Fabry and Hesse summarized the challenges, which give room for the process development and improvement: (i) the starting materials as well as the products are explosive, thus, safe handling could be an issue; (ii) purity and morphology of the deposits require continuous improvement; (iii) recycling and reutilization of by-products require permanent attention; of particular relevance are closed loop technologies; (iv) improvement of energy efficiency is also very important, as over 90% of the applied power is lost (Fabry and Hesse 2012). The seed rods are highly electrically resistant; therefore, high power supply is necessary to preheat the rods to the temperatures, at which resistivity of silicon is effectively reduced. Note also that the process operates in batches.

Bernreuter argues that only fluidized bed reactor (FBR) technology could "grab some percentage points" from the market share dominated by the Siemens technology (Bernreuter 2020). In terms of chemistries, the FBR technology is similar to the Siemens process. As in the Siemens technology, the FBR uses the trichlorosilane ($HSiCl_3$) as a precursor gas, and this is decomposed by physical vapor deposition (PVD) on silicon substrates at 1000 °C. The major difference between both technologies lies in the reactor configuration. The FBR technology uses the reactor, which is based on a tube through which the precursor gas and hydrogen are injected, and a fluidized bed is formed at the reactor bottom. The seed particles are fed through the top of the reactor. Granular silicon is produced and collected at the reactor bottom. Another feed gas used in FBR technology is monosilane (SiH_4). This gas decomposes at 650–700 °C, that is, at much lower temperature when compared with that needed to decompose $HSiCl_3$. It is estimated that the monosilane decomposition in the FBR process consumes about 10 times less electricity when compared with the energy, which is needed to decompose trichlorosilane in the Siemens process. Another FBR advantage is the possibility to produce granules, which can be withdrawn from the reactor continuously, whereas, in the Siemens technology, the process has to be interrupted to withdraw the produced silicon bars. The produced silicon is further

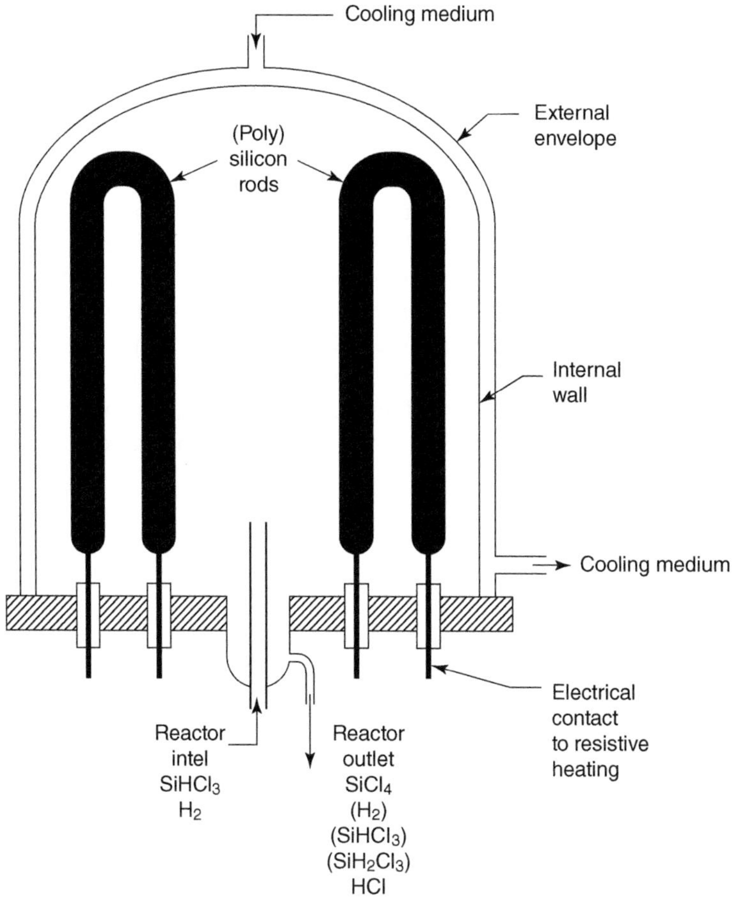

Figure 5.1 Schematic configuration of the Siemens reactor with the silicon rods, on which CVD of Si conversion occurs along with the inlet of the precursor ($SiHCl_3/H_2$) and outlet of the reaction products ($SiCl_4$, HCl). The silicon rods are electrically heated to ~1100 °C. Not shown are preheaters inside the reactor. Source: Reproduced with permission of publisher from (Ceccaroli and Lohne 2003). Copyright © John Wiley.

discharged into a crucible for melting. Granular polysilicon has a higher packing density than the chunks prepared from the bars produced by the Siemens process. Mixing of the produced granules with silicon chunks provide an opportunity to increase the charge weight by 30% (Bernreuter 2020). The application of the FBR will also be discussed below when introducing the ethyl process.

5.3 The Union Carbide Process

The specificity of the Union Carbide process lies in the application of monosilane (SiH_4) as a precursor for silicon formation by PVD. The process is named after the Union Carbide Corporation, which is a chemical corporation owned by the Dow

Chemical Company in the USA. To produce the precursor, MG-Si is used as a starting material, which is hydrochlorinated with STC in hydrogen atmosphere. The process begins with the reaction, which yields the TCS:

$$3SiCl_4 + 2H_2 + Si \rightarrow 4SiHCl_3 \tag{5.15}$$

This is an endothermic reaction with an enthalpy at 25 °C of 37 kJ mol^{-1} (Breneman 1987). Output of this reaction at the unit molar ratio of the reactants is about 20% of TCS when the temperature is 500 °C and a pressure of 35 bars are applied (Ciftja et al. 2008). The product is separated by distillation; the non-reacted STC is recycled back into the reactor for TCS production. The TCS is purified before hydrochlorination using rectification procedures. The high purity compound is further treated using quaternary ammonium ion-exchange resin or zeolite (a class of micro-porous aluminosilicate minerals) and amine-type catalysts, which promote formation of monosilane (Bailey et al. 1958; Litteral 1978, Kirsch and Laurent 1985):

$$2SiHCl_3 \rightarrow H_2SiCl_2 + SiCl_4 \tag{5.16}$$

$$2SiH_2Cl_2 \rightarrow H_3SiCl + HSiCl_3 \tag{5.17}$$

$$2SiH_3Cl \rightarrow \mathbf{SiH_4} + H_2SiCl_2 \tag{5.18}$$

The boiling points (T_b) of the produced compounds differ considerably: $T_{b\text{-}H_2SiCl_2} = 8.6°C$, $T_{b\text{-}HSiCl_3} = 31.8°C$, $T_{b\text{-}SiCl_4} = 57.6°C$, $T_{b\text{-}SiH_4} = -112°C$. This makes it possible to separate the products by distillation. The silicon precursor SiH_4 is purified by cryogenic distillation at very low temperature, as its boiling point is far below zero degrees Celsius. The high-quality precursor is thermally decomposed on heated silicon seed rods in a bell-jar reactor

$$SiH_4 \rightarrow 2H_2 + Si \tag{5.19}$$

The above reaction yields pure silicon at relatively low temperatures (370–450 °C); the method, therefore, is referred to as a "cold" conversion. This feature is advantageous when compared to the Siemens decomposition process, which demands much more energy in the reactor operation (~1100 °C). Another advantage of the Union Carbide process is that the decomposition of SiH_4 does not yield any corrosive by-products (5.19). Also, the efficiency of SiH_4 pyrolysis is very high, such that nearly complete decomposition can be attained; the process runs in a closed-loop (Ceccaroli and Lohne 2003). Among the disadvantages of this method is the high cost of distillation. Other detrimental factors include explosiveness and pyrophoricity of the precursor as well as the formation of silicon dust, which impedes the CVD process.

5.4 The Ethyl Process

The primary idea of the ethyl process is utilization of silicon fluoride (SiF_4) as a starting material rather than the MG-Si, which is used in the Siemens and the Union Carbide technologies. The Ethyl process was named after the Ethyl Corporation, which is now an owned subsidiary of NewMarket Corporation (USA) with almost one hundred years' experience in chemical manufacturing.

The SiF_4 precursor is produced from CaF_2, which is readily available as a by-product from the phosphate fertilizer – apatite. Apatite is a phosphate mineral-rich in OH^-, F^-, and Cl^- referred to as hydroxyapatite, fluorapatite, and chlorapatite with the formulas $Ca_{10}(PO_4)_6(OH)_2$, $Ca_{10}(PO_4)_6F_2$, and $Ca_{10}(PO_4)_6Cl_2$, respectively. Apatite raw material also contains silicates. When apatite is exposed to the concentrate sulfuric acid, calcium fluoride (CaF_2) is produced, which yields hydrofluoric acid (HF). The latter converts the silicates present in the raw material into $(SiF_6)^{2-}$, which further is converted into SiF_4 when drying at elevated temperatures. Note that large quantities of SiF_4 are available as a by-product in the fertilizer industry; therefore, application of this compound as a starting material for silicon production is economically effective.

The SiF_4 can be hydrogenated using several reactions, e.g. with sodium aluminum hydride (sodium alanate) (Marlett 1986)

$$SiF_4 + NaAlH_4 \rightarrow SiH_4 + AlNaF_4 \tag{5.20}$$

or lithium aluminum hydride (lithium alanate)

$$SiF_4 + LiAlH_4 \rightarrow SiH_4 + AlLiH_4 \tag{5.21}$$

The metal hydrides (alanates) are produced according to the following reactions

$$H_2 + Na + Al \rightarrow AlNaH_4 \tag{5.22}$$

$$H_2 + Li + Al \rightarrow AlLiH_4 \tag{5.23}$$

The byproduct – metal tetrafluoride – can be used as a valuable raw material in aluminum electrolysis.

The produced SiH_4 is purified by cryogenic distillation as already mentioned in the Union Carbide process. Chemistry of the silicon decomposition in the Ethyl process is analogous as that in the Union Carbide process: it bases on thermal decomposition of monosilane on silicon, which starts already at 370 °C (5.19), which is much lower than that in the Siemens process (~1100 °C). The distinctive feature, however, in the Ethyl process is the application of seed particles of silicon in a FBR. The product is obtained in the form of granules with an opportunity to tune the product particle sizes and forms. An advantageous feature is also that decomposition of the monosilane does not yield any corrosive by-products. Also, the efficiency of monosilane pyrolysis is high.

The methods discussed above are classified as chemical purification routes. There are also the methods, which are classified as metallurgical routes, which involve obtaining SoG-Si directly from metallurgical silicon. Braga et al. reviewed the principles as well as advantages and disadvantages of each alternative (Braga et al. 2008). The authors examined the developments of various companies worldwide: Wacker Chemie AG (Germany), Tokuyama Corporation (Japan), REC Silicon (headquartered in Norway with U.S based plants), Chisso Corporation (Japan), Elkem ASA (Norway), SINTEF (Norway), Energy Research Centre of the Netherlands

(ECN), Kawasaki Steel Corporation (Japan), Norwegian University of Science, and Technology (NTNU). The metallurgical route proposes the purification of MG-Si without chlorosilanes and hydrochloric acid, which are toxic and corrosive. Chlorosilanes are also explosive in the presence of water and hydrochloric acid. Also, the emission of chlorine is a challenge. The authors also claimed that the consumption of energy in metallurgical route is estimated to be by one-quarter lower when compared to the Siemens process. Fabry and Hesse summarized that in WACKER version of the Siemens process, a closed-loop is achieved; the process bases on only three main chemical reactions, which involve STC, TCS, HCl, and H_2 (Fabry and Hesse 2012). Hyper-pure silicon is obtained by this route, while the by-products chlorine and hydrogen are recycled. The capital expenditure (CapEx) of the process is also economically justifiable. The authors concluded that the Siemens process will maintain dominating positions "in the long term" due to maturity.

5.5 Electrorefining in Molten Salts

Molten salt electrolysis has the potential to separate silicon from metallic and non-metallic impurities. Electrochemical refining attracted considerable attention while attempting to develop technology of SoG-Si production from MG-Si feedstock, which would be less costly and a more energy-efficient alternative compared to the Siemens purification process. Another motivating factor of the research is an increased demand in technologies for waste handling of PV modules. Electrochemical treatment appears promising in reducing pollution by the hazardous metals contained in the modules. Note that major principles of electrorefining of liquid alloys were reported already in 1925 for Al refining known as the Hoopes process (Hoopes et al. 1925; Frary 1925).

The electrorefining bases on two major concepts: (i) Si made as anode of electrochemical cell dissolves at certain potential in molten salt electrolyte rather than the more noble impurities of the anode; (ii) Si is deposited on a cathode at certain potential rather than the less noble impurities of the molten electrolyte. Thus, by adjusting electrolysis conditions, for instance, the cell voltage, it is possible to upgrade MG-Si to higher purity. The principle is attractive as a relatively simple one in terms of technical operation and the opportunity to avoid the emission of hazardous volatile compounds. Note that two-electrode configuration is typically used to purify metals; however, higher purity could be achieved when using a three-electrode configuration. Such approach makes it possible to precisely control anode and cathode potentials with respect to a reference electrode enabling better control of the separation of impurities (Tao 2013).

An early work by Monnier and Giacometti reported the electrochemical Si refining in a solution of SiO_2 in cryolite (Na_3A1F_6) at 1000 °C (Monnier and Giacometti 1964). The authors demonstrated that silicon can be purified either from solid Si or from molten Si–Cu. Rather high purity of silicon was achieved by the electrochemical treatment in the range from 99.7–99.99%. Electrolytic refining of silicon and germanium was patented in 1965 (Monnier and Barakat 1965).

De Mattei et al. studied electrodeposition of silicon at temperatures above its melting point (De Mattei et al. 1981). They performed electrowinning of molten silicon at 1427 °C in $BaO–SiO_2–BaF_2$ melt. The percentage purity of 99.97% was obtained for the material. Spherical or ellipsoidal droplets have been formed during the processing, which solidified with a grain size of about 1 mm. We discuss subsequently (Section 15.4) that formation of Si micro-spheres in molten salt electrolytes could occur even at the temperatures below Si melting point due to the influence of metallic impurities: the melting point of contaminated Si could be significantly lower than that of pure Si.

Olson and Carleton (1981) as well as Sharma and Mukherjee (1986) lowered the bath operating temperature to 750 °C using KF–LiF-based electrolytes.

Johansen et al. proposed a high-temperature method for electrolytic refining of metals whose melting point is above ~1000 °C using either two or three layers configuration of the cell (Johansen et al. 2007). The two-layer cell comprises an upper molten electrolyte layer, which is reefed to as anode, and a lower molten alloy layer, which acts as a cathode. The latter is prepared from an alloy, which contains the metal to be purified, and at least one metal, which is nobler than the metal to be purified. The three-layer cell contains, in addition, an intermediate layer, which is prepared from a molten electrolyte having a density between the density of the upper and lower layers. Oxide-based electrolytes are employed: one electrolyte contains the oxide of the metal to be refined, and another oxide electrolyte has a melting point below the operating temperature of the process.

Olsen and Rolseth proposed a three-layer purification method, in which high melting point fluoride process was applied at the operating temperature above the melting point of silicon (1412 °C) (Olsen and Rolseth 2010). The electrochemical cell comprised the Si–Cu alloy at the bottom, which acted as an anode, the $CaF_2–BaF_2$ electrolyte at intermediate layer, and liquid Si at the top acted as a cathode, on which purified Si is deposited. Immiscibility and differences in densities of the liquid phases ensured the integrity of the layers: 3.5 g cm^{-3} of Si–Cu anode, 2.8 g cm^{-3} of $CaF_2–BaF_2$ electrolyte, and 2.57 g cm^{-3} of liquid Si. The method demonstrated high degree of purification removing all contaminants except for boron (B). According to the authors, this was caused by the similarity of thermodynamic properties of B and Si. Boron, however, shows higher affinity to specific transition metals and this could be exploited in the electrorefining process as electro-catalyzed precipitation of transition metal borides occurs at the cathode-electrolyte interface (Olsen et al. 2010). Precipitation of the following intermetallic compounds was detected after electrolysis: $FeB–Fe_2B$, $TiB_2–Ti_3B_4$, AlB_x.

Reduction of boron and phosphorous down to 2.4–4.3 ppmw, respectively, was achieved using the three-layer approach with liquid Al and Cu–Si electrodes in $K_3AlF_6–LiF–K_2SiF_6$ electrolyte at temperature below Si melting point, that is, at 950 °C (Lai et al. 2010). The anode was solidified from hypereutectic solution of Cu in MG-Si. The Cu-Si phase has been used as a filter for the electrochemical purification. High current densities produced Si particles embedded in the electrolyte, which aggregated into balls with diameter of ~2 cm of high purity.

Massot and coworkers achieved recovery of Si from Si–4.7Fe alloy by selective Si electro-dissolution in molten NaF–KF at 850 °C (Massot et al. 2013). The current efficiencies close to 100% and purity higher than 99.99% have been attained. Voltammetry showed that the dissolution potential of the common impurities, such as Ti, Mn, Cr, Ni, Mo, and Cu, was more anodic than that of Fe. Thus, the specific case of Si–Fe is an opportunity to extend the Si recovery approach to other metals, as well.

The fluoride-based electrolytes are very corrosive, and there are difficulties in removing the melt residue from Si deposits. To avoid such complications, attempts were undertaken to prepare the electrolyte from $CaCl_2$, which is well soluble in water, inexpensive, and nontoxic. MG-Si was dissolved in $CaCl_2$, NaCl, and CaO melt at 850 °C, and the possibility of electrorefining was demonstrated (Kongstein et al. 2007). Si anode, however, became passive during the polarization due to surface SiO_2 formation. Pulsed current was applied to overcome the problem. Nevertheless, significant passivation after ~24 hours of electrolysis has been observed.

A liquid anode prepared by alloying MG-Si with Cu was applied to prevent the anode passivation problems. Using of liquid Si–Cu electrode has been proposed in the early work of Monnier and Giacometti (1964). Cai et al. demonstrated an opportunity to use Si–Cu alloy as a purification filter, as concentrations of impurities (B, P, and Fe) in the Cu–Si alloy were higher than those in Si (Cai et al. 2012). When selecting an appropriate polarization, the elements more noble than Si (e.g. Cu, B, P) will not dissolve anodically, while less noble elements (e.g. Ca, Al) will not electrodeposit cathodically. By doing so, the content of boron was reduced from 36.0 to 0.27 ppmw and that of phosphorus from 25.0 to 0.89 ppmw. Such result was close to SoG-Si purity. Content of Fe, Al, and Ti was reduced by more than one order of magnitude as well. However, the content of Ca and Cu in the electrodeposit was higher than that in MG-Si. The authors suggested directional solidification process to remove these impurities, as their segregation coefficients in silicon are very small.

Oishi et al. proposed a refining concept, which is based on application of both the cathode and the anode in a liquid state (Oishi et al. 2016). Figure 5.2 outlines

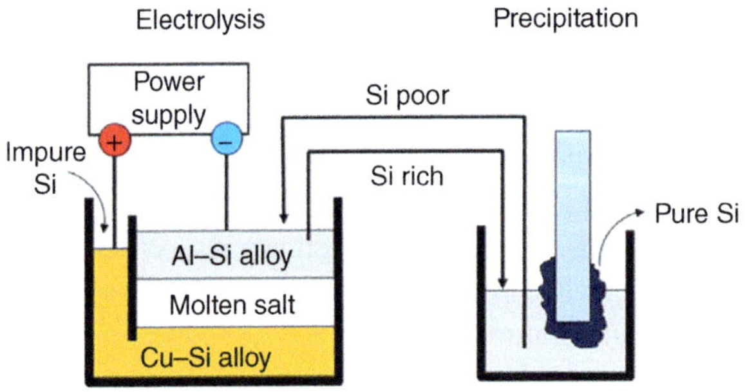

Figure 5.2 Conceptual drawing of the Si purification using Al–Si liquid cathode, Cu–Si liquid anode, molten salt electrolyte, and impure Si feedstock. Source: Oishi et al. (2016)/IOP Publishing / CC BY-4.0.

their conceptual drawing of such process. Low purity Si is added to liquid Cu–Si anode and Si is anodically dissolved along with the impurities less noble than Si. The Si and more noble impurities are deposited onto a liquid cathode like Al–Si. The Si-enriched liquid is transferred to another vessel where Si is obtained by precipitation, and the remaining liquid alloy is reused as a cathode material. The Al–Si cathode ensures proper separation between the melt and the cathode product at relatively low temperature (1000–1300 K). Continuous operation is possible, which is similar to that used in Al purification in the Hoopes cell. MG-Si and Si cells from commercial PV modules were used as feed material and purification has been attained, although the results did not meet the standard of SoG-Si. The treatment was in particular effective for Ag, which is used as the electrode on the front side of Si PV cell. The Ag concentration in recovered Si was reduced by four orders of magnitude with all the silver remaining in the liquid anode.

The energy consumption of Si purification using liquid Si–Cu anode and $CaCl_2$–SiO_2 melt was estimated as high as ~2 kWh kg^{-1} (Cai et al. 2012). The consumption in Siemens process is much higher: 65–160 kWh kg^{-1} (Fabry and Hesse 2012; De Wild-Scholten and Alsema 2004). These figures indicate potential of the electrorefining for industrial production of SoG-Si. However, industrially mature processes are still pending.

5.6 Zone Refining, Acid Leaching

High purity of silicon can be achieved by the zone refining method (Pfann 1952; Keck and Golay 1953; Pfann 1957; Pfann 1962). This technique has been introduced in the early 1950s of the last century to industrially produce important semiconductors such as germanium and silicon of extreme purity (Pfann 1952). The research began with germanium, which was urgently needed in ultrapure condition for transistor industry. Thereafter, zone refining has been applied to silicon, which became a dominant material in transistor industry. Zone refining is also applicable to a variety of other crystalline materials: metals, various organic, and inorganic compounds. The method is applied not only to purification purposes but also to precise control of materials composition.

The method bases on thermal induction of a short molten zone in a relatively long solid charge of the material to be purified; this small melted zone travels through the solid charge redistributing the solutes (impurities) in the charge. Two traveling liquid–solid interfaces take the impurities in and mix them with the liquid zone content by convection and diffusion. When freezing, a solid freezes out with different composition from that of the liquid phase. When traversing the liquid zone, the impurities are driven along the charge segregating at the end of the charge. Various distributions could be achieved depending on the size and number of zones, the rate of travel, as well as specimen configuration. The impurities, which lower the freezing point of the substrate material, travel with the molten zone to the end of charge. On the contrary, the impurities, which increase the freezing point, travel in the opposite direction accumulating at the beginning of the charge.

A crucible-free approach is typically applied when application of crucible is undesirable due to reaction of the liquid silicon with the crucible material, such as graphite or glass. For instance, liquid silicon actively reacts with the crucible materials and it becomes polycrystalline. The crucible-free approach employs the float zone arrangement when the specimen to be treated is prepared in the form of a bar and the liquid zone is induced using a heat radiator supplied with an induction coil.

The purification degree is increased, applying series of the molten zone traversing. When increasing the number of cycles, the charge end is gradually enriched by impurities; that is, impurity gradient is formed along the charge. A limiting distribution is reached, which cannot be exceeded by subsequent zone passes. Back-and-fort passes through the solid charge lead to the elimination of deviations from a uniform composition; thus, the molten zone equalizes the areas of high and low concentration of constituents.

While the zone refining is effective with respect to most impurities in silicon, some of them, for instance, boron is difficult to remove by this method. Restricted diameter of the charge (the ingot) poses another limitation in terms of large-scale applications.

Chemical refining of MG-Si is considered as a cost-effective and energy-efficient alternative of SoG-Si production when compared to the Siemens process. Metallic impurities are not homogeneously distributed within MG-Si, a major part of them concentrates at certain sites of polycrystalline structure. While most of the metals are highly soluble in liquid silicon, their solubility in the solid phase is much lower. Thus, when solidifying, the major part of impurities precipitate locally as silicide and intermetallic compounds at grain boundaries or interstitial positions of Si crystalline structure (Juneja and Mukherjee 1986; Ciftja et al. 2008). If solidification of MG-Si is rapid, the impurities can also be captured inside the grains. Rapid solidification makes it possible to obtain homogeneous product with even distribution of all components (Santos et al. 1990; Ciftja et al. 2008).

Mechanical crashing or pulverization of MG-Si leads to breakage at the grain boundaries exposing the contaminants at the surface of the produced particles. This opens up an opportunity to remove the contaminants by chemical leaching because most of the impurities readily dissolve in acids. Rather aggressive, hazardous, and corrosive media are applied to dissolve the contaminants: *aqua regia* ($HNO_3 + HCl$, optimal molar ratio 1 : 3), HF, H_2SO_4, etc. Prolonged treatment, which could take over 10 hours, as well as elevated temperature and high pressure is recommended for the leaching (Ciftja et al. 2008). Note that it is difficult to remove by acid leaching such impurities as boron, oxygen, and carbon. For instance, boron and phosphorous were removed by a more complicated route, which included the slag (Na_2O–CaO–SiO_2) treatment of molten silicon, solidification, and crashing using a ball mill and leaching in HCl–HF mixture (Meteleva-Fischer et al. 2012).

Various companies of silicon production have developed acid leaching processes: Elkem, Union Carbide, Alcoa, Heliotronic, Siemens, etc. The Elkem process is assumed to be mostly technologically viable with the industrial application (Ciftja et al. 2008).

References

Acker, J., Röver, I., Röwer, G., and Bohmhammel, K. (2000). Formation of silicides in the system metal-silicon-chlorine-hydrogen: consequences for the synthesis of TCS from silicon and dihydrogen chloride. In: *Silicon for the Chemical and Solar Industry V* (ed. H.A. Øye, H.M. Rong, L. Nygaard, et al.), 121–133. Tromso: Norwegian University of Science and Technology (NTNU).

Bailey, D.L., Shafer, P.W., and Wagner, G.H. (1958). Disproportionation of chlorosilanes on amine-type catalyst. US patent 2, 834, 648.

Bernreuter, J. (2020). Polysilicon production processes. Why the Siemens process is still the prevailing technology to manufacture polysilicon. https://www.bernreuter.com/polysilicon/production-processes (accessed 22 April 2022).

Braga, A.F.B., Moreira, S.P., Zampieri, P.R. et al. (2008). New processes for the production of solar-grade polycrystalline silicon: a review. *Sol. Energy Mater. Sol. Cells* 92: 418–424. https://doi.org/10.1016/j.solmat.2007.10.003.

Breneman, W.C. (1987). High purity silane and silicon production. US Patent 4676967A.

Cai, J., Luo, X., Haarberg, G.M. et al. (2012). Electrorefining of metallurgical grade silicon in molten $CaCl_2$ based salts. *J. Electrochem. Soc.* 159: D155–D158. https://doi.org/10.1149/2.041203jes.

Ceccaroli, B. and Lohne, O. (2003). Solar grade silicon feedstock. In: *Handbook of Photovoltaic Science and Engineering* (ed. A. Luque and S. Hegedus), 153–204. Wiley.

Ciftja, A.; Engh, T. A.; Tangstad, M. (2008). Refining and recycling of silicon: a review. Norwegian University of Science and Technology. https://www.researchgate.net/publication/267552614_Refining_and_Recycling_of_Silicon_A_Review/link/54732d5f0cf24bc8ea19c1ab/download.

De Mattei, R.C., Elwell, D., and Feigelson, R.S. (1981). Electrodeposition of silicon at temperatures above its melting point. *J. Electrochem. Soc.* 128: 1712–1714. https://doi.org/10.1149/1.2127716.

De Wild-Scholten, M. and Alsema, E. (2004). Towards cleaner solar PV: environmental and health impacts of crystalline silicon photovoltaics. *Refocus* 5: 46–49. https://doi.org/10.1016/S1471-0846(04)00225-2.

Dropka, N., Hoang, D.L., Kürschner, U. et al. (2006). Kinetic studies on TCS synthesis. In: *Silicon for the Chemical and Solar Industry VIII* (ed. H.A. Øye, H. Brekken, T. Foosnaes and L. Nygaard), 167–180. Trondheim: Norwegian University of Science and Technology (NTNU).

Fabry, L. and Hesse, K. (2012). Crystalline silicon feedstock preparation and analysis. In: *Semicond. Semimet.*, vol. 87, 185–261. https://doi.org/10.1016/B978-0-12-388419-0.00007-8.

Frary, F.C. (1925). The electrolytic refining of aluminum. *Trans. Am. Electrochem. Soc.* 47: 275–286.

Habashi, F. (ed.) (1997). *Handbook of Extractive Metallurgy*, vol. 4, 4861–1984. Weinheim: Wiley-VCH.

Hoopes, W., Frary, F.C., and Edwards, J.D. (1925). Electrolytic refining of aluminum. US Patent 1, 534, 318.

Jäger-Waldau, A. (2012). The photovoltaic industry. In: *Practical Handbook of Photovoltaics*, 2e (ed. A. McEvoy, T. Markvart and L. Castaner), 1153–1172. Waltham (USA): Academic Press https://doi.org/10.1016/B978-0-12-385934-1.00050-7.

Johansen, K., Sadoway, D.R., Myhre, B. et al. (2007). Method for electrolytic production and refining of metals. US patent, publication US20070215483A1.

Juneja, J.M. and Mukherjee, T.K. (1986). A study of the purification of metallurgical grade silicon. *Hydrometallurgy* 16: 69–75. https://doi.org/10.1016/0304-386X(86)90052-6.

Keck, P.H. and Golay, M.J.E. (1953). Crystallization of silicon from a floating liquid zone. *Physiol. Rev.* 89: 1297–1297. https://doi.org/10.1103/PhysRev. 89.1297.

Kirsch, W.B. and Laurent, S.M. (1985). Purification of halide. US patent 4, 557, 921.

Kongstein, O.E., Wollan, C., Sultana, S., and Haarberg, G.M. (2007). Electrorefining of silicon in molten calcium chloride. *ECS Trans.* 3: 357–361. https://doi.org/10.1149/1.2798679.

Lai, Y.-Q., Jia, M., Tian, Z.-L. et al. (2010). Study of the morphology evolution and purification of electrorefined silicon. *Metall. Mater. Trans. A* 41: 929–935. https://doi.org/10.1007/s11661-009-0154-1.

Litteral, C.J. (1978). Disproportionation of chlorosilane. US patent 4113845A, publication 1978-09-12.

Marlett, E.M. (1986). Process for production of silane. US patent 4, 632, 816.

Massot, L., Bieber, A.L., Gibilaro, M. et al. (2013). Silicon recovery from silicon-iron alloys by electrorefining in molten fluorides. *Electrochim. Acta* 96: 97–102. https://doi.org/10.1016/j.electacta.2013.02.065.

Meteleva-Fischer, Y.V., Yang, Y., Boom, R. et al. (2012). Slag treatment followed by acid leaching as a route to solar-grade silicon. *JOM* 64: 957–967. https://doi.org/10.1007/s11837-012-0383-4.

Monnier, R. and Barakat, D. (1965). Dual cell refining of silicon and germanium. US Patent 3, 219, 561.

Monnier, R. and Giacometti, J.C. (1964). Recherches sur le raffinage électrolytique du silicium. *Helv. Chim. Acta* 47: 345–353. https://doi.org/10.1002/hlca.19640470202.

Noda, S., Tanabe, K., Yahiro, T. et al. (2004). Reaction of Si with HCl to form chlorosilanes: time dependent nature and reaction model. *J. Electrochem. Soc.* 151: C399–C404. https://doi.org/10.1149/1.1737386.

Oishi, T., Koyama, K., and Tanaka, M. (2016). Electrorefining of silicon using molten salt and liquid alloy electrodes. *J. Electrochem. Soc.* 163: E385–E389. https://doi.org/10.1149/2.0391614jes.

Olsen, E. and Rolseth, S. (2010). Three-layer electrorefinng of silicon. *Metall. Mater. Trans. B* 41: 295–302. https://doi.org/10.1007/s11663-010-9362-8.

Olsen, E., Rolseth, S., and Thonstad, J. (2010). Electrocatalytic formation and inactivation of intermetallic compounds in electrorefining of silicon. *Metall. Mater. Trans.* 41: 752–757. https://doi.org/10.1007/s11663-010-9400-6.

Olson, J.M. and Carleton, K.L. (1981). A semipermeable anode for silicon electrorefining. *J. Electrochem. Soc.* 128: 2698–2699. https://doi.org/10.1149/1.2127340.

Pfann, W.G. (1952). Principles of zone-melting. *JOM* 194: 747–753. https://doi.org/10.1007/BF03398137.

Pfann, W.G. (1957). Techniques of zone melting and crystal growing. *Solid State Phys.* 4: 423–521. https://doi.org/10.1016/S0081-1947(08)60158-7.

Pfann, W.G. (1962). Zone melting. *Science* 135 (3509): 1101–1109. https://doi.org/10.1179/095066057790149822.

Santos, I.C., Gonçalves, A.P., Santos, C.S. et al. (1990). Purification of metallurgical grade silicon by acid leaching. *Hydrometallurgy* 23: 237–246. https://doi.org/10.1016/0304-386X(90)90007-0.

Sharma, I.G., and Mukherjee, T.K. (1986) A study on purification of metallurgical grade silicon by molten salt electrorefining. *Metall. Mater. Trans. B*, 17, 395–397. https://doi.org/10.1007/BF02655087

Sirtl, E. and Reuschel, K. (1964). Übber die Reduktion von Chlorsilanen mit Wasserstoff. *Z. Anorg. Allg. Chem.* 332: 113–123. https://doi.org/10.1002/zaac.19643320302.

Tao, M. (2013). Impurity segregation in electrochemical processes and its application to electrorefining of ultrapure silicon. *Electrochim. Acta* 89: 688–691. https://doi.org/10.1016/j.electacta.2012.10.148.

6

Silicon Electrowinning and Electrodeposition of Thin Layers

6.1 Electrodeposition in Molten Fluoride, Chloride, and Oxide Electrolytes

Electrodeposition of thin Si layers can be used to replace multiple processes of Si wafer fabrication, such as production of metallurgical- grad (MG)-Si, upgrading to solar-grade (SoG)-Si, ingot casting, and slicing. Traditional manufacturing includes highly energy-intensive purification of silicon, e.g. by the Siemens process. The production cost also increases due to significant material loss in the form of sawing dust, which is inevitable when sawing the ingot into wafers. Juzeliūnas and Fray recently reviewed the electrodeposition of thin Si layers from molten salt electrolytes (Juzeliūnas and Fray 2020).

Molten fluorides are prospective electrolytes for Si electrodeposition. A widely studied composition is a fluorosilicate (Na_2SiF_6 and K_2SiF_6) dissolved in molten FLiNaK salt – a ternary eutectic mixture composed of LiF–NaF–KF whose melting point is around 454 °C.

Fluorosilicic acid (H_2SiF_6) is a viable low-cost feedstock of silicon since this acid is a byproduct of phosphate fertilizer production from rocks, in which silicates occur together with phosphates (Elwell and Rao 1988; Xu and Haarberg 2013). Sodium or potassium salt (Na_2SiF_6 and K_2SiF_6) can be produced when neutralizing H_2SiF_6 with KF, KOH, or NaOH:

$$H_2SiF_6 + 2KF \rightarrow K_2SiF_6 + 2HF \tag{6.1}$$

$$H_2SiF_6 + 2KOH \rightarrow K_2SiF_6 + 2H_2O \tag{6.2}$$

$$H_2SiF_6 + 2NaOH \rightarrow Na_2SiF_6 + 2H_2O \tag{6.3}$$

Historical development of silicon electrodeposition from molten fluoride electrolytes was reviewed by Elwell and Rao (1988). The review shows that silicon

Silicon: Electrochemistry, Production, Purification and Applications, First Edition. Eimutis Juzeliūnas.

electrodeposition dates back to the mid-nineteenth century when DeVille[1] claimed in 1854 that the production of silicon by electrolysis is an impure melt of $NaAlCl_4$, and Ullik reported in 1865 the deposition of silicon from a solution of K_2SiF_6 in KF.

Cohen and Huggins studied the electrodeposition of silicon to solar cell fabrication in the 1970s (Cohen and Huggins 1976; Cohen 1977). Epitaxial and polycrystalline continuous films were electrodeposited from molten fluorides (K_2SiF_6, LiF, KF, and KF + HF) at 750 °C on Si, Ag, W, and Nb substrates using pure Si as an anode. The coatings were characterized as dense, coherent, and well adherent to Si as well as various metal substrates. Frazer and Welch showed the possibility of depositing silicon from silica dissolved in molten cryolite (Na_3AlF_6) (Frazer and Welch 1977).

Rao, Elwell, and Feigelson conducted in 1980s a series of experiments on electrowinning and electrocoating of silicon from molten fluoride systems (Rao et al. 1980, 1981a,b, 1982; Elwell and Rao 1982, 1984, 1988). They deposited coherent silicon films on a graphite substrate at temperatures around 850 °C in the LiF–KF melt with 8–14 M percentage of K_2SiF_6 fraction. They also assessed the semiconductor properties of the electrodeposits. Although the obtained samples were undoped, they exhibited n-type conductivity. The resistivity of the samples was $\rho \leq 3\,\Omega\,\text{cm}$, the carrier concentration was 10^{17} atoms per cm^{-3}, and their mobility was $100\,\text{cm}^2\,\text{V}^{-1}\,\text{s}^{-1}$ (Rao et al. 1982). The carrier mobility was increased by laser annealing with the power around 11 W. The authors claimed that the produced material is promising to meet the specifications of the SoG-Si layer for solar cell applications.

Controversial conclusions have been drawn regarding the Si reduction mechanism in fluoride-based electrolytes. Elwell and Rao applied stationary and rotating electrodes and concluded that electron transfer is the rate-controlling step of deposition (Elwell and Rao 1982). They also pointed out that diffusion and a chemical reaction, probably a dissociation of Si_nF_{2n+2}, cannot be neglected. Cai et al. deposited Si on steel and concluded the one-step, four-electron mechanism (Cai et al. 2011). The electron transfer was sufficiently rapid so that the process was controlled solely by diffusion. Electrochemical studies of Si reduction on Ag in KF–LiF–K_2SiF_6 at 550 and 800 °C led to a conclusion that the process is diffusion controlled with a two-step electron transfer ($Si^{4+} \rightarrow Si^{2+} \rightarrow Si$) (Osen et al. 2010; Haarberg et al. 2013). Formation of the ions of intermediate valence by a disproportionation reaction $Si + Si^{4+} \rightarrow Si^{2+}$ was also discussed (Rao et al. 1980, 1981a; Suzuki et al. 2019). This reaction would lead to silicon dissolution and, in turn, to non-coherent deposits. Formation of the ions of intermediate valence in LiF–KF–NaF electrolytes was also concluded from voltammetry studies: cyclic

1 Henri-Étienne Sainte-Claire Deville was born in 1818 in the island St. Thomas in Danish Virgin Islands. He studied medicine and received a degree in Paris in 1843. From 1845 to 1851, he was a professor of chemistry and a dean at the University of Besançon. In 1851, he was appointed professor of chemistry at the École Normale Supérieure in Paris. In 1853, he received professorship at the Sorbonne (the University of Paris). He worked on metallurgical processes and invented a process for obtaining aluminum from aluminum dichloride using sodium as a reducing agent. He obtained in a pure form silicon, magnesium, titanium, and platinum. https://www.britannica.com/biography/Henri-Etienne-Sainte-Claire-Deville

voltammetry (Boen and Bouteillon 1983), chronopotentiometry (De Lepinay et al. 1987), and square wave voltammetry (Osen et al. 2010; Haarberg et al. 2013). However, other authors, who performed voltammetry measurements in similar electrolyte systems, did not reach any conclusion concerning the existence of Si^{2+} (Elwell and Rao 1982; Bieber et al. 2012; Sakanaka and Goto 2015). The existence of Si^{2+} was also not concluded in the BaF_2–CaF_2 electrolyte with SiO_2 as a silicon source (Hu et al. 2013). Obviously, the reaction mechanisms need further investigations.

In spite of the cost and energy attractiveness, Si deposition from fluoride electrolytes still remains at an early stage of practical development. The problems encountered relate to the highly aggressive hydrofluoric acid (HF), difficulty in impurity control, severe Si pitting, as well as staining and clouding of the silicon surface. Diffusion of silicon into the substrate could also be an issue. Extensive diffusion of constituents of Inconel (Ni–Cr-based superalloy) and formation of the Fe_3Si layer when depositing Si on steel have been reported (Cai et al. 2011; Stern and McCollum 1985).

Silicon dioxide, also known as silica, is a very attractive material to be used as a precursor for silicon electrodeposition. This compound, most commonly found as quartz, is the second most abundant mineral in the Earth's crust. As a feedstock, this mineral is inexpensive and widely available in a rather pure form.

The oxide has been proposed as a source of silicon in the BaO–SiO_2–BaF_2 molten electrolyte (De Mattei et al. 1981a,b). The principal feature of this invention is operation at the temperatures above the silicon melting point, that is, in excess of 1420 °C, so that the silicon is electrodeposited in a liquid form. One component of the bath is an alkaline earth or alkali oxide. Barium oxide (BaO) has been proposed as a preferred component, which is decomposed at elevated temperatures from barium carbonate. It is assumed that other alkaline earth oxides, such as CaO or SrO, or alkali metal oxides, such as Na_2O, Li_2O, or K_2O, may be used instead of BaO. It is also assumed that SiO_2 reacts with BaO and forms barium silicate ($BaSiO_3$), which is electrochemically reduced when applying an appropriate voltage between an anode and a cathode placed into the melt. Another component of the molten electrolyte is a fluoride, preferably of the same metal as the oxide, for instance, BaF_2. This compound reduces the melting point of the bath, that is, its operating temperature, and the viscosity of the electrolyte, which facilitate dissolution and electrodeposition processes. Silica was melted with barium carbonate ($BaCO_3$) and barium fluoride (BaF_2). The production route includes the following steps: (i) preparation of a charge in certain proportions of SiO_2, $BaCO_3$, and BaF_2; (ii) heating up to about 400 °C in vacuum to remove moisture; (iii) supply of inert gas into the furnace, which is heated to about 1450 °C; $BaCO_3$ decomposes to BaO and CO_2, which is removed by the inert gas flow; (iv) graphite anode and cathode are inserted into the melt and voltage is applied; and (v) silicon is reduced in the form of droplets, which are removed from the bath and solidified. The current efficiency of 15% was attained with the silicon yield rate about 0.03 g h^{-1} of approximately 99.98% purity.

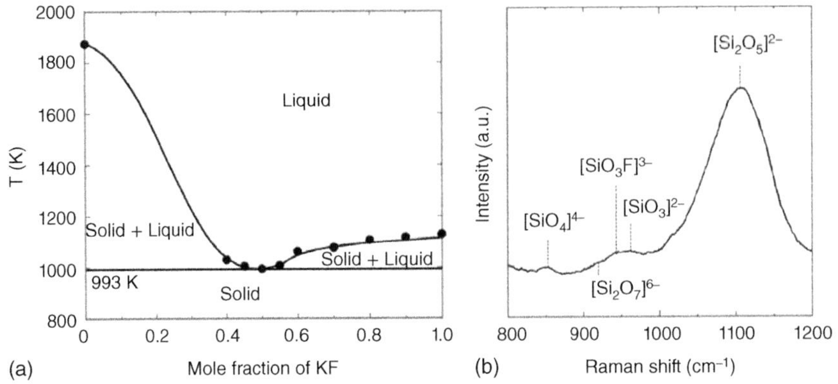

Figure 6.1 (a) Phase diagram of the SiO_2–KF binary system as determined by TG-DTA when heating continuously at a rate of 10 K min^{-1}. The phase diagram was obtained plotting the temperatures of endothermic peaks identified by the DTA; (b) Raman spectrum of the SiO_2–KF system obtained at the eutectic composition (50 : 50 mol%) and a temperature of 1073 K. Source: From (Suzuki et al. 2019), which is an open-access article distributed under the terms of the Creative Commons Attribution 4.0 License (CC BY). http://creativecommons.org/licenses/by/4.0/ (© The Authors, 2019).

Silica is soluble in other fluorides such as LiF–NaF–KF. The simplest electrolyte can be prepared by dissolving silica in molten potassium fluoride (KF). The phase diagrams and eutectic composition of this binary system have been studied by several authors (Konakov and Pivovarov 2003; Suzuki et al. 2019). The authors pointed out that within the composition region (0–0.33 mol% SiO_2), the characteristic of the interaction between the components is nearly ideal. Figure 6.1a shows the results of thermogravimetry and differential thermal analysis (TG-DTA), which revealed the interaction between silica and fluoride. The phase diagram indicates a single eutectic type with the eutectic composition at 50 mol% KF and temperature 993 K. Note that this temperature is lower by 990 K when compared to the melting point of single SiO_2.

Raman spectroscopy provides important information about silicon coordination structure in the molten salt electrolyte. Figure 6.1b shows the result obtained by high-temperature Raman spectroscopy at 1073 K in the range from 800 to 1200 cm^{-1} for the eutectic composition of the KF–SiO_2 system, which is at equal molar ratio (50 : 50 mol%). The wide band with the maximum at 1100 cm^{-2} indicates $[Si_2O_5]^{2-}$ in the electrolyte. This ion dominates over other ions; the intensity of the peak at 1100 cm^{-1} is much larger when compared to other peaks. These are observed at 850, 920, 950, and 960 cm^{-1} and can be assigned to $[SiO_4]^{4-}$, $[Si_2O_7]^{6-}$, $[SiO_3F]^{3-}$, and $[SiO_3]^{2-}$, respectively. The number of non-bridging oxygen atoms is 1 for $[Si_2O_5]^{2-}$, 4 for $[SiO_4]^{4-}$, 3 for $[Si_2O_7]^{6-}$, and 2 for $[SiO_3F]^{3-}$. Thus, a variety of species is produced when dissolving SiO_2 in the KF melt at the eutectic composition: the bond breakage of the Si–O–Si structure occurs, which leads to the formation of oxyfluoride and silicate ions.

The melting point is even lower when the KF electrolyte is supplied with LiF or LiF + NaF. Thus, the melting point of the LiF–KF system at a concentration ratio of 51.0 : 19.0 mol%, respectively, is 765 K and that of LiF–NaF–KF at 46.5 : 11.5 : 42.0 mol% is 730 K. The Raman spectra of these systems with the SiO_2 additive are similar to those of the KF–SiO_2 system (Suzuki et al. 2019). Such similarity implies that fluoride anions play a major role in the dissolution of SiO_2. Another finding of the authors lies in the important role played by oxygen ions (O^{2-}) in the SiO_2 dissolution process. Li_2O was added to the electrolyte as an oxygen source, which led to an increase in the formation of tetrahedron species $[SiO_3F]^{3-}$. A conclusion was drawn such that O^{2-} ions could cleave the bonds of silica or silicate ions and increase the Si precursor concentration in the electrolyte. Indeed, the reduction current increases when adding Li_2O into the electrolyte. While Si layers with a thickness of only a few micrometers were deposited in the Li_2O-free electrolyte, the layers over 100 μm were obtained when adding 5.0 mol% Li_2O to the electrolyte. The current efficiency was also enhanced by Li_2O at certain potentials from ~13% to ~50%. Note that the solubility of SiO_2 was enhanced by O^{2-} ions also in molten chloride electrolytes (Yang et al. 2017; Zou et al. 2017b). Note also that only Si traces were electrodeposited from the SiO_2 feedstock in the fluoride-based molten electrolytes, such as MgF_2–NaF, KF–BaF_2, KF–CaF_2, and KF–MgF_2 (Elwell 1981).

Zhuk reported the electrodeposition of Si coatings from the KF–KCl electrolyte with K_2SiF_6 additive as a silicon source (Zhuk et al. 2017). The author deposited polycrystalline silicon on graphite, glassy carbon, silver, and tungsten substrates. No intermediate phases were detected by Raman spectroscopy at the Si–C and Si–Ag interfaces.

Chloride molten salts have been recognized as promising electrolytes to deposit high-purity silicon. In particular, $CaCl_2$ has attracted considerable attention as an environmentally friendly, water-soluble, cheap, and widely available salt. Molten $CaCl_2$ exhibits rather a wide electrochemical window. The salt is soluble in water; the rest of it can be easily removed from the product by washing in water. Molten chloride requires higher operating temperatures when compared to fluoride; the $CaCl_2$ electrolyte operates typically at temperatures ~850 °C or higher. The Si layer was deposited in the molten $CaCl_2$ electrolyte at 850 °C using SiO_2 nano- or micro-sized particles as the precursors (Cho et al. 2012a,b; Yang et al. 2016). The authors added colloidal SiO_2 nanoparticles (NPs) (5–15 nm) to $CaCl_2$ and melted at 850 °C. Deposition of a crystalline discontinuous Si layer on Mo foil, n-type Si wafer, and glassy carbon was demonstrated. It is important that SiO_2 dissolves in the electrolyte due to interaction with the oxygen ions generated from CaO, which is added to the electrolyte or is inevitably present as an impurity. As a result, silicates are formed (e.g. SiO_3^{2-}, SiO_4^{4-}, and $Si_2O_7^{6-}$), which can further be electrochemically reduced to Si (Zou et al. 2017). The authors added $CaSiO_3$ to the $CaCl_2$–CaO electrolyte as a silicon precursor.

Yang et al. concluded that the reduction rate increased when using finer silica granules. They explained this effect by a more favored diffusion of released oxygen ions (O^{2-}) from smaller Si shells. Also, the effect of adding of pure Si particles from several micrometers to 1 mm on the electrochemical reduction rate of SiO_2 granules was reported (Zhong et al. 2018b). The question, however, regarding the deposition mechanism surfaces is: In which form the silicon precursors are present and reduced – the SiO_2 particles or the silicates formed due to SiO_2 particle dissolution? Both processes are distinctively different by electrochemical nature. The first route will be determined by direct electrochemical conversion of SiO_2 to Si releasing O^{2-}; the second route includes silica dissolution, silicate production, diffusion to electrode, and electrochemical reduction. Significant advantages have been achieved using calcium chloride electrolytes to deposit photosensitive silicon layers. These processes will be discussed in detail in the subsequent chapter, which is devoted to electrochemical formation of photoactive p–n junction.

While K_2SiF_6 and SiO_2 precursors have been most widely studied, an attempt to apply other precursors has also been undertaken. Thus, silicon tetrachloride ($SiCl_4$) was used to deposit silicon thin films in the LiCl–KCl eutectic melt electrolyte (Matsuda et al. 1996). The boiling point of this compound is 57.65 °C; it is almost insoluble in the electrolyte, so the silicon deposition occurred at the three-phase interface, which is composed of the Si substrate, the molten electrolyte, and the precursor in a gaseous phase. Electrochemical oscillation was detected as the deposition reaction progresses. The study shoved a principal opportunity to deposit silicon at the three-phase interface. Furthermore, the $SiCl_4$ precursor was applied in molten KF–KCl electrolyte (Yasuda et al. 2017a). Over 80% $SiCl_4$ dissolution efficiency was achieved in this melt. The authors deposited a silicon film from high-purity $SiCl_4$ dissolved in the KF–KCl eutectic composition

$$SiCl_4 + 6F^- \rightarrow SiF_6^{2-} + 4Cl^- \tag{6.4}$$

$$SiF_6^{2-} + 4e^- \rightarrow Si + 6F^- \tag{6.5}$$

However, the deposited film morphology was smooth only near to the substrate. High chloride concentration due to reaction (6.4) was concluded to be the main reason for the coarse silicon structure. Nevertheless, the approach is attractable in terms of purity, as high-purity precursor obtained by distillation can be used. Note that the well-known Siemens purification process employs gaseous species as well (HCl_4 and $HSiCl_3$). A specific feature of KF salt is its high solubility in water. The salt has much higher solubility when compared to other alkali and alkaline earth fluorides. For instance, the solubility of fluorides as grams per 100 g of water (298 K) is as follows: KF – 101.6, LiF – 0.13, NaF – 0.15, MgF_2 – 0.13, and CaF_2 – 0.0016.

The $SiCl_4$ precursor has also been applied to deposit silicon at low temperatures in non-aqueous solutions (Zeile et al. 2006; Liu et al. 2010). This precursor was also used for the preparation of Si–C composites for negative electrodes in Li-ion batteries by direct reduction with various alkali metal graphite intercalation compounds (Cahen et al. 2008).

The important electrolyte composition consisted of a mixture of molten chloride–fluoride salts (Boiko et al. 1985; Frolenko et al. 1992; Kuznetsova et al. 2009; Chemezov et al. 2012; Zaykov et al. 2015; Zhuk et al. 2017). Deposition of a photoresponsive n-type silicon film was also recently attained from KCl–KF–K_2SiF_6 at 650 °C (Peng et al. 2018). Even a more complex composition with the addition of potassium iodide (KCl–KF–KI) was studied using K_2SiF_6 (Laptev et al. 2020) or

$SiO_2 + K_2SiF_6$ (Zaykov et al. 2014) as Si sources. Laptev et al. deposited Si nanofilms under galvanostatic conditions; the films shoved good adhesion to a substrate made of glassy carbon or tungsten. The authors analyzed the mechanism of Si phase growth and concluded that the process being controlled by diffusion or charge transfer kinetics is the most important phenomenon of the deposition process. The authors also applied neutron transmutation doping (NTD) in order to produce a stable phosphorous isotope ^{31}P in the deposited films (Isakov et al. 2020). To this end, the Si films of 5 and 15 μm thickness were irradiated by a neutron flux density of $1.46 \times 10^{12}\ cm^{-2}\ s^{-1}$, duration of 168 hours, and temperature of 318 K. After the NTD processing, the number of the doping atoms ^{31}P was estimated in the film to be as high as 1.1×10^{14}.

6.2 Substrate Materials for Silicon Electrodeposition

Silver has widely been used as a cathode material to deposit the Si layer because Si easily nucleates on Ag and it does not dissolve in the substrate (Cohen and Huggins 1976; Cohen 1977; Rao et al. 1980; Boen and Bouteillon 1983; De Lepinay et al. 1987; Osen et al. 2010; Bieber et al. 2012). Dense, coherent, and adherent Si layers up to 3 mm thickness were deposited on the Ag substrate using K_2SiF_6 in fluoride melts (Rao et al. 1980). Pure Si layers up to 300 μm thickness were deposited on Ag in LiF–KF and LiF–NaF–KF eutectic mixtures, while using K_2SiF_6 as an additive (De Lepinay et al. 1987). Instantaneous nucleation with diffusion-controlled nucleus growth on Ag was determined using NaF–KF in the temperature range of 820–950 °C (Bieber et al. 2012). In recent years, Maeda et al. and Yasuda et al. obtained adherent, compact, and smooth Si films on the Ag substrate using the K_2SiF_6 precursor in a water-soluble KF–KCl molten salt (Maeda et al. 2014; Yasuda et al. 2016). A freestanding Si film was obtained when the Ag–Si system was washed in water. Other authors dissolved SiO_2 in the LiF–NaF–KF electrolyte at 600 °C and reduced silicon on silver electrode (Sakanaka and Goto 2015). In this work, an inert anode made of hercynite ($FeAl_2O_4$) instead of common graphite was used. The anode was prepared by sintering of Fe_2O_3 and Al_2O_3 powders.

The importance of the silver substrate has been emphasized when attempting to deposit photoresponsive Si films (Cho et al. 2012). The authors added colloidal SiO_2 NPs (5–15 nm) to $CaCl_2$ and melted the mixture at 850 °C. A discontinuous Si layer on Mo foil was deposited. The deposits were highly contaminated by carbon (up to 11 at.%) and did not exhibit a photocurrent. The situation changed substantially when the molybdenum substrate was replaced by silver. The authors electrodeposited the photoactive Si layers on a silver substrate (Cho et al. 2012). Si deposition on silver occurred by forming Si–Ag liquid droplets followed by supersaturation of Si in the droplet and precipitation of Si and segregation of impurities. The authors claimed 99.9 a.% purity of the Si product deposited on silver. The deposit consisted of micrometer-sized platelets or clusters of crystals, not a film. It showed some photopotential and photocurrent responses to xenon lamp irradiation in an aprotic solvent (acetonitrile containing ethyl viologen), whose reduction potential was within the band gap of Si. Note that the vapor–liquid–solid (VLS) deposition of Si from $SiCl_4$ gas is based on metal-catalyzed silicon growth as well. Precious metals (Au and Ag) are used as the growth agents in the VLS deposition.

Graphite is an attractive cathode material, which is stable in an inert atmosphere at high temperatures; it is also inexpensive and widely available. Rao et al. deposited well-adherent Si on graphite from LiF–KF–K_2SiF_6 at $745 \pm 5\,°C$ (Rao et al. 1981). A coherent Si layer of 0.2–0.5 mm thickness was deposited with a nodular or dendritic structure up to several millimeters in thickness. The impurity level (Cu, Fe, and Ni) was less than 0.005%. Boen and Bouteillon have deposited on a graphite cathode uniform Si layers up to 1 mm thickness by pulsed current electrolysis (Boen and Bouteillon 1983). The deposits exhibited semiconductor properties: resistivity was less than $1\,\Omega\,cm$ with the lifetime of free carriers $\sim 30\,\mu s$. Carleton et al. found instantaneous Si nucleation followed by three-dimensional growth on vitreous carbon and graphite (Carleton et al. 1983).

Crystalline silicon has also been studied as a cathode substrate material (Cohen 1976). The author obtained smooth epitaxial films on such substrate. The electrolytic growth on silicon on a parent substrate is of importance when seeking to form functional silicon surface architectures. A dense silicon layer was also deposited on the molybdenum substrate in BaF_2–CaF_2–SiO_2 melts at 1300 °C (Hu et al. 2013).

A variety of substrate materials has been studied in a KF- and KCl-mixed electrolyte with K_2SiF_6 as a silicon source (Zhuk et al. 2017). The authors electrodeposited Si continuous films on graphite, glassy carbon, silver, nickel, and tungsten substrates. They confirmed by Raman spectroscopy the absence of intermediate phases at the Si–C and Si–Ag interfaces; however, formation of two separate phases Si and Ni_2Si has been concluded when depositing silicon on the nickel substrate.

The above overview demonstrates the great efforts made by several groups to electrodeposit silicon thin layers with desired structural properties and functionalities. A great variety of experimental setups, such as electrolyte compositions, Si precursors, electrodes, and substrates, has been proposed and studied in detail. Nevertheless, technologically viable solutions on the industrial scale are still pending. An important feature to achieve is high-purity silicon, which is required for solar cell or electronic applications. Cost-effectiveness and energy efficiency are other challenging goals for researchers and engineers.

6.3 Electrodeposition of Photoactive Silicon and p–n Junction

Crystalline silicon is leading in solar cell production and, most likely, this material will maintain such position in the future photovoltaic market. The cost-effectiveness of solar energy utilization is a major issue; this is predicted to a large extent by the cost of silicon wafer production. A strategy for fabricating low-cost solar cells has been proposed, which focuses on the electrochemical deposition of thin photoresponsive Si layers and formation of p–n junction (Cho et al. 2012; Yang et al. 2016; Peng et al. 2018; Zou et al. 2019). Electrodeposition of thin Si layers could replace multiple processes of Si wafer fabrication, such as production of MG-Si, upgrading to SoG-Si, ingot casting, and slicing. Traditional manufacturing includes highly energy-intensive purification of silicon, e.g. by Siemens process. Production cost also increases due to significant material loss in the form of sawing dust, which is inevitable when sawing the ingot into wafers.

Recently, significant progress has been achieved in the demonstration of electrodeposition of crystalline silicon for low-cost photovoltaic applications (Zou et al. 2019). The electrodeposition process has been performed in calcium chloride–oxide electrolytes at 850 °C. Calcium chloride is attractive because of features such as environment friendliness, water solubility, low cost, and wide availability. Deposition of a photoresponsive *n*-type silicon film was also attained in a molten $KCl–KF–K_2SiF_6$ electrolyte (Peng et al. 2018). The operation temperature in this electrolyte was 650 °C, which is by 200° lower than that applied in the chloride–oxide electrolyte. Such temperature lowering relates directly to energy efficiency of the deposition process.

Cho et al. proposed the formation of a photoresponsive Si layer by direct reduction of silica NPs as a precursor (Cho et al. 2012a,b). Direct reduction of SiO_2 NPs in molten $CaCl_2$ is questionable. The $SiO_2–CaCl_2$ phase diagram shows that the double salt congruently melts at 798 °C (Wang and Morita 2016). Eutectic reaction occurs at 745 °C and ~27 mol% $CaCl_2$. The melting of NPs could occur even at lower temperatures due to well-known size-dependent depression of the melting point. Silica–silicate conversion is also possible:

$$SiO_2 + CaO \rightarrow CaSiO_3 \tag{6.6}$$

$$CaSiO_3 \rightarrow Ca^{2+} + SiO_3{}^{2-} \tag{6.7}$$

It is worth mentioning that CaO is inevitably present in molten $CaCl_2$ due to hygroscopic properties of the salt. Hydrolysis of $CaCl_2$ leads to the formation of CaO, which has always been observed in practice even in a prudently pretreated molten electrolyte. Thus, melting and dissolution of nanosilica in a $CaCl_2$ electrolyte is very likely. Calcium oxide (CaO) was intentionally added to the $CaCl_2$ electrolyte in order to facilitate SiO_2 dissolution and subsequent Si reduction (Zou et al. 2017).

The p-type silicon was deposited from the molten $CaCl_2$ electrolyte with a SiO_2 precursor (Cho et al. 2012a,b). The p-type conductivity was attained without adding to the electrolyte any common p-dopant, such as boron or aluminum. The p-type conductivity has been confirmed by the cathodic photocurrents, which are typical of a p-type photoelectrode. Titanium, which was a highest trace impurity in the SiO_2 precursor, was concluded to be a p-type dopant in the deposit. The Ti-to-Si ratio was determined by secondary ion mass spectroscopy; the Si/Ti count ratio was measured as high as 2500 : 1. Boron and aluminum were not detected in the deposit.

A photoresponsive nanostructured Si layer was obtained from silica powder in molten $CaCl_2$ at 900 °C (Xiao et al. 2013). The reduction mechanism was considered in terms of cathodic silica electrodeoxidation, carbon (di)oxide formation on graphite anode, and silicate formation followed by cathodic reduction to Si. The Si dispersed in chloroform exhibited an absorbance peak in the ultraviolet and visible zone and showed photoluminescence emission maximum at about 308 nm.

A rather thick macroporous Si structure (~10 μm) was obtained on the p-Si substrate in the cathodic region of Ca deposition in molten $CaCl_2$ (Juzeliūnas et al. 2016). The photoactivity of the structure was studied in a 0.1 M KCl + 0.05 M methyl viologen solution under white light irradiation. The surface exhibited enhanced photoactivity when compared with that of pristine wafer, which was evident from much higher photopotential and photocurrent responses. The effect was explained by internal light trapping by the porous structure.

Dense and uniform photoresponsive silicon films were electrodeposited from a molten KCl–KF–K_2SiF_6 salt mixture at 650 °C (Peng et al. 2018). Tin-doped films were formed by adding to the electrolyte minor amounts of tin (0.020–0.035 wt%). The films exhibited n-type conductivity and generated a photocurrent, which was about 38–44% of that generated by commercial Si wafer.

Zou et al. formed Si films and p–n junctions by electrochemical reduction of silicates (6–7) using $CaSiO_3$ and SiO_2 NPs as precursors (Zou et al. 2017). The basic principle of the electrochemical formation of Si p–n junction is shown in Figure 6.2. Calcium oxide was added to the electrolyte to convert SiO_2 into silicates, e.g. ${SiO_3}^{2-}$ and ${SiO_4}^{4-}$. Subsequent reduction of silicates yielded the Si layers 5–50 μm in thickness. The authors demonstrated that pulsed current deposition favors the formation of dense and smooth silicon films. The faradaic efficiency for the silicon film formation was evaluated as high as ~60%; the remaining 40% efficiency was attributed to the formation of silicon powder. Different film morphologies could be obtained when tuning the current densities, potentials, or pulse conditions, whereas a mirror finish was attained by surface polishing. The conductivity of n-type was attained when co-depositing Sb from Sb_2O_3, while the p-type conductivity originated from the Al non-controllably released from a quartz crucible. The p–n junction exhibited photocurrent and photovoltage responses; the authors qualified the photoperformance as modest.

Zou et al. recently reported another advantageous study, which supports a promising strategy for low-cost silicon solar cell production by means of molten salt electrochemistry (Zou et al. 2019). The authors proposed an electrochemical process with the claimed purity of silicon close to 6N obtained from silicon dioxide in molten calcium chloride. An important feature of the proposed approach is use of calcium oxide as an intermediate for ionization of silicon dioxide and, in turn, formation of silicate ions, from which silicon is reduced. It is worth noting that calcium oxide shows considerable solubility in molten calcium chloride. A distinctive difference of the approach from that outlined in Figure 6.2 is avoiding silicate as a starting material; thus, SiO_2, CaO, and $CaCl_2$ were used as input materials.

The authors consider that the overall dissolution–reduction process occurs according to the following reactions (Figure 6.3):

$$x\mathrm{SiO_2} + y\mathrm{CaO}(\mathrm{Ca^{2+}}, \mathrm{O^{2-}}) \rightarrow \mathrm{Ca}_y\mathrm{Si}_x\mathrm{O}_{(2x+y)}\ \left(y\mathrm{Ca^{2+}}, \mathrm{Si}_x\mathrm{O^{2y-}}_{(2x+y)}\right) \tag{6.8}$$

$$\mathrm{Si}_x\mathrm{O^{2y-}}_{(2x+y)} + 4xe^- \rightarrow x\mathrm{Si} + (2x+y)\mathrm{O^{2-}} \tag{6.9}$$

The Si layers of 10 nm thickness were deposited onto graphite as a substrate (Zou et al. 2019). Graphite was also used as an anode material. The purity of the deposits was studied by glow discharge mass spectrometry (GDMS), which showed

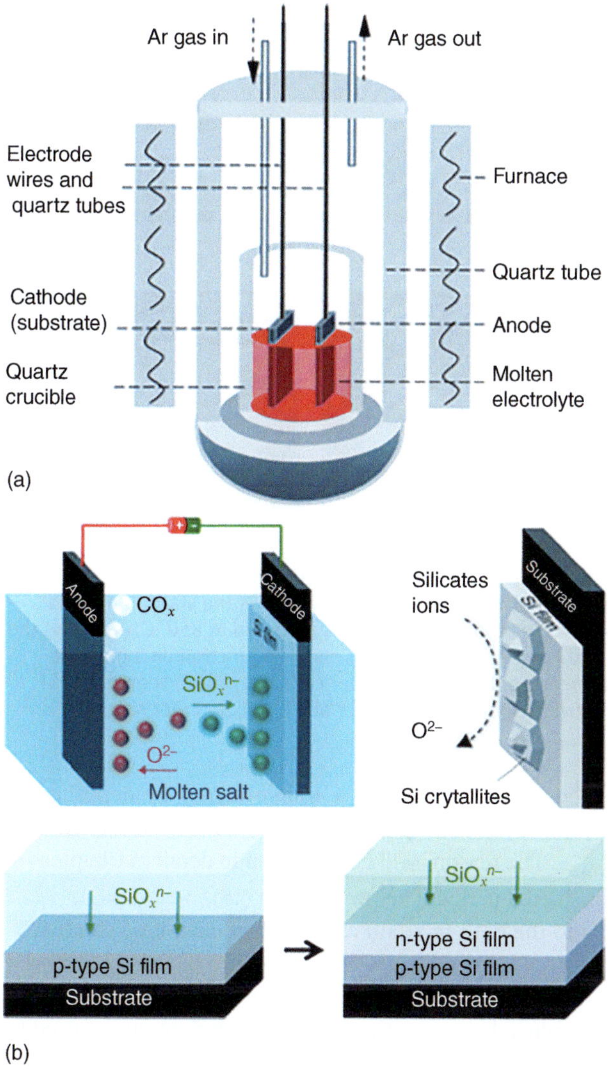

Figure 6.2 Basic principle of electrochemical formation of the Si p–n junction: (a) Molten salt cell with graphite plates as the cathode and anode and a molten electrolyte. The electrolyte composition was typically 100 g of $CaCl_2$, 1.0 g of $CaSiO_3$, 1.8 g of CaO, and 1.4 g of SiO_2. The reagents were poured into a closed quartz crucible and melted at 850 °C in an argon atmosphere. (b) Outline of two-step formation of n-type and p-type films. Source: Reprinted with permission from Zou et al. (2017b). Copyright 2017, American Chemical Society.

Ti, Cu, Ni, Cr, and Fe having the most impact on solar cell efficiency. It has been concluded that the impurity contents were below the tolerable threshold. However, other impurities, such as P, B, Ca, and C, were not analyzed. In particular, carbon and calcium are important in terms of contamination. Carbon could be present due to reduction on the cathode of CO_x, which originates from the interaction

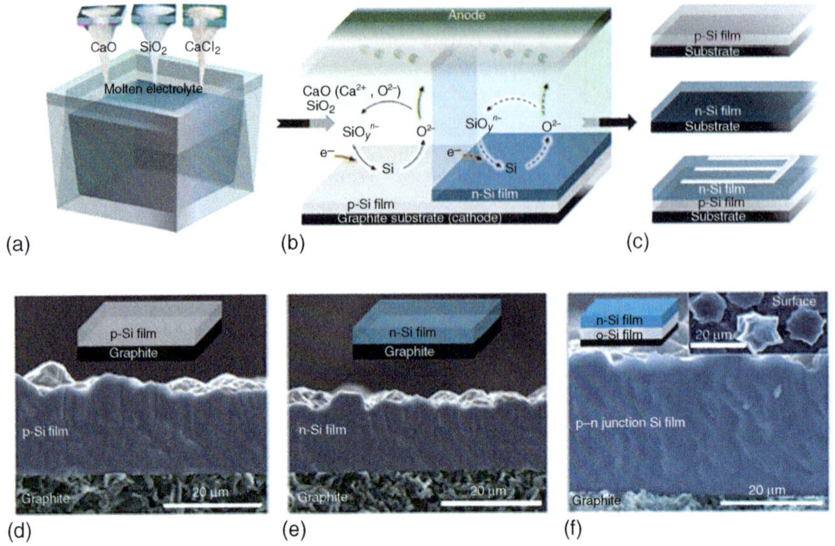

Figure 6.3 Schematic representation of the Si electrodeposition (a–c) and SEM images of the cross sections of the electrodeposited Si layers (d–f): (a) shows the starting materials CaO, SiO_2, and $CaCl_2$, from which the molten electrolyte is prepared at 850 °C; (b) shows the electrodeposition of p-type and n-type layers as well as the p–n junction on a graphite substrate; (c) outlines the schematic of the possible solar cell; (d, e, and f) show the p-Si, n-Si, and p–n junction, respectively. Source: From Zou et al. (2019), Springer Nature. Licensed under CC BY 4.0.

of the applied graphite anode with oxygen (O^{2-}). Calcium may be deposited at underpotential conditions, that is, at more positive polarizations than the calcium equilibrium potential $E^0_{Ca/Ca^{2+}}$. The reasons will be discussed in detail in Chapter 16. Purity concerns will be considered in detail below (Section 6.5).

Zou et al. added dopants to the bath to attain either p-type or n-type conductivity of the silicon layer (Zou et al. 2019). Electrodeposition of p-Si can be achieved when alumina or boric anhydride powders are added to the molten salt electrolyte. For deposition n-Si, calcium phosphate or antimony oxide powders can be used to provide P or Sn as a dopant. The electrodeposited films exhibited the photocurrents, which were up to 50% of those induced by a commercial silicon wafer.

The electrodeposition of p–n junction was demonstrated as a proof of concept: the solar devices constructed from the prepared films showed the power conversion efficiency to be around 3.1%. The approach proposed the route on how to greatly reduce the capital cost and energy consumption for solar cell production and other applications.

6.4 Electrodeposition of Silicon from Ionic Liquids and Organic Solvents

Ionic liquids (ILs) also called "room-temperature molten salts" are the materials composed of ionic species: an organic cation and a complex anion that are liquid

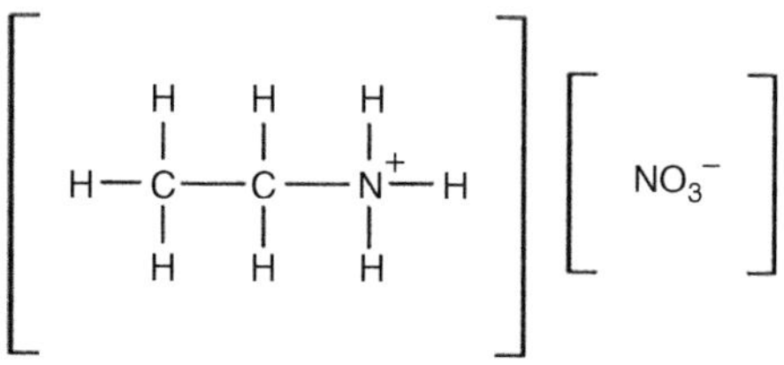

Figure 6.4 Structural formula of the ethylammonium nitrate ($C_2H_8N_2O_3$) – the first ionic liquid, which was synthesized by Paul Walden in 1914.

at relatively low temperatures (arbitrarily below 100 °C (Endres et al. 2008)). According to Beyersdorf et al. an incredible number 10^9 to 10^{18} of ILs, binary and ternary mixtures, are – theoretically – achievable (Beyersdorf et al. 2008). The cations typically are bulky asymmetric compounds, which could be combined in the IL with a great variety of anions, such as inorganic ions (e.g. halides, hexafluorophosphate PF_6^-, and tetrafluoroborate BF_4^-) or complex organic species such as methylsulfonylimides or lactates. The wide range of combinations allows great diversity of ILs with a wide choice of physical and chemical properties for specific applications. The number of ILs is assumed to be as high as one million or higher when taking only the simplest IL compositions into consideration; a much higher number is assumed for binary and ternary compositions.

The beginning of the history of ILs is linked to Paul Walden's[2] report published in 1914 on the synthesis of ethylammonium nitrate (Walden 1914). After one century, more precisely, during the past two decades, the ILs developed to a major scientific area with an exponential increase in the number of publications from a few in 1996 to over 5000 in 2016 (Lei et al. 2017). The authors claim that such annual growth rate exceeds the rate of other popular scientific areas (Figure 6.4).

The primary advantage of ILs for electrochemical applications is a wide electrochemical window (up to 6 V), which makes it possible to supply a higher polarization to electrode than in the aqueous solutions limited by water decomposition. Therefore, electroreduction of electronegative metals (Al, Mg, Ta, Nb, Mo, and W) could be attained in ILs. The other advantages of ILs for electrochemical applications include good electrical conductivity and possibility to avoid hydrogen evolution owing to the aprotic nature of most IL-based electrolytes.

It is widely recognized that ILs provide high health and safety standards owing to insignificant vapor pressure, non-flammability, and biodegradability. The ILs due to low vapor pressure are being considered as environment-friendly replacements for industrial volatile compounds. However, there is some paradigm shift observed from the oft-quoted "nonvolatile, non-flammable, and air and water stable" to that of "distinctly volatile, flammable, and unstable" (Lei et al. 2017). The toxicity of the ILs to aquatic organisms is also a concern (Pham et al. 2010). The authors

2 Paul Walden (Pauls Valdens; Paul von Walden) was born in 1863 near Cēsis (ger. Wenden) in Latvia. He started chemical engineering studies in 1882 in the Riga Technical University and received his doctorate in 1891 at the University of Leipzig. Paul Walden and Wilhelm Ostwald published their first joint work in 1887, which focused on electrical conductivity depending on the molecular weight of salts. The interested reader can find further details on Walden's research and biography in relevant biographical literature. A brief overview written by Walden's contemporary P. Günter could be found in Angew. Chem. 1933, 46, 497. https://onlinelibrary.wiley.com/doi/epdf/10.1002/ange.19330463002

emphasized the importance of the knowledge of IL behavior in the terrestrial environment, in particular, the processes of microbial degradation, sorption, and desorption in soil and aquatic milieu as possible recipients of ILs. Amde et al. addressed the environmentally relevant issues of ILs as well (Amde et al. 2015). The authors discussed the management aspects of potential hazards, their environmental behavior, and fate effects. Nevertheless, ILs remain to be desirable materials as "green" alternatives to volatile organic solvents. Due to tuning opportunities, ILs are also referred to as "designer solvents."

Economically, applications of the ILs have limitations due to rather high production cost of these materials. In this respect, methods of recycling and reuse of ILs are important. These methods were properly surveyed (Mai et al. 2014).

Various works have been published in the field of ILs in recent decades. Interested readers are referred to comprehensive reviews published in a topical issue of *Chemical Reviews* devoted to ILs Lei et al. (2017). The reviews published in this issue cover recent achievements in numerous fields related to IL research and applications; to mention a few: nanoconfined ILs, fuel production in ILs, catalytic applications, selective oxidation, extraction and separation processes for bioactive compounds, vibrational spectroscopy, chemical stability, and biological activity, applications in pharmaceutics and medicine as well as in energy storage and conversion. Application of ILs as anticorrosion agents to protect metals and alloys has also been recently reviewed (Verma et al. 2021). The properties and applications of ILs have also been surveyed in earlier reviews (Seddon 2003; Endres et al. 2008; Olivier-Bourbigou et al. 2010; Amde et al. 2015). The electrodeposition of Si in IL electrolytes was also reviewed (Juzeliūnas and Fray 2020).

Most studies of silicon electrodeposition from the ILs were based on the bis(trifluoromethylsulfonyl)amide [TFSA] (sometimes called as – imide [bis(trifluoromethylsulfonyl)imide, TFSI]). Many authors constituted the electrolyte using $[TFSA]^-$ as an electrolyte anion: (Katayama et al. 2001; Zein El Abedin et al. 2004; Nishimura et al. 2008; Martinez et al. 2010; Fournier and Favier 2011; Komadina et al. 2013; Tsuyuki et al. 2015; Park et al. 2013; Zhang et al. 2016; Shah et al. 2017; Thomas et al. 2018; Tsuyuki et al. 2018).

The molecular structure in Figure 6.4 shows how atoms are arranged in the $[TFSA]^-$. Apart from $[TFSA]^-$, studies have been performed using other anions, such as trifluoromethylsulfonate $[TFO]^-$ or tris(pentafluoroethyl)-trifluorophosphate $[FAP]^-$ (Pulletikurthi et al. 2013). The molecular structure of these anions is shown in Figure 6.5. Inorganic anions, such as chloride (Nishimura et al. 2009) or hexafluorophosphate (PF_6^-) (Shah et al. 2017), have also been applied.

A variety of organic cations have been used in the synthesis of IL electrolytes for Si deposition: 1-ethyl-3-methylimidazolium [EMIm] (Katayama et al. 2001), *N*-butyl-*N*-methylpyrrolidinium [BMPyrr] (Zein El Abedin et al. 2004; Martinez et al. 2010; Park et al. 2013; Thomas et al. 2018), *N*-trimethyl-*N*-hexylammonium [TMHA] (Nishimura et al. 2008; Komadina et al. 2013; Tsuyuki et al. 2018), *N*-ethyl-*N*-methylpyrrolidinium [EMPyrr] (Nishimura et al. 2009), 1-butyl-3-methylpyridinium [BMPy] (Martinez et al. 2010; Zein El Abedin et al. 2004; Park et al. 2013; Thomas et al. 2018), tributyl(methyl)ammonium [TBMA] (Zhang et al. 2016), and 1-butyl-3-methylimidazolium [BMIm] (Shah et al. 2017; Zhao et al. 2018). The molecular structure of the above organic cations is given in Figure 6.5.

Silicon tetrachloride ($SiCl_4$) is a typical precursor introduced into an IL electrolyte for Si deposition. $SiCl_4$ is a colorless volatile liquid, whose boiling point is rather low $T_b = 57.65\,°C$ (330.80 K). We discussed in previous chapters the application of this compound in industrial production of highly pure silicon. Other Si precursors included the hexafluorosilicate (SiF_6) (Katayama et al. 2001) and the silicon tetrabromide ($SiBr_4$) (Martinez et al. 2010).

Katayama et al. succeeded in depositing only a very thin layer of silicon from 1-ethyl-3-methylimidazolium hexafluorosilicate at an elevated temperature up to 90 °C (Katayama et al. 2001). The deposit, however, was completely converted to silica when exposed to ambient air. Voltammetry showed possibility to deposit silicon from $SiCl_4$ in 1-butyl-1-methylpyrrolidinium bis(trifluoromethylsulfonyl) amide, the air- and water-stable IL (Zein El Abedin et al. 2004). However, only a thin layer, few hundred nanometers in thickness, composed of 50–200 nm spherical crystallites was deposited on the gold substrate. Si nanoislands, which agglomerate into larger formations, were deposited and observed by scanning tunneling microscopy (STM) on the Au(111) substrate (Borisenko et al. 2006). NPs of silicon in the range of 100–500 nm were deposited on copper from the IL as well as propylene carbonate as the electrolyte solvent (Schmuck et al. 2010). Purity of the deposits appeared to be the main source of concern, particularly due to alloying of NPs with a foreign metal (Au and Cu) substrate. A large part of deposited silicon was found oxidized, even though the experiments have been performed in a glove box in order to avoid contact with air.

Salman et al. electrodeposited silicon-germanium (Si_xGe_{1-x}) composition from 1-butyl-1-methylpyrrolidinium bis(trifluoromethylsulfonyl)amide [BMPyrr] [TFSA], which is an air- and water-stabile IL (Al-Salman et al. 2008). $GeCl_4$ and $SiCl_4$ were used as germanium and silicon precursors. The Si_xGe_{1-x} deposit showed a color change during electrodeposition from red to blue. The authors attributed this change to a quantum size effect.

The electrolysis in ILs was performed at room temperature as well as at higher temperatures elevated up to 150 °C (Nishimura et al. 2009). Various metals have been used in the studies referred to above as the substrates for Si electrodeposition: Ag, Pt, Au, Ni, Al, Si, Cu, and Ti. The influence of Ga and InGa substrates on the structure of Si electrodeposits was revealed (Zhao et al. 2021). Carbon was also applied as a substrate (Fournier and Favier 2011; Shah et al. 2017).

The Si electrodeposition from ILs is a rather limited process; in most cases, only submicron films were deposited. Electrochemical quartz crystal microbalance (EQCM) is a useful tool for the identification of the deposition process. We will consider this method in detail in Chapter 7, when discussing silicon electrode passivation with oxide films. Here, it was only mentioned that this method allows determining the electrode mass change with nanogram resolution *in situ* and in real time. Also, the simultaneous combination of a mass measurement with electrochemical (potential, current) measurement is possible. Tsuyuki et al. applied the EQCM method for Si electrodeposition from ILs

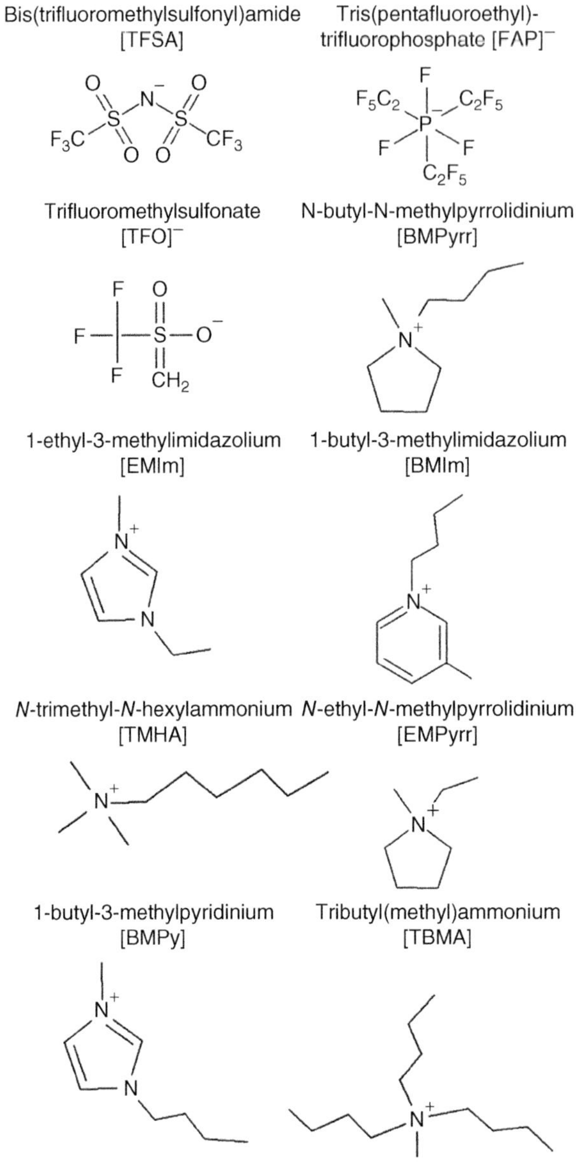

Figure 6.5 Structural formulas of the cations and anions used for IL synthesis for Si electrodeposition. The anion bis(trifluoromethylsulfonyl)amide [TFSA] was most widely used. This compound with the empirical formula $(CF_3SO_2)_2N$ is also called bis(trifluoromethylsulfonyl)imide [TFSI], bistriflamide, or bistriflimide. It is also referred to as bis(trifluoromethylsulfonyl)imide (NTF), Tf_2N, NTf_2, and TFSI (Endres et al. 2008, p. xxi).

(Tsuyuki et al. 2015). They studied the deposition mechanism in trimethyl-*n*-hexylammonium bis(trifluoromethylsulfonyl)imide [TMHA][TFSI] with $SiCl_4$ as a precursor. The method allowed detecting of the $SiCl_4$ reduction as well as current efficiencies at various electrode potentials. The efficiency was evaluated to be nearly 100%, assuming a four-electron reduction reaction. Also, EQCM identified a reduction peak of the Si electrodeposition and several elementary steps of intermediate formation prior to the Si deposition (Tsuyuki et al. 2017). In conclusion, the authors supported the EQCM data using the data obtained by the X-ray photoelectron spectroscopy (XPS) and Raman spectroscopy. Note that XPS and Raman spectroscopies indicated the amorphous structure of the deposited Si with Si—Si bonds. They obtained crystalline Si films only after annealing in an Ar gas.

Recently, an innovative approach was proposed by Tsuyuki et al., which is based on the formation of Si thin films in ILs under light irradiation conditions (Tsuyuki et al. 2018). Similar to the above studies, the photoprocess has been studied in the [TMHA][TFSI] electrolyte using $SiCl_4$ as the Si source. A possibility was also studied to electrodeposit doped Si films using $AlCl_3$ as a p-type dopant. The light irradiation enhanced Si electrodeposition and facilitated the development of smoother layers. Nevertheless, the films were largely contaminated by oxygen, which the authors attributed to exposure of the film to air or oxidation during annealing when attempting to remove the impurities stemming from ILs.

The electrolytes prepared using organic solvents are also characterized by a wide electrochemical window, which is sufficient to deposit Si. Organic solvents have low ionic conductivity; therefore, supporting electrolytes are applied to increase the conductivity of the deposition bath. Agrawal and Austin deposited amorphous Si from trichlorosilane ($SiHCl_3$, $T_b = 31.8\,°C$) dissolved in propylene carbonate (PC, $C_4H_6O_3$, and $T_b = 240–243\,°C$) containing tetrabutylammonium chloride (TBAC and $[CH_3(CH_2)_3]_4NCl$) as the supporting electrolyte (Agrawal and Austin 1981). Experiments have been carried out at 35–45 °C using various substrates: Pt, Ti, n-Si, and indium tin oxide (ITO). The deposits contained hydrogen-bonded compounds SiH_2 or SiH.

Lee and Kröger reported that no Si deposit was obtained using ethyl alcohol as a solvent and SiF_4 as a precursor (Lee and Kröger 1982). The authors deposited a mixture of silicon, carbon, and fluorine from K_2SiF_6 dissolved in acetone with the addition of HF. Silicic acid was dissolved in ethylene glycol and formamide mixture with HF and the deposition has been studied in the temperature interval from 27 to 85 °C (Rama Mohan and Kröger 1982). Infrared reflection spectra identified the deposition of a mixture of Si, F, C, N, and O. Solutions of tetraethyl orthosilicate (TEOS) in acetic acid, propylene carbonate, and 1-chloropropane were also studied (Takeda et al. 1981). Deposition of the Si thin film from TEOS in acetic acid appeared promising; however, the deposits contained a large fraction of silica.

Gobet and Tannenberger used the following Si precursors: $SiHCl_3$, $SiCl_4$, $SiBr_4$, $Si(CH_2CH_3)_4$, $Si(OCH_2CH_3)_4$, $Si(OCH_2CH_3)_4$, and $Si[N(CH_3)_2]_4$ (Gobet and Tannenberger 1988). The precursors were dissolved in tetrahydrofuran. Tetrabutylammonium bromide (TBAB) and $LiClO_4$ were used as supporting electrolytes.

Electrodeposition has been performed on Pt, Au, Ni, Cu, glassy carbon, and ITO substrates. Thin layers up to 0.25 μm were deposited, which were largely contaminated by carbon (8 at.%), oxygen (8 at.%), and chlorine (1–5 at.%).

Nicholson performed an extensive study using various supporting electrolytes: tetraethylammonium chloride (TEAC), tetrapropylammonium chloride (TPAC), tetrabutylammonium chloride (TBAC), LiCl, and corresponding bromides (tetraethylammonium bromide [TEAB], tetrapropylammonium bromide [TPAB], TBAB, and LiBr) (Nicholson 2005). Solvents included PC, tetrahydrofuran (THF), acetonitrile (AN), dimethyl sulfide (DMS), and 1-methyl naphthalene. Silicon precursors were $SiCl_4$ and $SiBr_4$. The deposited layers were about 100 nm in thickness and contained large fractions of oxygen and carbon. Minor amounts of hydrogen, chlorine, and phosphorus were also found. Nevertheless, the films showed photoresponsiveness. Low concentrations of $AlCl_3$ were added to the bath to provide p-doping and to form n–p junction on n-type Si(100) wafer. The photoeffects were attributed to pure and non-oxidized deposits close to the substrate.

A porous deposit was obtained in the PC solvent using $SiCl_4$ as a precursor and TBAC as a supporting electrolyte (Nishimura and Fukunaka 2007). Raman spectroscopy performed *in situ* identified deposition of Si. The deposit, however, was very active: it oxidized immediately in the air.

Si deposition has also been studied on Ni, Ag, Pt, Au, and glassy carbon from TBAC ($[CH_3(CH_2)_3]_4NCl$), acetonitrile (CH_3CN, $T_b = 82\,°C$), and THF (C_4H_8O, $T_b = 66\,°C$) baths (Munisamy and Bard 2010). The authors used trichlorosilane ($SiHCl_3$, $T_b = 31.8\,°C$) and Si tetrachloride ($SiCl_4$, $T_b = 57.65\,°C$) as silicon precursors. The deposits apart from silicon contained up to 35 at.% carbon. Silicon was found largely in the oxidized state of SiO_2. Other contaminations included N (up to 9 at.%) and Cl (1–2 at.%).

The studies referred to the above showed that the quality of electrodeposited silicon is poor due to its oxidation and incorporation of the species from the organic solvent as well as the electrolyte. The growth is largely prevented by the deposited Si, which isolates the electrode. Further growth can occur as a result of structural rearrangements, diffusion of the precursors through the defects, and to the conductivity sites produced by deposited carbon, if any. No solution was found to the problem of the Si oxidation. The films were of higher quality at the coating–substrate interface; however, the purity zone was only several tens of nanometers thick. Further investigations are also needed on the control of hydrogen content.

6.5 Purity Concerns and Solutions

Purity is the major performance indicator of silicon to be applied for solar and electronic devices. Electrochemical production of solar-grade silicon (SoG-Si) or higher purity was and remains a challenge for researchers.

Pre-electrolysis is typically used to reduce the level of impurities in molten salt electrolytes. There are different approaches depending on the presence or absence of a Si precursor in the electrolyte. If a precursor is present, the pre-electrolysis potential (E) is selected more positive than that of Si deposition ($E^0_{Si/Si^{4+}}$).

In such case, the impurities whose reduction potential is greater than $E^0_{Si/Si^{4+}}$ will be removed. On the contrary, the impurities will not be removed with reduction potential less than $E^0_{Si/Si^{4+}}$. When SiO_2 or silicate is used as a Si source, there is an opportunity to pre-purify the precursor-free electrolyte. Removal of impurities, whose reduction potential is greater than that of the electrolyte cations (Ca^{2+}, Na^+, K^+, and Li^+), is possible. Effectiveness of pre-electrolysis in fluoride electrolytes has been reviewed (Xu and Haarberg 2013).

Purity of 99.8 at.% was achieved in early attempts of electrochemical production of SoG-Si from SiO_2 (Yasuda et al. 2007). The reduced product was melted under vacuum at 1773 K. The ingot was contaminated by C (10 100 ppm by weight), Ca (14 000 ppm), Mn (11 000 ppm), and Fe (31 000 ppm). Other contaminants were Ag, Mo, Ni, Cu, Cr, Ti, and Al. All the impurities originated mainly from a stainless steel vessel (Fe, Cr, Mn, Ni, and Mo), carbon anodes (C), Mo contact wire, and the electrolyte residue (Ca and Cl). Chlorine gas produced on the anode reacted with the stainless steel reactor as well as with the lead wire and contaminated the silicon. The p–n samples were contaminated by W, which originated from the electrode leads, whereas Al contamination was released from the quartz crucible (Zou et al. 2017). The purity was increased when using Si contacting electrodes instead of the foreign metal electrode (Yasuda et al. 2013; Akpinar et al. 2017). The electroreduced silicon was upgraded to SoG-Si by subsequent zone melting purification (Sun et al. 2015; Zhong et al. 2018a).

Contamination by calcium originates either from the molten salt entrapped in the surface structure or the calcium metal deposited at negative polarization. Calcium silicides can form when exceeding the equilibrium potential $E < E^0_{Ca/Ca^{2+}}$ or even under conditions of underpotential deposition (UPD), that is, in some region of $E > E^0_{Ca/Ca^{2+}}$ (Xiao et al. 2010; Xiao and Wang 2014; Nohira et al. 2003; Gao et al. 2005; Yasuda et al. 2005a,b,c). The calcium content was significantly reduced by leaching in the following solutions: $NH_4OH/H_2O_2/H_2O$, $HF/H_2O_2/H_2O$, and $HCl/H_2O_2/H_2O$ (Yasuda et al. 2013). The calcium concentration dropped down from 84 000 to 120 ppm when water was replaced by $HCl/H_2O_2/H_2O$ as a washing medium. Chloride contamination decreased accordingly: from 82 000 to 76 ppm The approximate balance between the calcium and chloride concentrations implies that the contamination originated mainly from the residue of the electrolyte.

An advantage in terms of purity suggests Si electrodeposition using a gaseous Si precursor. Principal opportunity of Si reduction from $SiCl_4$ gas was reported by Matsuda et al. (1996). They used Si electrode in the LiCl–KCl eutectic melt and succeeded to deposit Si only near the electrolyte–$SiCl_4$ boundary, as the gas was almost insoluble in the electrolyte. Over 80% $SiCl_4$ dissolution efficiency was achieved in the KF–KCl melt (Yasuda et al. 2017a). The authors deposited a silicon film from high-purity $SiCl_4$ dissolved in KF–KCl eutectic composition.

$$SiCl_4 + 6F^- \rightarrow SiF_6{}^{2-} + 4Cl^- \quad (6.10)$$

$$SiF_6{}^{2-} + 4e^- \rightarrow Si + 6F^- \quad (6.11)$$

However, the deposited film morphology was smooth only near to the substrate. High chloride concentration due to the reaction (6.10) was concluded to be the main reason for the coarse silicon structure. Note that the well-known Siemens purification process employs a gaseous precursor as well ($HSiCl_3$).

A process using liquid alloy Si–Zn cathode has been proposed for SoG-Si production in molten $CaCl_2$ (Yasuda et al. 2017b). The molten salt suppressed the evaporation of Zn metal. Reduction of SiO_2 on a Zn cathode proceeded at $E < 1.45\ V_{Ca/Ca^{2+}}$, which is much more positive than the silica reduction on the Si substrate. Micrometer-sized Si particles were obtained by precipitation when cooling down the Si–Zn system.

Highly pure silicon was obtained by SiO_2 reduction in the form of Si powder and subsequent forming of ingots by the floating zone method (Zhong et al. 2018a). The purity of the produced silicon was higher than that of SoG-Si. The minority carrier lifetimes in the ingot were 2 orders of magnitude shorter when compared to 10N Si.

The vast majority of the studies of electrochemical SiO_2 reduction are performed using carbon-based anodes (Table 9.1). This raises the question about silicon contamination by carbon. The oxygen ions released from silica or silicates

$$SiO_2 + 4e^- \rightarrow Si + O^{2-} \quad (6.12)$$

$$SiO_3{}^{2-} + 4e^- \rightarrow Si + 3O^{2-} \quad (6.13)$$

can react with the carbon anode producing CO_x

$$C + xO^{2-} \rightarrow CO_x \quad (6.14)$$

The carbon oxide can be reduced back to C on the Si cathode with a possible carbonization reaction (Zou et al. 2017a; Juzeliūnas et al. 2018).

$$Si + C \rightarrow SiC \quad (6.15)$$

The carbon impurity level was found to be over 10 000 ppm when deoxidizing silica using the carbon anode (Yasuda et al. 2007). The contamination was reduced to 1000–6800 ppm when the deoxidized sample was treated in HF, HCl, H_2O_2, and H_2O mixtures (Yasuda et al. 2013). A large carbon content of up to 11 at.% was determined for Si deposited using SiO_2 NPs as the precursor (Cho et al. 2012a). Electrodeposited photoactive surfaces were also largely contaminated by carbon (Cho et al. 2012b, Angew. Chem, 51, p. 12740). Moreover, electrolysis of the SiO_2 layer using graphite anode yielded a carbon–carbide structure (Juzeliūnas et al. 2018). The structure exhibited distinctive photoresponse, which was much more pronounced when compared to the p-Si wafer. Thus, photoactivity due to carbon contamination appears to be important in photoelectrochemical studies of electroreduced Si. Nevertheless, the carbon content in the produced n–p films was not determined (Zou et al. 2017; Yang et al. 2017).

The creation of cost-effective and non-corroding substitutes of graphite anodes in avoiding the contamination of reduced silicon by carbon as well as CO_x emission poses a challenge for future research. In a different application, Jiao and Fray used calcium ruthenate and solid solutions of calcium ruthenate and calcium titanate successfully in the electrodeoxidation of TiO_2 (Jiao and Fray 2010). It was shown that calcium-ruthenate-based anodes endure the applied electrical field with oxygen production on the anode.

References

Agrawal, A.K. and Austin, A.E. (1981). Electrodeposition of silicon from solutions of silicon halides in aprotic solvents. *J. Electrochem. Soc.* 128: 2292–2296. https://doi.org/10.1149/1.2127237.

Akpinar, B., Erdogan, M., Akduman, B., and Karakaya, I. (2017). The use of silicon wafer barriers in the electrochemical reduction of solid silica to form silicon in molten salts. *ECS Trans.* 77: 2011–2016. https://doi.org/10.1149/07711.2011ecst.

Al-Salman, R., Zein El Abedin, S., and Endres, F. (2008). Electrodeposition of Ge, Si and Si_xGe_{1-x} from an air- and water-stable ionic liquid. *Phys. Chem. Chem. Phys.* 10: 4650–4657. https://doi.org/10.1039/b806996b.

Amde, M., Liu, J.-F., and Pang, L. (2015). Environmental application, fate, effects, and concerns of ionic liquids: a review. *Environ. Sci. Technol.* 49: 12611–12627. https://doi.org/10.1021/acs.est.5b03123.

Beyersdorf, T., Schubert, T.J.S., Weltz-Biermann, U. et al. (2008). Synthesis of ionic liquids. In: *Electrodeposition from Ionic Liquids* (ed. F. Endres, A.P. Abbot and D.R. MacFarlane), 15–46. Wiley-VCH Verlag GmbH & Co.

Bieber, A.L., Massot, L., Gibilaro, M. et al. (2012). Silicon electrodeposition in molten fluorides. *Electrochim. Acta* 62: 282–289. https://doi.org/10.1016/j.electacta.2011.12.039.

Boen, R. and Bouteillon, J. (1983). The electrodeposition of silicon in fluoride melts. *J. Appl. Electrochem.* 13: 277–288. https://doi.org/10.1007/BF00941599.

Boiko, O.I., Delimarskii, Y.K., and Chernov, R.V. (1985). Electrochemical reduction of silicon(IV) from fluoride-chloride melt. *Ukr. Khim. Zh. (Russ. Ed.)* 51: 385–390.

Borisenko, N., Zein El Abedin, S., and Endres, F. (2006). In situ STM investigation of gold reconstruction and of silicon electrodeposition on Au(111) in the room temperature ionic liquid 1-butyl-1-methylpyrrolidinium bis(trifluoromethylsulfonyl) imide. *J. Phys. Chem. B* 110: 6250–6256. https://doi.org/10.1021/jp057337d.

Cahen, S., Janot, R., Laffont-Dantras, L., and Tarascon, J.M. (2008). Chemical reduction of $SiCl_4$ for the preparation of silicon-graphite composites used as negative electrodes in lithium-ion batteries. *J. Electrochem. Soc.* 155: A512–A519.

Cai, Z., Li, Y., and Tian, W. (2011). Electrochemical behavior of silicon compound in LiF–NaF–KF–Na_2SiF_6 molten salt. *Ionics* 17: 821–826. https://doi.org/10.1007/s11581-011-0582-y.

Carleton, K.L., Olson, J.M., and Kibbler, A. (1983). Electrochemical nucleation and growth of silicon in molten fluorides. *J. Electrochem. Soc.* 130: 782–786. https://doi.org/10.1149/1.2119803.

Chemezov, O.; Apisarov, A.; Isakov, A.; Zaikov, Yu. (2012) Structure silicon deposits obtained by electrolysis SiO_2 in the chloride-fluoride melts. L. Zhang, J. A. Pomykala, A. Ciftja https://doi.org/10.1002/9781118359341.ch58

Cho, S.K., Fan, F.R.F., and Bard, A.J. (2012a). Formation of a silicon layer by electroreduction of SiO_2 nanoparticles in $CaCl_2$ molten salt. *Electrochim. Acta* 68: 123–127. https://doi.org/10.1016/j.electacta.2012.01.008.

Cho, S.K., Fan, F.R.F., and Bard, A.J. (2012b). Electrodeposition of crystalline and photoactive silicon directly from silicon dioxide nanoparticles in molten $CaCl_2$. *Angew. Chem. Int. Ed.* 51: 12740–12744. https://doi.org/10.1002/anie.201206789.

Cohen, U. (1976) Electrodeposition of polycrystalline silicon from a molten fluoride bath and product. US Patent No 3, 983,012.

Cohen, U. (1977). Some prospective applications of silicon electrodeposition from molten fluorides to solar cell fabrication. *J. Electron. Mater.* 6: 607–643. https://doi.org/10.1007/BF02660341.

Cohen, U. and Huggins, R.A. (1976). Silicon epitaxial growth by electrodeposition from molten fluorides. *J. Electrochem. Soc.* 123: 381–383. https://doi.org/10.1149/1.2132829.

De Lepinay, J., Bouteillon, J., Traore, S. et al. (1987). Electroplating silicon and titanium in molten fluoride media. *J. Appl. Electrochem.* 17: 294–302. https://doi.org/10.1007/BF01023295.

De Mattei, R.C., Elwell, D., and Feigelson, R.S. (1981a). Electrodeposition of silicon at temperatures above its melting point. *J. Electrochem. Soc.* 128: 1712–1714. https://doi.org/10.1002/chin.198148039.

De Mattei, R.C., Elwell, D., and Feigelson, R.S. (1981b). Electrodeposition of molten silicon. US Patent, No 4, 292,145.

Elwell, D. (1981). Electrowinning of silicon from solutions of silica in alkali metal fluoride/alkaline earth fluoride eutectics. *Solar Energy Mater.* 5: 205–210. https://doi.org/10.1016/0165-1633(81)90032-0.

Elwell, D. and Rao, G.M. (1982). Mechanism of electrodeposition of silicon from K_2SiF_6-flinak. *Electrochim. Acta* 27: 673–676. https://doi.org/10.1016/0013-4686(82)85058-5.

Elwell, D. and Rao, G.M. (1984). Time dependence of sell voltage during electrolysis of K_2SiF_6–KF–LiF melts. *Surf. Technol.* 23: 67–72. https://doi.org/10.1016/0376-4583(84)90077-3.

Elwell, D. and Rao, G.M. (1988). Electrolytic production of silicon. *J. Appl. Electrochem.* 18: 15–22. https://doi.org/10.1007/BF01016199.

Endres, F., Abbot, A.P., and MacFarlane, D.R. (ed.) (2008). *Electrodeposition from Ionic Liquids*, 387. Weinheim. https://doi.org/10.1007/s10008-009-0821-6: Wiley-VCH. ISBN: 978-3-527-31565-9.

Fournier, C. and Favier, F. (2011). Zn, Ti and Si nanowires by electrodeposition in ionic liquid. *Electrochem. Commun.* 13: 1252–1255. https://doi.org/10.1016/j.elecom.2011.08.031.

Frazer, E.J. and Welch, B.J. (1977). A galvanostatic study of the electrolytic reduction silica in molten cryolite. *Electrochim. Acta* 22: 1179–1182. https://doi.org/10.1016/0013-4686(77)80058-3.

Frolenko, D.B., Martemjanova, Z.S., Valeev, Z.I., and Baraboshkin, A.N. (1992). Structure of silicon deposits obtained by electrolyzing mixed chloride-fluoride melts. *Sov. Electrochem.* (translated from Elektrokhimiya) 28: 1427–1434.

Gao, P., Jin, X.B., Wang, D.H. et al. (2005). A quartz sealed Ag/AgCl reference electrode for $CaCl_2$ based molten salts. *J. Electroanal. Chem.* 579: 321–328. https://doi.org/10.1016/j.jelechem.2005.03.004.

Gobet, J. and Tannenberger, H. (1988). Electrodeposition of silicon from nonaqueous solvent. *J. Electrochem. Soc.* 135: 109–112. https://doi.org/10.1149/1.2083227.

Haarberg, G.M., Famiyeh, L., Martinez, A.M., and Osen, K.S. (2013). Electrodeposition of silicon from fluoride melts. *Electrochim. Acta* 100: 226–228. https://doi.org/10.1016/j.electacta.2012.11.052.

Hu, Y., Wang, X., Xiao, J. et al. (2013). Electrochemical behavior of silicon(IV) ion in BaF_2–CaF_2–SiO_2 melts at 1573 K. *J. Electrochem. Soc.* 160: D81–D84. https://doi.org/10.1149/2.038303jes.

Isakov, A., Khvostov, S., Kinev, E. et al. (2020). Neutron transmutation doping of thin silicon films electrodeposited from the KF–KCl–KI–K_2SiF_6 melt. *J. Electrochem. Soc.* 167: 082515. https://doi.org/10.1149/1945-7111/ab933c.

Jiao, S. and Fray, D.J. (2010). Development of an inert anode for electrowinning in calcium chloride-calcium oxide melts. *Metall. Mater. Trans. B* 41: 74–79. https://doi.org/10.1007/s11663-009-9281-8.

Juzeliūnas, E. and Fray, D.J. (2020). Silicon electrochemistry in molten salts. *Chem. Rev.* 120: 1690–1709. https://doi.org/10.1021/acs.chem.rev.9b00428.

Juzeliūnas, E., Coxon, P.R., Fray, D.J. et al. (2016). Photoelectrochemical activity of macroporous silicon formed in molten calcium chloride. *Chemija* 27: 11–16.

Juzeliūnas, E., Fray, D.J., Kalinauskas, P. et al. (2018). Electrochemical synthesis of photoactive carbon-carbide structure on silicon in molten salt. *Electrochem. Commun.* 90: 6–10. https://doi.org/10.1016/j.elecom.2018.03.003.

Katayama, Y., Yokomizo, M., Miura, T., and Kishi, T. (2001). Preparation of novel fluorosilicate salt for electrodeposition of silicon at low temperature. *Electrochemistry* 69: 834–836. https://doi.org/10.5796/electrochemistry.69.834.

Komadina, J., Akiyoshi, T., Ishibashi, Y. et al. (2013). Electrochemical quartz crystal microbalance study of Si electrodeposition in ionic liquid. *Electrochim. Acta* 100: 236–241. https://doi.org/10.1016/j.electacta.2012.07.043.

Konakov, V.G. and Pivovarov, M.M. (2003). Interaction between components in the KF–SiO_2 system. *Glass Phys. Chem* 29: 69–74. https://doi.org/10.1023/A:1022313911036.

Kuznetsova, S.V., Dolmatov, V.S., and Kuznetsov, S.A. (2009). Voltammetric study of electoreduction of silicon complexes in a chloride-fluoride melt. *Russ. J. Electrochem.* 45: 742–748. https://doi.org/10.1134/S1023193509070052.

Laptev, M.V., Isakov, A.V., Grishenkova, O.V. et al. (2020). Electrodeposition of thin silicon films from the KF–KCl–KI–K_2SiF_6 melt. *J. Electrochem. Soc.* 167: 042506. https://doi.org/10.1149/1945-7111/ab7aec.

Lee, C.H. and Kröger, F.A. (1982). Cathodic deposition of amorphous alloys of silicon, carbon, and fluorine. *J. Electrochem. Soc.* 129: 936–942. https://doi.org/10.1149/1.2124069.

Lei, Z., Chen, B., Koo, Y.-M., and MacFarlane, D.R. (2017). Introduction: ionic liquids. *Chem. Rev.* 117: 6633–6635. https://doi.org/10.1021/acs.chemrev.7b00246.

Liu, W., Ma, Q.P., Guo, J.Y. et al. (2010). The comparison research of two working electrodes in electrochemical reduction of silicon tetrachloride. *Adv. Mater. Res.* 97–101: 32–35. https://doi.org/10.4028/www.scientific.net/AMR.97-101.32.

Maeda, K., Yasuda, K., Nohira, T. et al. (2014). A new electrodeposition process of crystalline silicon utilizing water-soluble KF–KCl molten salt. *ECS Trans.* 64: 285–291. https://doi.org/10.1149/06404.0285ecst.

Mai, N.L., Ahn, K.A., and Koo, Y.-M. (2014). Methods for recovery of ionic liquids – a review. *Process Biochem.* 49: 872–881. https://doi.org/10.1016/j.procbio.2014.01.016.

Martinez, A.M., Osen, K.S., Kongstein, O.E. et al. (2010). Electrodeposition of silicon thin films from ionic liquids. *ECS Trans.* 25: 107–118. https://doi.org/10.1149/1.3318509.

Matsuda, T., Nakamura, S., Ide, K. et al. (1996). Oscillatory behavior in electrochemical deposition reaction of polycrystalline silicon thin films through reduction of silicon tetrachloride in a molten salt electrolyte. *Chem. Lett.* 25: 569–570. https://doi.org/10.1246/cl.1996.569.

Munisamy, T. and Bard, A.J. (2010). Electrodeposition of Si from organic solvents and studies related to initial stages of Si growth. *Electrochim. Acta* 55: 3797–3803. https://doi.org/10.1016/j.electacta.2010.01.097.

Nicholson, J.P. (2005). Electrodeposition of silicon from nonaqueous solvents. *J. Electrochem. Soc.* 152: C795–C802. https://doi.org/10.1149/1.2083227.

Nishimura, Y. and Fukunaka, Y. (2007). Electrochemical reduction of silicon chloride in a non-aqueous solvent. *Electrochim. Acta* 53: 111–116. https://doi.org/10.1016/j.electacta.2007.06.026.

Nishimura, Y., Fukunaka, Y., Nishida, T. et al. (2008). Electrodeposition of Si thin film in a hydrophobic room-temperature molten salt. *Electrochem. Solid-State Lett.* 11: D75–D79. https://doi.org/10.1149/1.2952190.

Nishimura, Y., Nohira, T., Morioka, T., and Hagiwara, R. (2009). Electrochemical reduction of silicon tetrachloride in an intermediate-temperature ionic liquid. *Electrochemistry* 77: 683–686. https://doi.org/10.5796/electrochemistry.77.683.

Nohira, T., Yasuda, K., and Ito, Y. (2003). Pinpoint and bulk electrochemical reduction of insulating silicon dioxide to silicon. *Nat. Mater.* 2: 397–401. https://doi.org/10.1038/nmat900.

Olivier-Bourbigou, H., Magna, L., and Morvan, D. (2010). Ionic liquids and catalysis: recent progress from knowledge to applications. *Appl. Catal., A* 373: 1–56. https://doi.org/10.1016/j.apcata.2009.10.008.

Osen, K.S., Martinez, A.M., Rolseth, S. et al. (2010). Electrodeposition of crystalline silicon films from alkali fluoride mixtures. *ECS Trans.* 33: 429–438. https://doi.org/10.1149/1.3484801.

Park, J., Lee, C.K., Kwon, K., and Kim, H. (2013). Electrodeposition of silicon from 1-butyl-3-methylpyridinium bis(trifluromethylsulfonyl) imide ionic liquid. *Int. J. Electrochem. Sci.* 8: 4206–4214. https://doi.org/10.1007/978-3-642-02723-9_117.

Peng, J.J., Yin, H.Y., Zhao, J. et al. (2018). Liquid-tin-assisted molten salt electrodeposition of photoresponsive n-type silicon films. *Adv. Funct. Mater.* 28: 1703551. https://doi.org/10.1002/adfm.201870194.

Pham, T.P.T., Cho, C.-W., and Yun, Y.-S. (2010). Environmental fate and toxicity of ionic liquids: a review. *Water Res.* 44: 352–372. https://doi.org/10.1016/j.watres.2009.09.030.

Pulletikurthi, G., Lahiri, A., Carstens, T. et al. (2013). Electrodeposition of silicon from three different ionic liquids: possible influence of the anion on the deposition process. *J. Solid State Electrochem.* 17: 2823–2832. https://doi.org/10.1007/s10008-013-2185-1.

Rama Mohan, T.R. and Kröger, F.A. (1982). Cathodic deposition of amorphous silicon from solutions of silicic acid and tetraethyl orthosilicate in ethylene glycol and formamide containing HF. *Electrochim. Acta* 27: 371–377. https://doi.org/10.1016/0013-4686(82)85009-3.

Rao, G.M., Elwell, D., and Feigelson, R.S. (1980). Electrowinning of silicon from K_2SiF_6-molten fluoride systems. *J. Electrochem. Soc.* 127: 1940–1944. https://doi.org/10.1149/1.2130041.

Rao, G.M., Elwell, D., and Feigelson, R.S. (1981a). Electrodeposition of silicon onto graphite. *J. Electrochem. Soc.* 128: 1708–1711. https://doi.org/10.1149/1.2123696.

Rao, G.M., Elwell, D., and Feigelson, R.S. (1981b). Electrocoating of silicon and its dependence on the time of electrolysis. *Surf. Technol.* 13: 331–337. https://doi.org/10.1016/0376-4583(81)90107-2.

Rao, G.M., Elwell, D., and Feigelson, R.S. (1982). Characterization of electrodeposited silicon on graphite. *Solar Energy Mater.* 7: 15–21. https://doi.org/10.1016/0165-1633(82)90092-2.

Sakanaka, Y. and Goto, T. (2015). Electrodeposition of Si film on Ag substrate in molten LiF–NaF–KF directly dissolving SiO_2. *Electrochim. Acta* 164: 139–142. https://doi.org/10.1016/j.electacta.2014.12.159.

Schmuck, M., Balducci, A., Rupp, B. et al. (2010). Alloying of electrodeposited silicon with lithium – a principal study of applicability as anode material for lithium ion batteries. *J. Solid State Electrochem.* 14: 2203–2207. https://doi.org/10.1007/s10008-008-0763-4.

Seddon, K.R. (2003). A taste of the future. *Nat. Mater.* 2: 363–365. https://doi.org/10.1038/nmat907.

Shah, N.K., Pati, R.K., Ray, A., and Mukhopadhyay, I. (2017). Electrodeposition of Si from an ionic liquid bath at room temperature in the presence of water. *Langmuir* 33: 1599–1604. https://doi.org/10.1021/acs.langmuir.6b03621.

Stern, K.H. and McCollum, M.E. (1985). Electrodeposition of silicon from molten salts. *Thin Solid Films* 124: 129–134. https://doi.org/10.1016/0040-6090(85)90255-X.

Sun, D.-Y., He, L.-W., Lian, J.-Q. et al. (2015). Process for solar grade silicon production by molten salt electrolysis and zone melting. *J. Mater. Eng.* 43: 73–78. https://doi.org/10.11868/j.issn.1001-4381.2015.04.013.

Suzuki, Y., Inoue, Y., Yokota, M., and Goto, T. (2019). Effects of oxide ions on the electrodeposition process of silicon in molten fluorides. *J. Electrochem. Soc.* 166: D564–D568. https://doi.org/10.1149/2.0441913jes.

Takeda, Y., Kanno, R., Yamamoto, O. et al. (1981). Cathodic deposition of amorphous silicon from tetraethylorthosilicate in organic solvents. *J. Electrochem. Soc.* 128: 1221–1224. https://doi.org/10.1149/1.2127597.

Thomas, S., Kowalski, D., Molinari, M., and Mallet, J. (2018). Role of electrochemical process parameters on the electrodeposition of silicon from 1-butyl-1-methylpyrrolidinium bis(trifluoromethanesulfonyl)imide ionic liquid. *Electrochim. Acta* 265: 166–174. https://doi.org/10.1016/j.electacta.2018.01.139.

Tsuyuki, Y., Huynh, T.A.P., Komadina, J. et al. (2015). Electrochemical quartz crystal microbalance, X-ray photoelectron spectroscopy, and Raman spectroscopy analysis of $SiCl_4$ reduction in ionic liquids. *Electrochim. Acta* 183: 49–55. https://doi.org/10.1016/j.electacta.2015.04.167.

Tsuyuki, Y., Fujimura, T., Kunimoto, M. et al. (2017). Analysis of cathodic reaction process of $SiCl_4$ during Si electrodeposition in ionic liquids. *J. Electrochem. Soc.* 164: D994–D998. https://doi.org/10.1149/2.0761714jes.

Tsuyuki, Y., Takai, H., Fukunaka, Y., and Homma, T. (2018). Formation of Si thin films by electrodeposition in ionic liquids for solar cell applications. *Jpn. J. Appl. Phys.* 57: 08RB11. https://doi.org/10.7567/JJAP.57.08RB11.

Verma, C., Alrefaee, S.H., Quraishi, M.A. et al. (2021). Recent developments in sustainable corrosion inhibition using ionic liquids: a review. *J. Mol. Liq.* 321: 114484. https://doi.org/10.1016/j.molliq.2020.114484.

Walden, P. (1914). Über die Molekulargrösse und elektricsche Leitfähigkeit einigergeschmolzenen Salze. *Bull. Acad. Imp. Sci. St.-Petersbourg.* 405–422.

Wang, Y. and Morita, K. (2016). Measurement of the phase diagram of the SiO_2–$CaCl_2$ system and liquid area study of the SiO_2–CaO–$CaCl_2$ system. *Metall. Mater. Trans.* 47B: 1542–1547. https://doi.org/10.1007/s11663-016-0641-x.

Xiao, W. and Wang, D. (2014). The electrochemical reduction process of solid compounds in high temperature molten salts. *Chem. Soc. Rev.* 43: 3215–3228. https://doi.org/10.1039/c3cs60327j.

Xiao, W., Jin, X.B., Deng, Y. et al. (2010). Rationalisation and optimization of solid state electro-reduction of SiO_2 to Si in molten $CaCl_2$ in accordance with dynamic three-phase interlines based voltammetry. *J. Electroanal. Chem.* 639: 130–140. https://doi.org/10.1016/j.elechem.2009.12.001.

Xiao, W., Jin, X.B., and Chen, G.Z. (2013). Up-scalable and controllable electrolytic production of photo-responsive nanostructured silicon. *J. Mater. Chem. A* 1: 10243–10250. https://doi.org/10.1039/C3TA11823A.

Xu, J. and Haarberg, G.M. (2013). Electrodeposition of solar cell grade silicon in high temperature molten salts. *High Temp. Mater. Proc.* 32: 97–105. https://doi.org/10.1515/htmp-2012-0045.

Yang, X., Yasuda, K., Nohira, T. et al. (2016). The role of granule size on the kinetics of electrochemical reduction of SiO_2 granules in molten $CaCl_2$. *Metall. Mater. Trans. B* 47: 788–797. https://doi.org/10.1007/s11663-015-0456-1.

Yang, X., Ji, L., Zou, X. et al. (2017). Toward cost-effective manufacturing of silicon solar cells: electrodeposition of high-quality Si films in a $CaCl_2$-based molten salt. *Angew. Chem. Int. Ed.* 56: 15078–15082. https://doi.org/10.1002/anie.201707635.

Yasuda, K., Nohira, T., and Ito, Y. (2005a). Effect of electrolysis potential on reduction of solid silicon dioxide in molten $CaCl_2$. *J. Phys. Chem. Solids* 66: 443–447. https://doi.org/10.1016/j.jpcs.2004.06.037.

Yasuda, K., Nohira, T., Ogata, Y.H., and Ito, Y. (2005b). Direct electrolytic reduction of solid silicon dioxide in molten LiCl–KCl–$CaCl_2$ at 773 K. *J. Electrochem. Soc.* 152: D208–D212. https://doi.org/10.1149/1.2042910.

Yasuda, K., Nohira, T., Amezawa, K. et al. (2005c). Mechanism of direct electrolytic reduction of solid SiO_2 to Si in molten $CaCl_2$. *J. Electrochem. Soc.* 152: D69–D74. https://doi.org/10.1149/1.1864453.

Yasuda, K., Nohira, T., Hagiwara, R., and Ogata, Y.H. (2007). Direct electrolytic reduction of solid SiO_2 in molten $CaCl_2$ for the production of solar grade silicon. *Electrochim. Acta* 53: 106–110. https://doi.org/10.1016/j.electacta.2007.01.024.

Yasuda, K., Nohira, T., Kobayashi, K. et al. (2013). Improving purity and process volume during direct electrolytic reduction of solid SiO_2 in molten $CaCl_2$ for the production of solar-grade silicon. *Energy Technol.* 1: 245–252. https://doi.org/10.1002/ente.201300001.

Yasuda, K., Maeda, K., Nohira, T. et al. (2016). Silicon electrodeposition in water-soluble KF–KCl molten salt: optimization of electrolysis conditions at 923 K. *J. Electrochem. Soc.* 163: D95–D99. https://doi.org/10.1149/2.0791603jes.

Yasuda, K., Maeda, K., Hagiwara, R. et al. (2017a). Silicon electrodepopsition in a water-soluble KF–KCl molten salt: utilization of $SiCl_4$ as Si source. *J. Electrochem. Soc.* 164: D67–D71. https://doi.org/10.1149/2.0641702jes.

Yasuda, K., Shimao, T., Hagiwara, R. et al. (2017b). Electrolytic production of silicon using liquid zinc alloy in molten $CaCl_2$. *J. Electrochem. Soc.* 164: H5049–H5056. https://doi.org/10.1149/2.0121708jes.

Zaykov, Y.P., Isakov, A.V., Zakiryanova, I.D. et al. (2014). Interaction between SiO_2 and a KF–KCl–K_2SiF_6 melt. *J. Phys. Chem. B* 118: 1584–1588. https://doi.org/10.1021/jp4086816.

Zaykov, Y.P., Zhuk, S.I., Isakov, A.V. et al. (2015). Electrochemical nucleation and growth of silicon in the KF–KCl–K_2SiF_6 melt. *J. Solid State Electrochem.* 19: 1341–1345. https://doi.org/10.1007/s10008-014-2729-z.

Zeile, J. V.; Janek, R. P.; Prudhomme, J.; Owens, G. L. (2006) Hope springs eternal: silicon films from the electrochemical reduction of silicon tetrachloride in nonaqueous solution. In: Abstracts of Papers of the American Chemical Society. 231, Meeting abstract: 523-INOR.

Zein El Abedin, S., Borissenko, N., and Endres, F. (2004). Electrodeposition of nanoscale silicon in a room temperature ionic liquid. *Electrochem. Commun.* 6: 510–514. https://doi.org/10.1016/j.elecom.2004.03.013.

Zhang, J., Chen, S., Zhang, H. et al. (2016). Electrodeposition of crystalline silicon directly from silicon tetrachloride in ionic liquid at low temperature. *RSC Adv.* 6: 12061–12067. https://doi.org/10.1039/C5RA23085C.

Zhao, Z., Liu, J., and Zhang, C. (2018). Preparation and characterization of electrochemically deposited silicon films in ionic liquid from polysilicon byproduct $SiCl_4$. *Int. J. Electrochem. Sci.* 13 (9): 8766–8774. https://doi.org/10.20964/2018.09.51.

Zhao, Z., Yang, C., Wu, L. et al. (2021). Preparation and characterization of crystalline silicon by electrochemical liquid–liquid–solid crystal growth in ionic liquid. *ACS Omega* 6: 11935–11942. https://doi.org/10.1021/acsomega.1c00304.

Zhong, M., Yasuda, K., Homma, T., and Nohira, T. (2018a). Purity and minority carrier lifetime in silicon produced by direct electrolytic reduction of SiO_2 in molten $CaCl_2$. *Electrochemistry* 86: 77–81. https://doi.org/10.5796/electrochemistry.17-00087.

Zhong, M., Yang, X., Yasuda, K. et al. (2018b). Effect of Si addition on the electrochemical reduction rate of SiO_2 granules in molten $CaCl_2$. *Metall. Mater. Trans. B* 49 (B): 341–348. https://doi.org/10.1007/s11663-017-1132-4.
Zhuk, S.I., Isakov, A.V., Apisarov, A.P. et al. (2017). Electrodeposition of continuous silicon coatings from the KF–KCl–K_2SiF_6 melts. *J. Electrochem. Soc.* 164: H5135–H5138. https://doi.org/10.1149/2.0171708jes.
Zou, X., Ji, L., Lu, X., and Zhou, Z. (2017a). Facile electrosynthesis of silicon carbide nanowires from silica/carbon precursors in molten salt. *Sci. Rep.* 7: 1–9, art. no. 9978. https://doi.org/10.1038/s41598-017-10587-5.
Zou, X., Ji, L., Yang, X. et al. (2017b). Electrochemical formation of a p–n junction of thin film silicon deposited in molten salt. *J. Am. Chem. Soc.* 139: 16060–16063. https://doi.org/10.1021/jacs.7b09090.
Zou, X., Ji, L., Ge, J. et al. (2019). Electrodeposition of crystalline silicon films from silicon dioxide for low-cost photovoltaic applications. *Nat. Commun.* 10: 5772. https://doi.org/10.1038/s41467-019-13065-w.

7

Photoelectrochemistry and Nanogravimetry of Si and Si-Oxide Electrodes

7.1 Topicality of Si Photoelectrochemical Research

The photoelectrochemical (PEC) studies of silicon are important in several aspects. The PEC measurements provide information, which correlates with the information of the solid-state cells, so that there is no prior need to design a solar cell that characterizes Si surface photoresponsiveness (Ye et al. 2011; Cho et al. 2012). PEC studies provide important characteristics of electrode–solution interface: dynamics of the surface photo- and electrochemical charging, capacitance of the space–charge region (SCR) and its dependence on the supplied energy (electrode potential), the zero-charge potential (flat band potential, E_{fb}), etc. The electronic properties of silicon are of primary importance in the development of efficient photovoltaic (PV) devices. The number of donors or acceptors in a semiconductor as well as the number of light-induced photocharges is obtained from the measurement of electrode capacitance vs. applied voltage.

PEC studies are also of importance in the development of light energy conversion systems. PEC systems enable light energy conversion into electrical energy and its storage in chemical bonds, such as in the case of solar-driven water splitting. The PEC system converts the energy of the incident photons to the exited states of charge carriers of a semiconducting electrode. The energy is further captured and converted to chemical bonds producing hydrogen or oxygen on the photocathode and photoanode, respectively. According to thermodynamic assumptions, the process requires a minimum potential difference $\Delta E = 1.23$ V between the cathode and the anode. However, because of high overpotentials of these reactions on semiconductor electrodes, there is a need to apply electrocatalysts in order to facilitate the water splitting process. Hydrogen energy conversion is expected to become the choice of the future energy produced from sustainable and renewable sources. The electrochemical conversion of hydrogen energy is relatively simple when compared to conventional fossil fuels.

The electronic properties of semiconductor electrodes are defined using the band model. It comprises the valence and the conducting bands separated by a bandgap

Silicon: Electrochemistry, Production, Purification and Applications, First Edition. Eimutis Juzeliūnas.

whose energy is given in electron volts (E_g). The width of the bandgap determines to a large extent the optical properties of the electrode. The energy for silicon is as high as $E_g = 1.1$ eV; its conduction band is almost vacant. Silicon electric conduction is possible because of thermal excitation of electrons. This property is at the heart of Si electrochemistry in high-temperature molten salts. When the temperature is increased, the electrons in the conduction band are produced, and their mobility is sufficiently high to respond to an electric field or, in other words, to support the conduction of the silicon electrode. The pure material acts as an intrinsic semiconductor where thermally excited electrons leave behind the holes in the valence band. Both charge carriers are in a state of equilibrium, which is determined by dissociation and recombination phenomena.

Semiconductors are characterized by density of charge carriers, which for Si is $\sim 1.4 \times 10^{10}$ cm^{-3} at ~25 °C. This is a rather low value when compared to doped silicon. The number of charge carriers in moderately or heavily doped semiconductors typically range from 10^{13} cm^{-3} to 10^{18} cm^{-3}, and the degenerately doped material contains above 10^{18} cm^{-3} of a dopant. Note also that the mobility of charge carriers is orders of magnitude higher when compared to the movement of ions in solutions (Bard 2001). The authors record the mobility of electrons in Si to be as high as $u_n = 1350$ cm^2 V^{-1} s^{-1} and that of holes $u_p = 480$ cm^2 V^{-1} s^{-1}.

When a semiconductor comes in contact with an electrolyte, charge transfer between both phases occurs. Thus, n-type Si will release its majority carriers – electrons – to the solution phase where they will be accepted by electroactive species, for instance, a redox couple (Red + ne$^-$ → Ox) introduced into the solution. As a result, an electrostatic equilibrium will be achieved with a positively charged semiconductor and a negatively charged adjacent solution phase. The excess charge in the solid phase does not reside at the solid–liquid junction; it is distributed within a much wider area known as a SCR. The thickness of the SCR is typically in the range of 50–2000 Å (Bard 2001, p. 753). The potential drop is localized within the semiconductor across the SCR rather than at the liquid junction as in the case of metallic electrodes, which form the electric double layer with the electrolyte.

The properties of the semiconductor–electrolyte junction could significantly change when a semiconductor surface is passivated, for instance, with an oxide layer. The oxide film can also form spontaneously during exposure to the atmosphere or to an aqueous liquid environment.

Passivation of silicon by thin oxide films is a promising approach to increasing the efficiency of PV devices. When an oxide film is applied, the effective surface recombination velocity (S_{eff}) and the interface state density of silicon are substantially reduced. This leads to substantial improvement in the efficiency of PV devices. In organic solar cells, an ultra-thin oxide interlayer reduces the recombination losses, improves the electron transport at the interface, and enhances the stability of the solar cells up to 80% of their initial efficiency after 750 hours in operation (Polydorou et al. 2018).

Another important area of application of oxides on Si is protection of the substrate from aggressive environment as well as improving its electrochemical stability. The electrochemical instability of semiconductors in aqueous electrolytes is an issue, which limits the development of efficient sunlight-driven hydrogen–oxygen generators from water. Application of a thin film material for the protection of photoelectrodes in solar fuel generators has been reviewed (Hu et al. 2015). The authors summarized a variety of metal and oxide films, which have been proposed for the protection of n-Si photoanodes applied for oxygen evolution. These include metals (Au, Pd, and Ni), oxides (indium tin [ITO], TiO_2, NiO_x, MnO_x, Fe_2O_3, CoO_x, $NiRuO_x$, and $NiCo_2O_4$), metal oxide systems (TiO_2/Ir, Pt–Pd/MnO_x, and Ti–Fe_2O_3), and oxide stacks (Cr_2O_3/TiO_2/SbO_x, ITO with RuO_2, TiO_2/Ni/NiO_x, Pt/SiO_x, CoO_x, and Fe_2O_3). The following layers were studied for the protection of hydrogen-evolving p-Si photocathodes: Pt, Ti, Ti/Ni, SiO_2, Al_2O_3, TiO_2, ZnO, $SrTiO_3$, Mo_3S_4, Mo/MoS_2, and CdS.

Passivation in semiconductor physics (including PV) and electrochemistry refers to, albeit related, different phenomena. Electronic passivation in PV addresses the reduction of interface state density (D_{if}) (unsaturated bonds) on the semiconductor surface, e.g. by chemical reduction applying passivating dielectrics (Cui et al. 2017a; Singh et al. 2016). Another important type is the field effect passivation, which relates to electrostatic shielding of the charge carriers from the interface by an internal electric field, e.g. the films having a negative (Al_2O_3 and $SiO_2 + HfO_2$) or positive (SiN_x) charge (Cui et al. 2017a; Oudot et al. 2018).

In electrochemistry, an additional component – electrolyte – plays an essential role. Particularly, specific adsorption of H^+ or OH^- ions on silicon is of importance. The adsorbed ions can charge the interface and create a field effect by analogy with that under gaseous (atmosphere) conditions. The interface is typically composed of the semiconducting substrate, the passive layer, and the electrolyte. The properties of the passive layer, e.g. electric conductivity, in gaseous and electrolyte environments can be different because of electrolyte penetration into the passive layer. Electrochemical passivation primarily refers to inhibition of the charge transfer from a semiconductor to an electroactive species in an electrolyte. To this end, the semiconductor is coated by a layer, which protects the substrate from aggressive environment and prevents electron transfer to the electrolyte. Another kind of passivation is the spontaneous formation of protective film from corrosion products. It is desirable that the passive layer would be able to reduce recombination velocity in the semiconductor, protect the Si/passive layer/electrolyte junctions against corrosion, and, at the same time, does not inhibit substantially the electron transfer to the electrolyte. However, in most cases, the passivating films mitigate Si corrosion at the expense of surface activity: the electrode becomes electrochemically inactive in certain potential regions. The dilemma is typically solved in favor of passivity because dissolution leads to continual destruction of the semiconductor and failure propagation throughout active regions (Hu et al. 2015).

Electrochemists are in the habit of evaluating electrode energy levels using a conventional scale that is in respect to a normal hydrogen electrode (NHE) as a reference electrode (RE). Data were derived from early works of determination of NHE on an absolute scale (that is in respect to vacuum), which scattered from −4.3 to −4.7 eV (Lohman 1967; Gerischer 1975; Nozik 1996). Typically, an average value of $E = -4.5$ eV for NHE is used (Bard 1979, 2003; Nozik 1996).

It is worth mentioning that the protected electrode contains a buried solid junction (e.g. Si-oxide), and the interface represents rather a PV cell than the PEC cell. Such electrode can be ideally polarizable: no Faradaic process occurs on the electrode during polarization if the conductivity of the passive layer is high. However, unlike in gaseous environments, the buried solid junction in the electrolyte can undergo transformations. Distortion of oxide surfaces by penetration of water and ions is a general phenomenon (Perram et al. 1974). Water incorporation into oxide could reduce its density, refraction index, or acoustic velocity (Davis and Tomozawa 1995). Proton and water penetration into oxide is of importance to understand the instability of dielectrics during operation of ion-sensitive field effect transistors (ISFETs). The presence of water- or hydrogen-related species (–H and –OH) in ISFETs could enhance interface state generation and charge trapping (Topkar and Lal 1995; Hazarika and Sharma 2014). These aspects will be discussed in more detail below when considering the nanogravimetric evidence of electrolytic transformations.

7.2 Basic Parameters: Photopotential, Photocurrent, and Photocapacitance

Figure 7.1 outlines a typical experimental setup used for the PEC studies of a silicon wafer with a passive layer. The illumination is supplied through the transparent optical window, which is typically made of quartz. The three-electrode cell comprises the working, the counter, and the reference electrodes. The Si wafer with a passive layer acts as a working electrode. The counter electrode is typically made of an inert metal (e.g. Pt). The silver/silver chloride (Ag/AgCl) or calomel (Hg/Hg_2Cl_2) electrodes are used as the reference electrodes. The electrodes are connected to a potentiostat, an instrument, which controls the cell voltage and measures the current density that flows across the working electrode, that is, through the cell between the working and the counter electrodes. The entire voltage drop in the

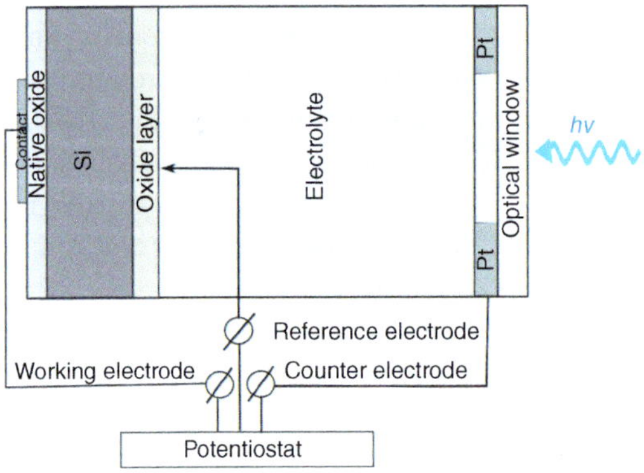

Figure 7.1 Schematic representation of the experimental setup typically used for the PEC studies of Si wafer covered with a passivating layer.

cell includes several contributions: the potential drop in the electrolyte, that within the passivating layer and the wafer, and the drop at the backside contact with a native oxide film.

When the semiconductor comes in contact with the electrolyte, a SCR in the solid phase forms and bending of the energy band at the solution–semiconductor interface occurs. In an n-type semiconductor, access holes are produced, and the created electric field has a direction, which drives the holes toward the electrolyte, whereas the produced electrons move toward the semiconductor bulk. A vice versa process occurs in a p-type semiconductor: the electrons move toward the solution and the holes toward the bulk of the solid. When the sample is irradiated, photons create electron–hole pairs once the light energy is greater than that of the bandgap. In an n-type semiconductor, minority carriers – holes – are produced, which, because of the space–charge field, move to the surface and could oxidize the species in the electrolyte. In contrast, in a p-type semiconductor, photoelectrons are produced, and the field in the SCR promotes the electron movement toward the surface with the opportunity of electroreduction reaction of species in the electrolyte. Thus, illumination induces a photocathodic current in a p-type semiconductor, and photoanodic current flows in an n-type semiconductor.

The typical photopotential variation of p-Si at an open circuit is shown in Figure 7.2. The electrode has been prepared from a p-Si wafer removing the native oxide in the hydrofluoric acid (HF) solution. The HF-treated Si surface is terminated by hydrogen in the form of mono-, di-, and tri-hydrides (Zhang 2004, pp. 58–59). Minor amounts of fluorine- and oxygen-terminated sites are also possible. Silicon oxide could also develop in the solution because of the replacement of the hydrogen termination by a hydroxyl one. We will denote such electrode as hydrogen terminated (Si–H) keeping in mind that this is a certain simplification. Figure 7.2 shows the PEC measurements performed using $\lambda = 505$ nm illumination, which is around the middle of the visible spectrum (~380 to ~750 nm). The applied photon energy (2.45 eV) is greater than the bandgap energy of Si ($E_{g,Si} = 1.1$ eV); it was sufficient to excite the Si photoelectrons to the conduction band. The PEC measurements have been performed in an acid perchlorate solution without adding any specific redox couple. The positive potential drift, which is marked as a baseline (E_{BL}), indicates gradual surface oxidation, that is, development of an oxide film. The thickness of the native oxide on silicon developed in solutions is typically less than 2 nm depending on the environmental conditions (Zhang 2004, pp. 58–59). The potential of the illuminated electrode in Figure 7.2 rises abruptly and reaches a quasi-steady-state value in respect to E_{BL}. In the dark, a vice versa behavior is observed: the potential drops sharply down and returns to an initial value, that is, E_{BL}.

When the p-Si surface comes in contact with the electrolyte, an energy band is bended negatively, producing an excess charge, which resides in the Helmholtz layer and within a much wider space-charged region. Under illumination, the minority carriers of the p-type semiconductor – electrons – are generated,

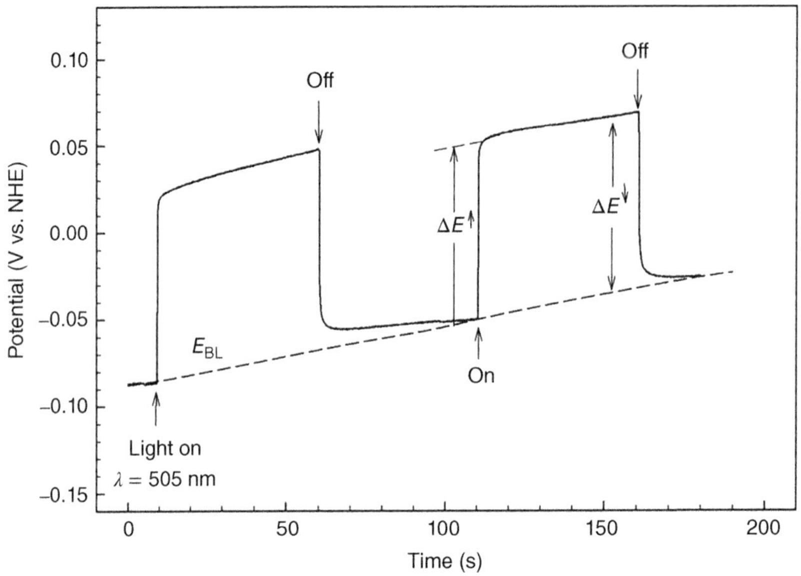

Figure 7.2 Photopotential variation at an open circuit in 1 M $NaClO_4$ (pH 3) of p-Si pretreated in HF solution. A p-type <100> single-crystal silicon polished wafer was used whose diameter was 5 cm, geometric area was 22.8 cm^2, thickness was 275 ± 25 μm, and resistivity was 10–20 $\Omega\,cm^{-1}$. The thickness of the native silica film, which formed spontaneously during polishing and exposing under ambient conditions, was estimated as $d = 0.8$ nm by spectroscopic ellipsometry. Irradiation of $\lambda = 505$ nm and 100 $mW\,cm^{-2}$ power has been applied. The light was chopped at one minute intervals. Source: Reproduced from (P. Kalinauskas et al. 2022), which is an open-access article distributed under the terms of the Creative Commons Attribution 4.0 License (CC BY). http://creativecommons.org/licenses/by/4.0 (© The Authors, 2022).

producing electron–hole (e^-/h^+) pairs so that holes arrive at the surface and charge it positively (Zhang 2004, p. 31). A vice versa process occurs when illumination is terminated: photoelectrons recombine and the electrode potential drops down to an initial value. Such conventional concept substantiates the experimental observations in Figure 7.2.

Figure 7.3 demonstrates the cathodic photocurrents induced on the p-Si photoelectrode. The photoresponse is due to a reaction of hydrogen reduction ($2H^+ + 2e^- \rightarrow H_2$) in an acid solution. The photocurrent grows exponentially with the cathodic bias and reaches some limit. The electrode is passive in the dark: the current drops to a negligible value when illumination is terminated. Thus, the dark electrode shows the properties of an ideally polarizable electrode.

Capacitance (C, $F\,cm^{-2}$) is one of the fundamental properties of the semiconductor–electrolyte interface, which characterizes the ability of the photoelectrode to store electrical charge. When measuring C as a function of applied voltage (E), relevant information is obtained about the type of the semiconductor (n–p) the photoelectrode is made from. Another fundamental property to obtain

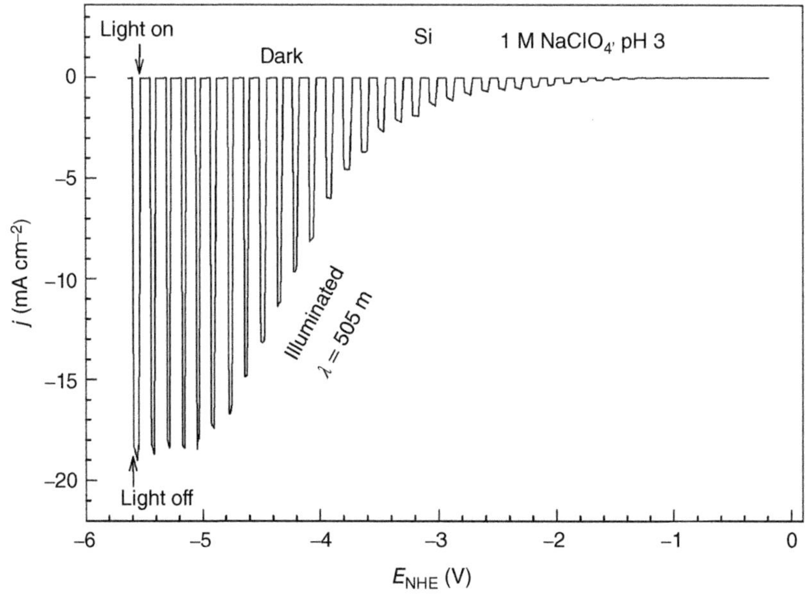

Figure 7.3 Photocurrents induced on p-Si in 1 M $NaClO_4$ (pH 3) in a wide range of cathodic polarizations. A polished p-type <100> silicon wafer was used to prepare the electrode whose thickness was 275 ± 25 μm and resistivity was 10–20 Ω cm^{-1}. The electrode contact was made on the backside of the wafer. Irradiation of $\lambda = 505$ nm and 100 mW cm^{-2} power has been applied. The light was chopped at one minute intervals. The entire voltage includes the drops in the electrolyte, within the Si wafer, and native oxide at the back contact. Source: Authors: P. Kalinauskas and E. Juzeliūnas.

is the flat-band potential (E_{fb}), which is the potential at which there is no charge depletion at the interface. The capacitance data are typically plotted C^{-2} vs. E that is in accordance with the Mott–Schottky relation:

$$C_{sc}^{-2} = \frac{2}{\varepsilon_0 \varepsilon N_{D/A} e} \left(E - E_{fb} - \frac{kT}{e} \right) \tag{7.1}$$

where $N_{D/A}$ stands for the number of donors or acceptors in the space–charge layer, k is Boltzmann's constant, ε is the dielectric constant of the semiconductor, ε_0 is the permittivity of free space, T is the absolute temperature, e is the charge of hole or electron, and E_{fb} is the flat band potential. E_{fb} indicates the potential, at which the energy band bending turns to zero. It can be determined when extrapolating the C^{-2}–E curve to $C^{-2} = 0$. The curve slope yields another important parameter – $N_{D/A}$, which characterizes the doping level of the dark electrode or the density of the photogenerated charges in the space–charge layer.

Figure 7.4 demonstrates the Mott–Schottky plots for the dark and illuminated p-Si–H electrode. Obviously, the E – C behavior under dark and illuminated conditions is different: the capacitance of the dark electrode decreases with negative polarization, whereas the capacitance of the illuminated electrode increases.

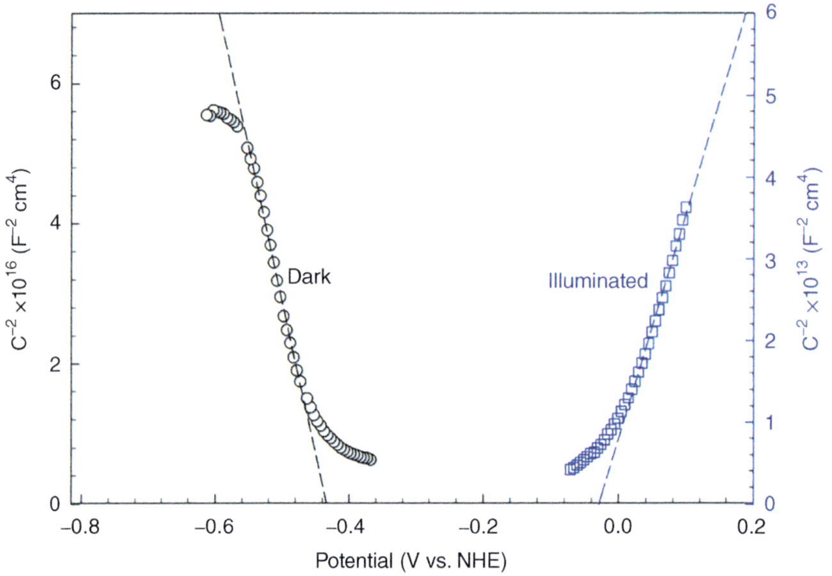

Figure 7.4 Mott–Schottky plots for the dark and illuminated p-Si-H electrode in a 1.0 M NaClO4 (pH 3) electrolyte obtained when polarizing the electrode in negative direction. Source: Authors: E. Juzeliūnas, P. Kalinauskas, and A. Grigucevičienė.

This indicates a decrease in the number of the positive charge carriers (holes) of the dark electrode because of the supply of negative charges (electrons) to the SCR. In contrast, photoelectrons are produced during illumination, and negative polarization increases the number of negative carriers. It is also obvious that the *C* values of the illuminated electrode are more than by 1 order of magnitude higher (6–10 nF cm^{-2} in the dark vs. 160–450 nF cm^{-2} under illumination). More examples of Mott–Schottky relation will be demonstrated when discussing the properties of the Si–HfO_2 electrode.

The PEC measurements enable rapid characterization of quality of Si films. Figure 7.5 shows the photocurrents generated on n-Si and p-Si layers, which were electrodeposited in a molten SiO_2–CaO–$CaCl_2$ electrolyte at 850 °C (Zou et al. 2019). The PEC measurements have been conducted in the electrolytes with redox agents: ethyl viologen diperchlorate ($EV(ClO_4)_2$) for p-Si and ferrocene $Fe(C_5H_5)_2$ for n-Si. One could see that the photocurrent density of the deposited silicon is about 40–50% of the currents generated on commercial wafers. This important achievement demonstrated significant progress in the electrochemical deposition of Si layers as a strategy to fabricate crystalline silicon films for low-cost PV applications.

The schematics of photogeneration of electrons and holes are also shown in Figure 7.5. When light is absorbed in p-Si, the energy of photoelectrons increases from valence to conduction band and a redox agent in the solution accepts these electrons. In n-Si, the holes are produced, which oxidizes the redox agent that is Fe^{2+} to Fe^{3+}.

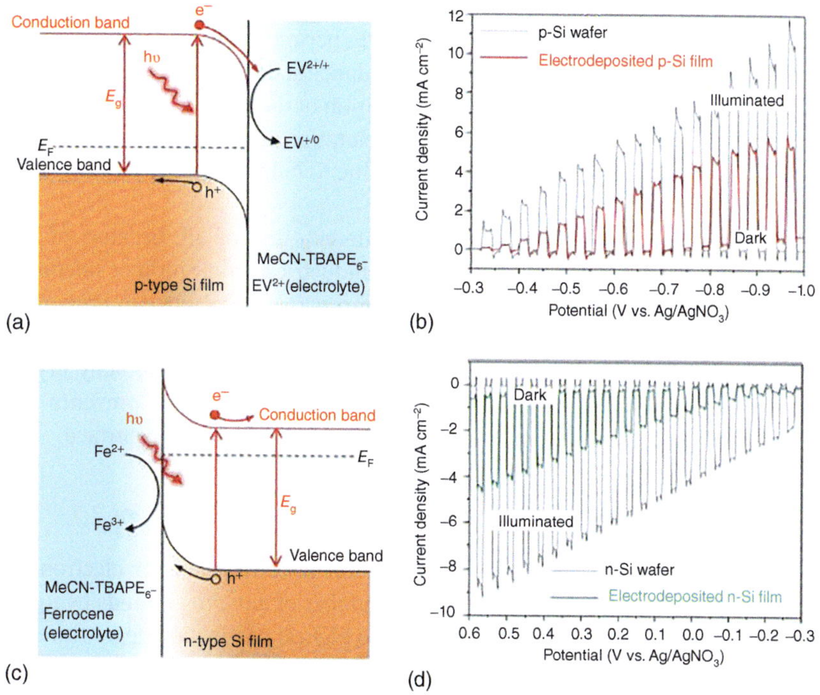

Figure 7.5 Schematic illustration of electron photoexcitation to the conduction band by illumination of a p-type semiconductor (a) as well as generation of holes in a n-type semiconductor (c) in contact with an electrolyte composed of the acetonitrile (CH_3CN) solution, 0.1 M tetrabutylammonium hexafluorophosphate ($TBAPF_6$) as the supporting electrolyte, and 0.05 M ethyl viologen diperchlorate ($EV(ClO_4)_2$) as the redox agent for p-Si and 0.05 M ferrocene $Fe(C_5H_5)_2$ for n-Si. Figures (b) and (d) show the photocurrents generated at different electrode polarizations by the p-type and n-type Si layers electrodeposited in a molten SiO_2–CaO–$CaCl_2$ electrolyte at 850 °C. Xenon lamp irradiation at 100 mW cm^{-2} was applied in a chopping on/off regime. Source: Reproduced from (Zou et al. 2019), which is an open-access article distributed under the terms of the Creative Commons Attribution 4.0 License (CC BY). http://creativecommons.org/licenses/by/4.0 (© The Authors, 2019).

7.3 Photoelectrochemical Features of the Si-Oxide Electrodes

Most of the technically important metals and alloys are spontaneously covered with a native oxide layer. An oxygen-containing layer may form in the electrolyte as well. Such layers typically exhibit semiconductor properties, for instance, photosensitivity, which is detectable as photopotentials and photocurrents. When irradiated, the metal oxide electrode absorbs the light and produces the photocurrent, which depends on various factors such as the wavelength, electrolyte composition, or electrode potential. PEC studies are also important in terms of exploitation of metals under solar irradiation. For instance, the phenomena of photoaccelerated or photoinhibited corrosion of zinc has been studied under polychromatic and laser illumination (Kalinauskas et al. 2001). Experiments using quartz crystal microbalance (QCM) and electrochemical impedance spectroscopy (EIS) indicated that light absorption does not induce decomposition of the oxide layer. A dual light effect on zinc corrosion was observed: photoacceleration at first stages and photoinhibition during prolonged exposures. This was due to light-assisted generation of a protective layer.

Silicon covered with an oxide layer is of interest, first of all, because of the importance of the systems in electronic and PV industries. Passivation of silicon by ultrathin oxide films is a typical approach to improving the electronic properties of silicon-based PV devices. The films reduce the number of unsaturated Si bonds (dangling bonds), which arise because of the disruption of the periodicity of the crystal structure at the surface (Zhang 2004). The surface oxide prevents the development of the recombination centers for minority carriers at the surface.

7.3.1 Si–SiO_2 Electrode

Silicon passivated with its oxide is very important in semiconductor electronics. Passivation by thermal SiO_2 could reduce the number of dangling bounds by up to 6 orders of magnitude, and this is regarded as the highest passivation degree among the semiconductors with the surface film (Michaelis 2008). The Si–SiO_2 interface, therefore, could serve as a reference to compare the photoresponsiveness of the Si passivated with ultrathin oxide layers.

Various methods of SiO_2 layer production on silicon are applied, such as wet or dry thermal oxidation, anodic oxidation of p-Si, chemical vapor deposition (CVD), or physical vapor deposition (PVD). These classical methods of Si oxidation have been properly surveyed (Wolf and Tauber 2000; Michaelis 2008). The thermally formed silica layer is regarded as a material of highest passivation capability (Michaelis 2008). Thermal oxidation is usually performed in the temperature range between 800 and 1200 °C. Oxygen penetrates at these temperatures into the Si lattice and forms the Si–SiO_2 interface with no other inhomogeneities so that the problems related to interface state densities caused by a mismatch in the lattice constants, or other geometric factors, are minimal.

We will limit our discussion of the Si-oxide electrodes with the p-type Si, which is an important photocathode for PEC hydrogen production. Figure 7.6 shows the PEC responses of a Si–SiO_2 electrode with a 10 nm oxide layer at the open circuit, that is, when no voltage to the electrode is supplied. The measurements have been performed using $\lambda = 505$ nm illumination, which is around the middle of the visible spectrum (~380 to 750 nm). The photon energy of this illumination is 2.45 eV, which is greater than the bandgap of silicon ($E_g = 1.1$ eV). Note also that the bandgap of dry amorphous SiO_2 is as high as $E_{g,SiO_2} = 7.64$ eV (Tan et al. 2005). The oxide can retain its high insulating properties in the electrolyte as well, particularly at the initial stages of exposure. Another important factor that determines the photopotential responses is film transparency (*Tr*). Evaluation based on the complex matrix form on the Fresnel equations at $\lambda = 505$ nm shows

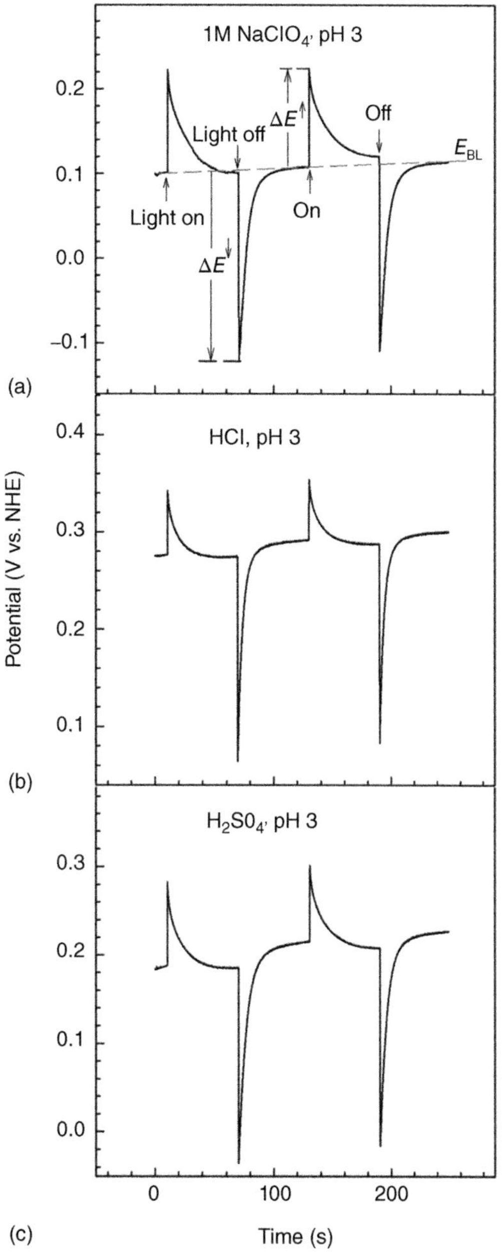

Figure 7.6 Photopotential variation at an open circuit for p-Si with 10 nm thermal oxide in 1 M $NaClO_4$, HCl, and H_2SO_4 solutions adjusted to pH 3. Illumination $\lambda = 505$ nm and $N = 50$ mW cm^{-2} has been chopped at one minute intervals. Source: Reproduced from (Kalinauskas et al. 2022), which is an open-access article distributed under the terms of the Creative Commons Attribution 4.0 License (CC BY). © The Authors, 2022. http://creativecommons.org/licenses/by/4.0.

that the 10 nm silica film is nearly completely transparent: $Tr_{SiO_2} = 0.995$ (http://www.filmetrics.com/reflectance-calculator).

The photopotential responses of the Si–SiO_2 electrode (Figure 7.6) are distinctively different from those of the electrode without a passive film (Figure 7.2). The photopotential of the passivated electrode increases abruptly ($\Delta E^{\uparrow}$) and, then, shifts slowly down, reaching the initial value in the dark (marked as a baseline, E_{BL}).

A vice versa variation is observed when illumination is terminated: the potential drops sharply down and, afterward, increases to the E_{BL} level. A distinctive feature is the prolonged time of potential relaxation (t_R), which takes around one minute. Another symptomatic feature is asymmetry of the curves when illumination is switched on and off: while the curve shapes are mirrored, their size is different: $|\Delta E^{\uparrow}| < |\Delta E^{\downarrow}|$. It is also important that the photopotential curves are similar in various electrolytes: perchlorate, sulfate, and chloride.

The sample exhibited similar photopotential features with a native oxide whose thickness was estimated by spectroscopic ellipsometry to be 0.8 nm (Figure 7.7). The curve shapes of this electrode are similar to those observed for the Si having 10 nm SiO_2 (Figure 7.6a) with the only difference being that the photopotential does not return to an initial value (E_{BL}) – it retains some residual value: $E_R = \Delta E^{\uparrow} - E_{BL} > 0$.

The above PEC measurements have been performed in the electrolytes without a specific redox couple. Under such conditions, the p-Si electrodes are nearly ideally polarizable; actually, no Faradaic process is induced during the illumination. The only electroactive species in the electrolyte are hydrogen ions. However, reduction of these ions is highly suppressed by the oxide: considerable photocurrents for the Si-oxide electrodes appeared at high cathodic polarization $E < -1.5\ V_{NHE}$. In the dark, no Faradaic process was observed even at such high polarization. Such cathodic behavior is typical of a p-type semiconductor.

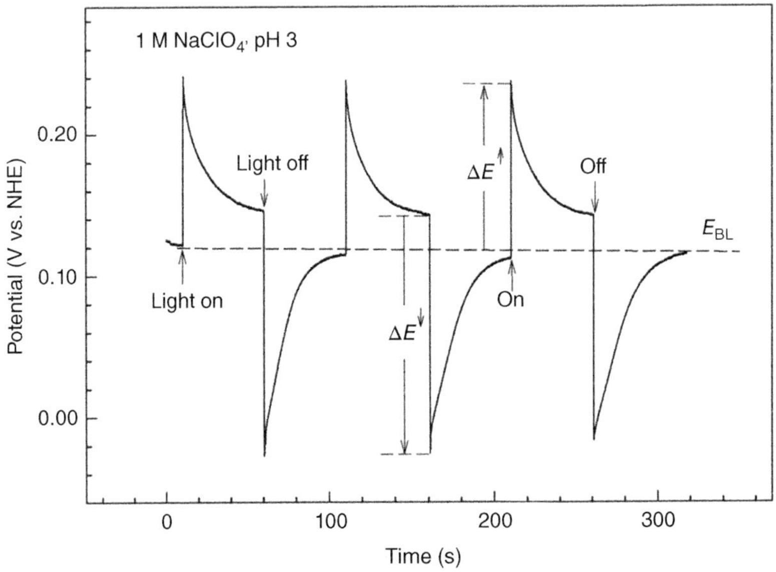

Figure 7.7 Photopotential variation at an open circuit for p-Si with a native oxide film (~0.8 nm) in 1 M $NaClO_4$ (pH 3). Illumination $\lambda = 505$ nm and $N = 50$ mW cm^{-2} has been chopped at one minute intervals. Source: Reproduced from (Kalinauskas et al. 2022), which is an open-access article distributed under the terms of the Creative Commons Attribution 4.0 License (CC BY). http://creativecommons.org/licenses/by/4.0 (© The Authors, 2022).

It is also important that the carrier concentration in Si wafer, produced through defined manufacturing processes (e.g. Czochralski process, CZ), varies locally – both axially and radially, and the produced wafers are specified to a certain resistivity range. It is difficult to prepare two identical Si electrodes with adequate reproducible characteristics. Therefore, the photocharacteristics can somewhat vary from sample to sample.

Several phenomena could be considered as the reasons for the observed specifics of the photoelectrode. One of them is adsorption of electrolyte ions, which counterbalances the photoinduced Si surface charging. There is movement of photoelectrons in the p-type semiconductor so that the Si surface at the interface becomes positively charged according to the reasons discussed above. The positive photocharging attracts negative electrolyte ions pursuing an electrostatic equilibrium. The process by nature is a non-specific adsorption when ions approach an interface because of the long-range Coulombic interactions and retain their solvent shell. Thus, the ions are separated from the interface by molecular layers (IUPAC 1997). As a result, the electrode potential shifts negatively. When the adsorbed anions completely compensate the photoinduced charge, the electrode potential returns to its initial value in the dark. The process should be relatively slow; it will include such steps as ion diffusion, partial dehydration, and adsorption. It is also worth mentioning that charge (electrons and holes) mobility in semiconductors is orders of magnitude higher than the charge (ions) mobility in solutions (Bard and Faulkner 2001, p. 52). When illumination is terminated, the electron–hole pairs recombine, the photoelectrons no longer exist, and the potential is determined solely by the ions, which have been attracted to the interface during the photocharging. As a result, the potential reduces abruptly ($\Delta E^{\downarrow}$). Thereafter, the potential returns to the initial (E_{BL}) level because of desorption of the photoadsorbed anions.

The ions are non-specifically adsorbed when they are attracted (or repulsed) solely by long-rage Coulombic interactions. For non-specific adsorption, chemical identities of ions could be ignored because the electric charge of the adsorbate plays the main role in such process and not its chemical nature. However, the photoresponses in Figure 7.6 actually do not depend on electrolyte concentration: the curve shapes are similar in the electrolytes when ion concentrations differ by 3 orders of magnitude. This indicates that the photoeffects relate rather to the solid phase and not to the solution phase.

Another concept to explain the relaxation behavior is based on the negative photocharging of the passivating film. As already discussed, the silicon surface becomes positively charged during the first instances of illumination because of photoelectron generation. One could assume that with illumination time, the photoelectrons, being isolated from the solution, penetrate the oxide film and balance the positive charge of the Si surface. As a result, the electrode potential slowly returns to the initial value in the dark. When illumination is terminated, electrons and holes recombine immediately, and the electrode potential remains determined by the charge in the oxide film as $\Delta E\downarrow$ indicates. This charge slowly ceases and the electrode potential returns to the initial value.

Measurement of electrode capacitance supports the above statement (Figure 7.8).

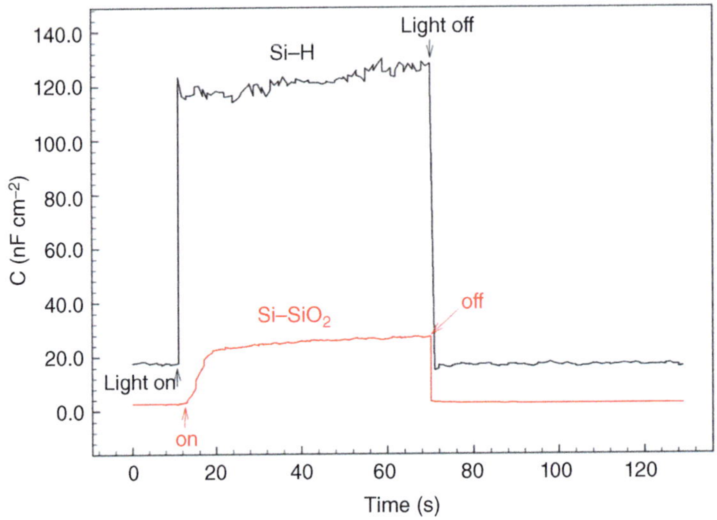

Figure 7.8 Capacitance variation when illuminating p-Si–H and p-Si with a thermal 10 nm SiO_2 film ($\lambda = 505$ nm and $N = 50$ mW cm^{-2}) in 1 M $NaClO_4$ (pH 3). Source: Reproduced from (Kalinauskas et al. 2022), which is an open-access article distributed under the terms of the Creative Commons Attribution 4.0 License (CC BY). http://creativecommons.org/licenses/by/4.0 (© The Authors, 2022).

When illuminated, the electrode photocapacitance increases rather slowly; this corresponds to the slow photopotential relaxation in Figure 7.6. When illumination is terminated, the photocapacitance immediately reduces, whereas the potential curve in Figure 7.6 shows prolonged relaxation. Such behavior implies fast discharging of Si because of the carriers' recombination, which is followed by much slower discharge of the oxide film. The observed asymmetry $\Delta E\uparrow < \Delta E\downarrow$ indicates that the interfacial charging lasts longer than the $\Delta E\uparrow$ indicates. The positive potential shift $\Delta E\uparrow$ is compensated from the very beginning of illumination because of the discussed electron transfer to the oxide film. Note that the slow photocharging is not observed for the Si–H electrode, whose capacitance is determined by the double layer at the Si–solution interface. In this case, there is no phase to be charged.

The Si–SiO_2 electrode capacitance measured in the dark are on the order of nanofarads (nF cm^{-2}). Such values are rather low when compared to metallic electrodes whose capacitances are typically on the order of microfarads. This great difference is because the number of electrons in metals is by far higher than the number of charge carriers in moderately or heavily doped semiconductors.

The potential of an illuminated sample with a native film does not completely relax to the initial value; it holds at some value $E_R > E_{BL}$ (Figure 7.7). This is an indication of some contribution of the positive photopotential of the Si/electrolyte interface as demonstrated for the Si–H electrode (Figure 7.2). The native oxide is very thin, and the electrolyte could penetrate the layer already during the first stages of exposure. Thus, E_R characterizes the Si surface protection by the oxide layer from the contact with the electrolyte.

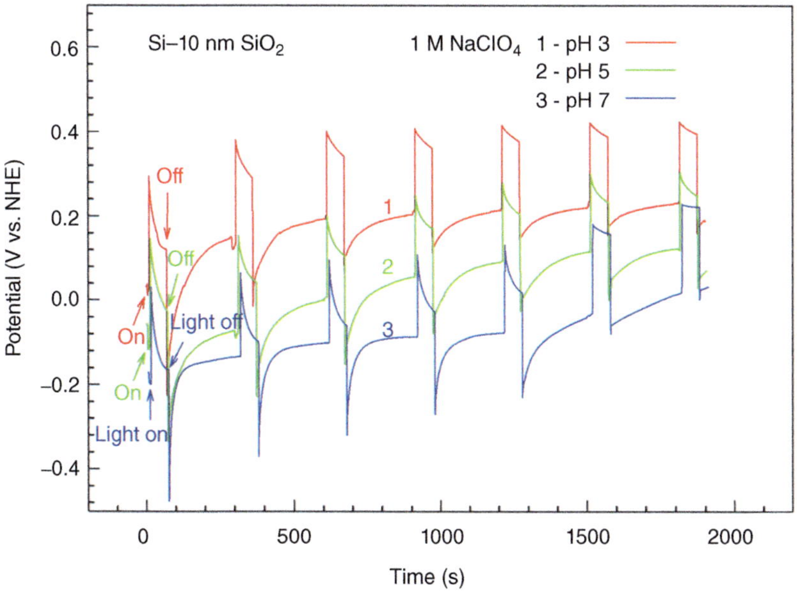

Figure 7.9 Potential variation when illumination has been chopped at an open circuit for p-Si with a 10 nm SiO_2 layer during half an hour exposure in a 1 M $NaClO_4$ solution adjusted to pH 3, 5, and 7. Irradiation was 505 nm wavelength and power was $N = 50\ mW\ cm^{-2}$. Source: Reproduced from (Kalinauskas et al. 2022), which is an open-access article distributed under the terms of the Creative Commons Attribution 4.0 License (CC BY). http://creativecommons.org/licenses/by/4.0 (© The Authors, 2022).

Additional information on the electrolytic transformation of the Si–SiO_2 interface provides the prolonged exposure experiments (Figures 7.9 and 7.10). Figure 7.9 compares the photoresponses determined during half an hour immersion at different pHs in the range from acid (pH 3) to neutral (pH 7). Qualitatively similar responses with distinctive relaxations are observed at the exposure beginning. With exposure time, the relaxations become less pronounced and finally they are not at all observed – the responses resemble those determined for the oxide-free silicon (Figure 7.2). Figure 7.9 demonstrates nearly rectangular responses, which appear when the exposure period is extended to one hour.

As for many semiconductors, the Si surface charge is determined by the specific adsorption of hydrogen or hydroxyl ions. The adsorption–desorption processes of these ions determine the potential of the Helmholtz layer and this obeys the relation (Zhang 2004, p. 13)

$$eV_H = B + kT\ln[H_3O^+] = B - 0.059\ \text{pH} \tag{7.2}$$

Thus, the electrode potential varies by −59 mV per pH unit according to the thermodynamic assumptions.

Experimentally, the $\Delta E-$pH dependence of the Si–SiO_2 electrode (Figure 7.11) varies from ca. $\Delta E/\Delta\text{pH} \sim -25$ mV at the beginning of immersion to $\Delta E/\Delta\text{pH} \sim -60$ mV at the final stages, which is in agreement with the above relation. The electrolyte penetration into the oxide at the beginning of exposure is unlikely.

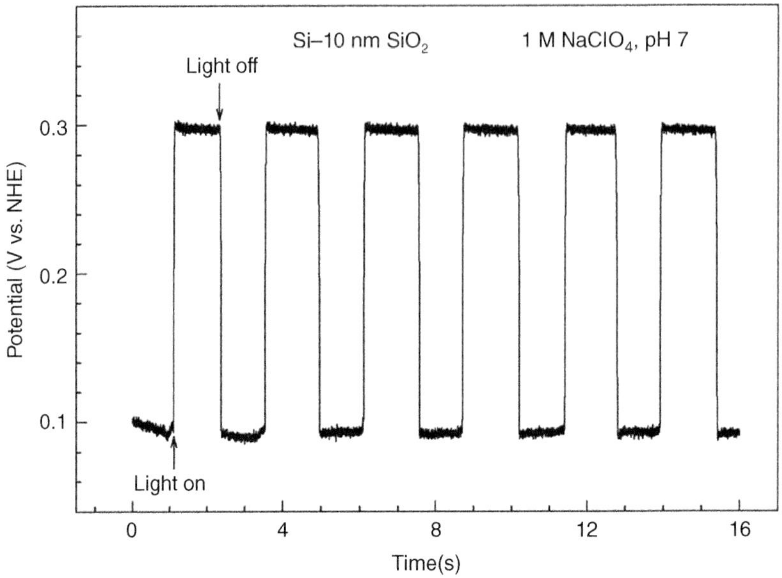

Figure 7.10 Photopotential steps when illumination has been chopped at an open circuit for p-Si with a 10 nm SiO_2 layer after one hour exposure in 1 M $NaClO_4$ (pH 7). Illumination of 505 nm wavelength and 50 mW cm^{-2} power have been applied. Source: Reproduced from (Kalinauskas et al. 2022), which is an open-access article distributed under the terms of the Creative Commons Attribution 4.0 License (CC BY). http://creativecommons.org/licenses/by/4.0 (© The Authors, 2022).

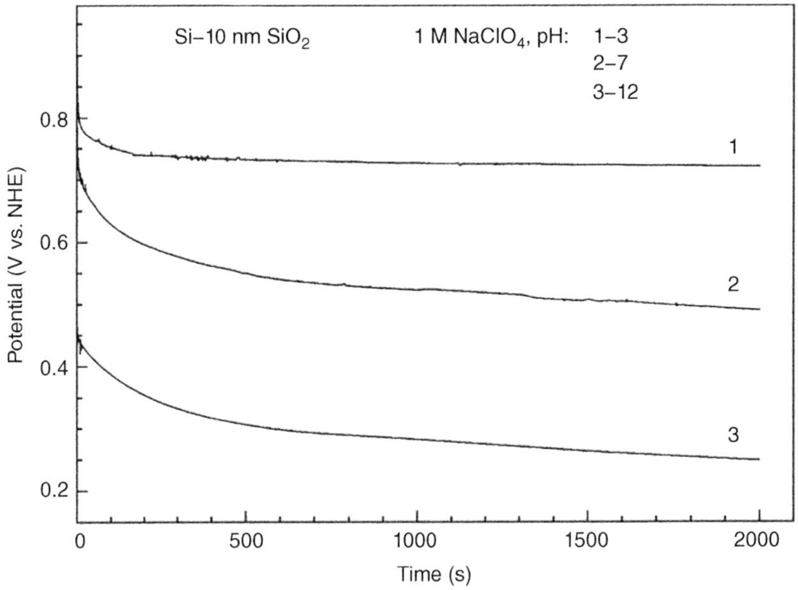

Figure 7.11 Variation of an open-circuit potential of p-Si with a thermal 10 nm SiO_2 layer in acid (pH 3), neutral (pH 7), and alkaline (pH 12) perchlorate electrolytes. Source: Authors: E. Juzeliūnas and P. Kalinauskas.

Nevertheless, the electrode is pH sensitive. With the exposure time, the electrolyte transforms the Si–SiO_2 interface into the Si–SiO_2 electrolyte.

Also, the potential of the dark electrode in Figure 7.9 varies by $\Delta E \sim -51$ mV per pH unit at the beginning of immersion and by $\Delta E \sim 55$ mV at the end of immersion. This is in satisfactory agreement with the condition $\Delta E/\Delta pH = -59$ mV, which is predicted by the specific adsorption of H^+ and OH^-. Again, the Si–SiO_2 interface transforms during the exposure into a hybrid structure of the Si–SiO_2–electrolyte.

An important feature of the Si–SiO_2 electrode is suppression of its photoresponsiveness by electrolyte during exposure. Thus, the initial photopotential ($\Delta E^{\uparrow}$) decreases from 0.29 to 0.19 V (34.5%) at pH 3, 0.26 to 0.19 V (26.9%) at pH 5, and 0.23 to 0.21 V (8.7%) at pH 7. Thus, the depression depends on the electrolyte acidity: the effect weakens when the pH is increased. The highly passive Si–SiO_2 interface degrades in the electrolyte, and this is more pronounced in an acidic environment. The gradual passivity fading leads to an increase of the surface recombination velocity as well as the number of Si surface states, at which charge transference occurs. This aspect is of importance when applying the PEC measurements to characterize the photoefficiency of PV electrodes. The residual potential E_R observed for the native oxide (Figure 7.7) is associated with the formation of the Si/electrolyte interface as well. The oxide layer is relatively thin (0.8 nm); the electrolyte could penetrate it even during the first stages of exposure. On the other hand, E_R provides a measure to evaluate the protection of the surface film from Si contact with the electrolyte.

Silicon oxide highly isolates the electron transfer from Si to electroactive species in an electrolyte. Figure 7.12 compares the photocurrents of the reaction $2H^+ + 2e^- \rightarrow H_2$ on the illuminated Si–H electrode with those on Si with a native and 10 nm thermal oxide. The 10 nm film reduces the photocurrent to negligible values, whereas the photocurrents on the Si–H electrode are substantial ($j \leq 18$ mA cm^{-2}). The curves are almost identical when the potential is swept in the negative and afterward in positive direction. This indicates that no reaction other than hydrogen reduction occurs during the cycling.

7.3.2 Si–HfO_2 Electrode

Hafnium oxide (hafnia, HfO_2) is an electric insulator with a bandgap of 5.3–5.7 eV (Bersch et al. 2008). It has high chemical stability and optical transparence and exhibits photoluminescence properties. Because of these features, hafnium oxide has the potential to be applied in a variety of technological fields: capacitors (Mangla and Gupta 2016), switching memories (memristors) (Bayerl and Lanza 2014; Verelli and Tsoukalas 2011; Shang et al. 2017; Rahaman et al. 2017; Kumar et al. 2017), field effect transistors (Müller et al. 2016; Marjani et al. 2016), heat mirrors (Ramzan et al. 2015), radiotherapy dose enhancement in the treatment of various types of cancers (Magiorella et al. 2012; Marill et al. 2014; Sherck and Won 2017), and transdermal drug delivery (Zhang et al. 2017a). Hafnium oxide is also widely used for the production of various sensors: biomedical (Ortiz-Dosal et al. 2017; Fohlerova and

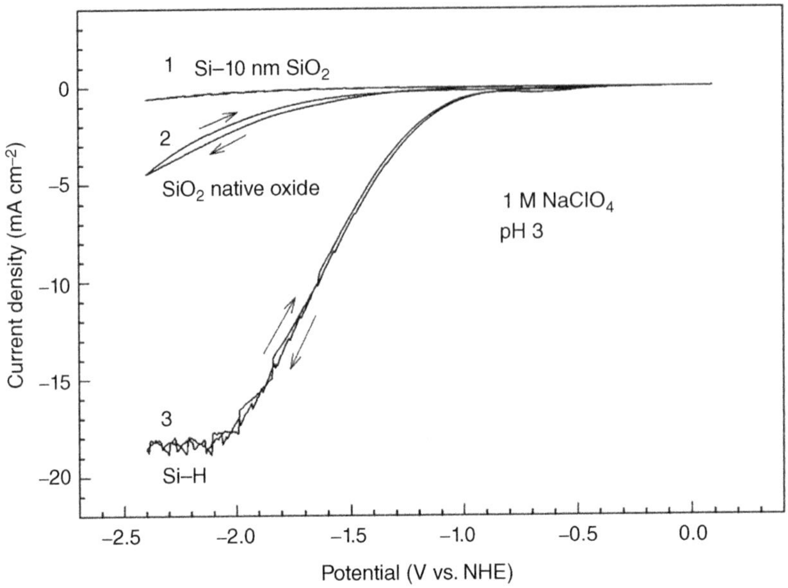

Figure 7.12 Cyclic cathodic potentiodynamic curves obtained for an illuminated p-Si–H electrode and Si with a native oxide (0.8 nm) and 10 nm thermal oxide. Illumination $\lambda = 505$ nm, $N = 50$ mW cm^{-2}, and potential sweep rate 100 mV s^{-1} have been applied. The curves are almost identical when cycling the potential in negative and positive directions. Source: Authors: E. Juzeliūnas and P. Kalinauskas.

Mozalev 2019; Jayaraman et al. 2015; Chen et al. 2010), carbon monoxide (Dave et al. 2018), and phosphopeptide (Huang et al. 2015).

Hafnium oxide is a highly resistive material, whose dielectric constant is at least four times higher when compared to conventional silica gate dielectrics (Chen et al. 2010). The Pourbaix diagram shows that hafnium forms a stable passive oxide over the entire potential-pH range (Pourbaix 1974). Because of high passivity, thin hafnia layers are potential candidates for anticorrosion protection of stainless steel (304 SS) (Li et al. 2019) and magnesium alloys (Staišiūnas et al. 2020). The samples prepared from AZ31 alloy (Mg–3Al–1Zn) with 30 and 60 nm HfO_2 films endured without visible corrosion indications at least 240 hours exposure, which indicates meeting of the requirements for the corrosivity conditions such as those in industrial and coastal areas with moderate salinity as well as chemical plants, swimming pools, coastal ship, and boat yards.

Another potential area of HfO_2 application is protection of silicon from aggressive environment as well as improving the electrochemical stability of the substrate (Staišiūnas et al. 2022). Several other oxides have been proposed for the protection of p-Si photocathodes for sunlight-driven water splitting: TiO_2 (Seger et al. 2013), ZnO (Sun et al. 2012), SiO_x (Esposito et al. 2013; Choi et al. 2014), Al_2O_3 (Choi et al. 2014), and $SrTiO_3$ (Ji et al. 2015).

The HfO_2 films have the potential to be applied in solar devices. Efficiency enhancement by hafnium oxide in organic solar cells has been reported (Polydorou et al. 2018). Another important field in solar energy research is electronic Si passivation by HfO_2 thin films in order to improve the Si surface properties by reducing the number of dangling bonds (Singh et al. 2016; Cui et al. 2017a,b). For instance, p-type Si with a 15 nm-thick hafnia film showed an effective surface recombination velocity as low as $9.9\,cm\,s^{-1}$, an interface state density of $3.6 \times 10^{10}\,cm^{-2}\,eV^{-1}$, and a positive charge density of $5 \times 10^{11}\,cm^{-2}$ (Cui et al. 2017a).

Hafnia films are obtained using various techniques: radio frequency (RF) and direct current (DC) magnetron sputtering (Verelli and Tsoukalas 2011), electrochemical anodizing (Fohlerova and Mozalev 2019), electron beam evaporation (Ramzan et al. 2015), ultrasonic spray pyrolysis (Mendoza et al. 2010), ultraviolet light (UV)-stimulated plasma anodizing (Kushitashvili et al. 2017), pulsed laser deposition (Plociennik et al. 2014), and electrochemical formation in organic electrolytes (Wang et al. 2017). Atomic layer deposition (ALD) is the most widely used method to form thin hafnia films and microstructures. Thermal, mechanical, electrical, optical, and structural properties of ALD-formed HfO_2 have been investigated and reviewed (Gaskins et al. 2017; Berdova and Liu 2016; Jogiaas et al. 2015).

The structural properties of the ALD HfO_2 films are typically studied using the grazing incidence X-ray diffractometry (GI-XRD) (Figure 7.13). The grazing incidence geometry of this method enables investigation of the crystallographic structure of thin films and reduces the influence of the silicon substrate. GI-XRD provided the structural data of HfO_2 and ZrO_2 thin films starting from ~9 nm in thickness (Park et al. 2017; Sharma et al. 2017).

The XRD broad peak in Figure 7.13 indicates the amorphous structure of the hafnia films. A similar peak was observed for amorphous HfO_2 nanoparticles with ~4 nm diameter prepared by the ammonia-catalyzed hydrolysis and condensation of hafnium(IV) *tert*-butoxide (Chaubey et al. 2012). The authors showed that the amorphous structure did not change during annealing the samples up to 400 °C; indications of the monoclinic HfO_2 phase appeared at 500 °C or higher. Another study showed the appearance of HfO_2 crystallinity with a monoclinic structure when the coating thickness has been increased to 80 nm (Staišiūnas et al. 2020). The thicker hafnia films are thermally unstable. For instance, the micrograph in Figure 7.14 shows that the 60 nm hafnia film on silicon delaminated and disintegrated into small islands when annealing the sample at 400 °C.

The crystallization phenomena of HfO_2 nanofilms have attracted considerable attention (Cui et al. 2017a; Singh et al. 2016; Mendoza et al. 2010; Kim et al. 2003; Fan et al. 2012; Zhang et al. 2019). The crystallinity of the films is important in terms of Si surface passivation. The films on silicon up to 35 nm in thickness were found predominantly amorphous with minor localized nucleation of crystals (Cui et al. 2017a; Singh et al. 2016). The amorphous structure yielded highly effective electronic passivation, whereas the film crystallization achieved by annealing at higher temperatures led to degradation of the effective surface recombination velocity.

Zhang et al. (2019) studied structural evolution and crystallization effects of HfO_2/Si interface depending on the post-annealing temperature. The authors concluded that during annealing, the oxygen diffuses from hafnia to silicon,

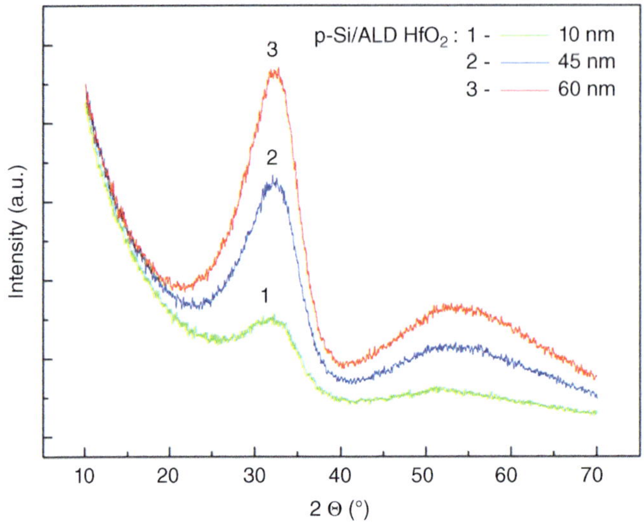

Figure 7.13 Grazing incidence X-ray diffraction (GI-XRD) patterns obtained for p-Si with ALD-deposited HfO_2 layers of different thicknesses: 10 nm, 45 nm, and 60 nm. Source: Reproduced from (Staišiūnas et al. 2022), which is an open-access article distributed under the terms of the Creative Commons Attribution 4.0 License (CC BY). http://creativecommons.org/licenses/by/4.0 (© The Authors, 2022).

Figure 7.14 Thermal surface fragmentation and delamination of the ALD-deposited 60 nm HfO_2 film (bright areas) on the Si substrate (dark areas) when annealed at 400 °C for 30 minutes. The actual width of the imaged area is 1.5 mm. Source: Adopted from (Kalinauskas et al. 2022), which is an open-access article distributed under the terms of the Creative Commons Attribution 4.0 License (CC BY). http://creativecommons.org/licenses/by/4.0 (© The Authors, 2022).

converting the Si surface to silica. At temperatures below 400 °C, both HfO_2 and SiO_2 layers showed an amorphous structure. At temperatures 450–550 °C, the HfO_2 layer transformed into a multiphase polycrystalline structure; the interfacial SiO_2 also became crystalline. At higher temperatures, both oxides transformed to a single-phase polycrystalline structures. The principal mechanism of temperature-dependent structural transformations at the HfO_2/Si interface is illustrated in Figure 7.15, as proposed by Zhang et al. (2019).

Thickness-dependent crystallization was also studied for the films obtained by ion-assisted RF deposition (Zhang et al. 2017b). The authors observed the formation of grains when the coating thickness was greater than 12 nm. The grain size and their density increased with the coating thickness. This phenomenon was explained in terms of Gibbs energy change of the atom clusters. The crystallization depends on the ratio between the cluster surface area and its volume. Thinner coatings have a smaller cluster size and higher surface-to-volume ratio; this results in positive change in Gibbs energy and suppression of the amorphous-to-crystalline transformation. In ALD deposition, such suppression is likely because of the deposition of extremely thin single layers using the layer-by-layer ALD approach (~0.13 nm/cycle in Figure 7.13).

The photopotential responses at the open circuit provide an insight into the PEC properties of the Si–HfO_2 system. Figure 7.16 shows the responses of the Si electrode with the ALD-deposited hafnia films of 10 and 60 nm thickness. In general, the photopotential variation of the Si–HfO_2 electrode is similar to that of the Si–SiO_2 electrode (Figures 7.6 and 7.7). The photopotential increases abruptly during the first instances of illumination ($\Delta E^{\uparrow}$);the increase lasts for tens of microseconds. The photopotential returns to the initial value rather slowly; the relaxation time is on the order of seconds. When illumination is terminated, a vice versa behavior is observed: the potential drops sharply down ($\Delta E^{\downarrow}$) and, afterward, it increases to the initial value in the dark marked as E_{BL}. Table 7.1 summarizes the light-induced potentials and the relaxation times. The data are also given for the Si–SiO_2 electrode thermally coated with a 10 nm silica layer, for comparison.

When repeating the illumination cycles, the potential does not return to the initial value; it retains some residual value $E_R = \Delta E^{\uparrow} - E_{BL} > 0$. The residual parameter is observed for a 10 nm film; this is not the case for the electrode with a thicker film (60 nm) whose potential returns to the value in the dark in both negative and positive courses. As already discussed, the parameter E_R indicates the transformation of silicon oxide interface into silicon oxide electrolyte because of electrolyte absorption by the oxide film. This phenomenon will be discussed in Section 7.4 when introducing the nanogravimetric study. Another symptomatic feature is asymmetry of the curves when illumination is switched on and off: while the curve shapes are mirrored, their size is different: $|\Delta E^{\uparrow}| < |\Delta E^{\downarrow}|$.

The photobehavior of the Si–HfO_2 electrode could be explained analogously as for the Si–SiO_2 electrode as discussed above. Under illumination, minority carriers – electrons – are produced and the Si surface is positively charged. Simultaneously, the charging on the oxide side by the produced electrons and/or

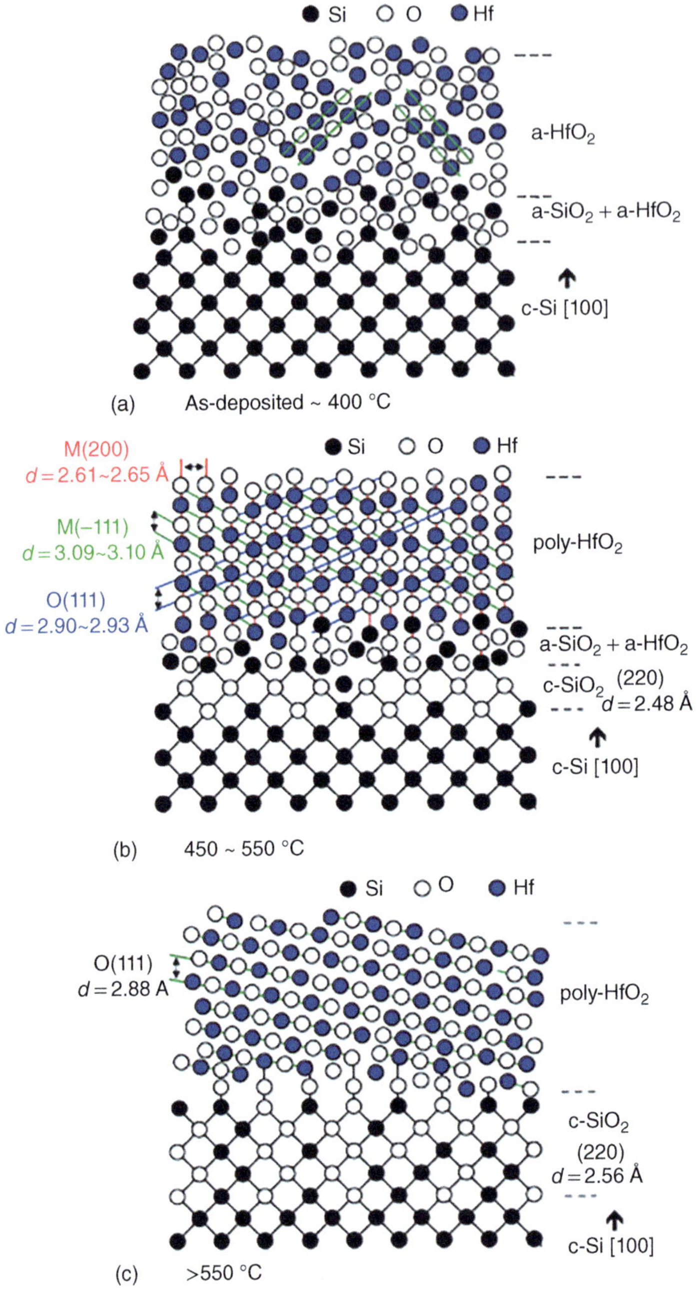

Figure 7.15 Diagram of the mechanism of HfO_2/Si interface transformation from amorphous (a-HfO_2 and a-SiO_2) to crystalline (c-HfO_2 and c-SiO_2) at different annealing temperatures. The diagram also shows the d-parameter and crystal orientations. Source: Reproduced from (Zhang et al. 2019), which is an open-access article distributed under the terms of the Creative Commons Attribution 4.0 License (CC BY). http://creativecommons.org/licenses/by/4.0 (© The Authors, 2019).

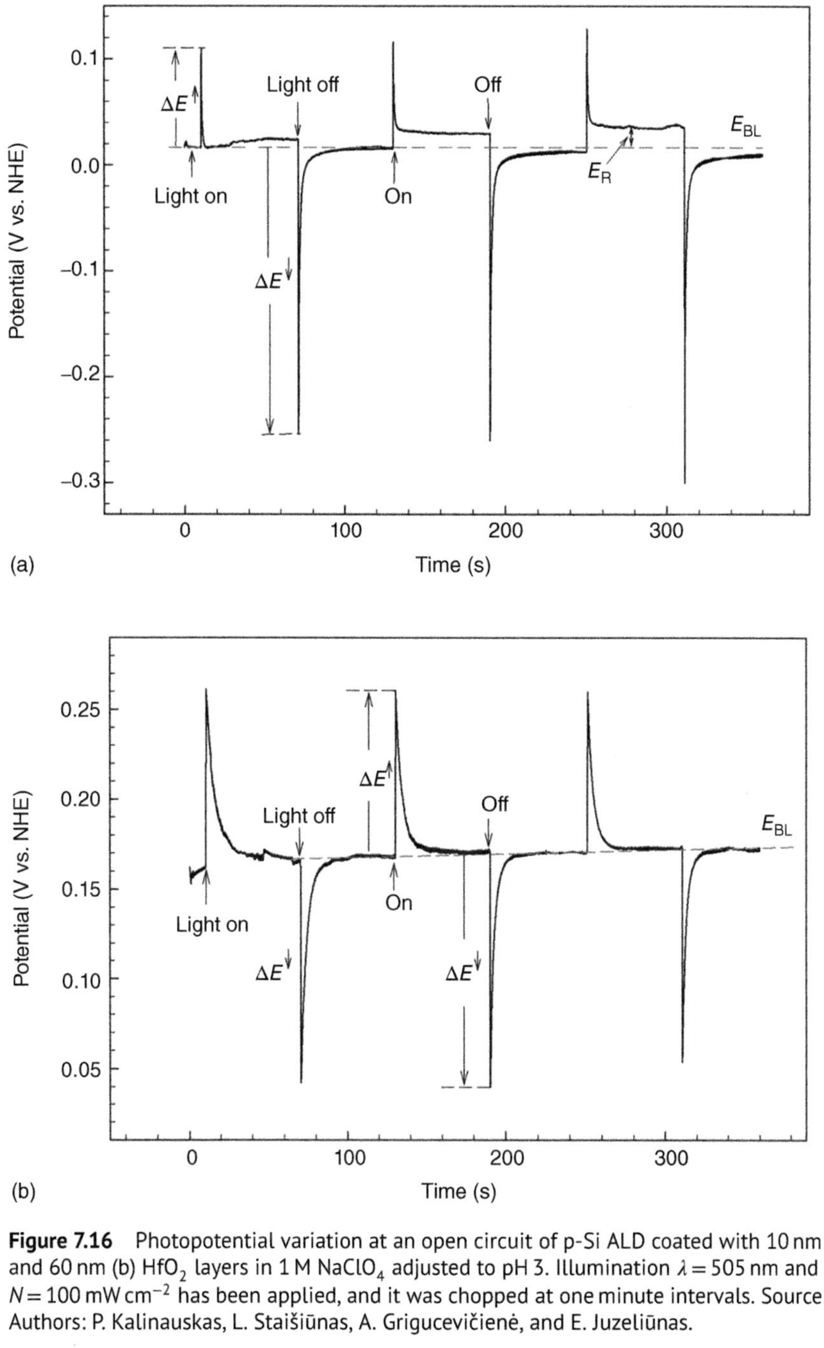

Figure 7.16 Photopotential variation at an open circuit of p-Si ALD coated with 10 nm (a) and 60 nm (b) HfO_2 layers in 1 M $NaClO_4$ adjusted to pH 3. Illumination $\lambda = 505$ nm and $N = 100$ mW cm^{-2} has been applied, and it was chopped at one minute intervals. Source: Authors: P. Kalinauskas, L. Staišiūnas, A. Grigucevičienė, and E. Juzeliūnas.

Table 7.1 The negative and positive potential jumps ($\Delta E^{\uparrow}$ and $\Delta E^{\downarrow}$) measured for Si with 10 nm HfO_2 and 10 nm SiO_2 layers when illumination is chopped as given in Figures 7.6 and 7.16.

Layer	$\Delta E^{\uparrow}$ mV	$\lvert\Delta E^{\downarrow}\rvert$ mV	E_R mV	t_R, s$\Delta E^{\downarrow}$	t_R, s$\Delta E^{\uparrow}$	*Tr*
10 nm HfO_2	101	276	0–20	6	18	0.972
60 nm HfO_2	93	127	—	16	11	0.667
10 nm SiO_2	118	223	0–14	42	38	0.995

Displayed are also the remnant potentials (E_R), the potential relaxation times (t, s), and transparency (*Tr*) of the layers.

at the Si-oxide boundary by the anions occurs. As a result, the positive interface is counterbalanced, and the electrode potential returns to its initial value in the dark. The process is relatively slow. When illumination is terminated, the photoelectrons recombine and the electrode potential reduces abruptly; the negative sheet of the charge, which counterbalanced the positive photocharge, determines the potential. With time, this charging ceases and the potential returns to the initial value in the dark. The schematic representation in Figure 7.17 illustrates the potential variation in the dark and under illumination.

Figure 7.18 shows capacitance variations of Si–H and Si–HfO_2 electrodes when illumination is applied and terminated. The photocapacitance of the coating-free sample (Si–H) remains constant during illumination, whereas that of the HfO_2-coated electrode indicates a slow extra-charging during illumination. Such behavior confirms that the Si–HfO_2 interface undergo some slow charging during the illumination.

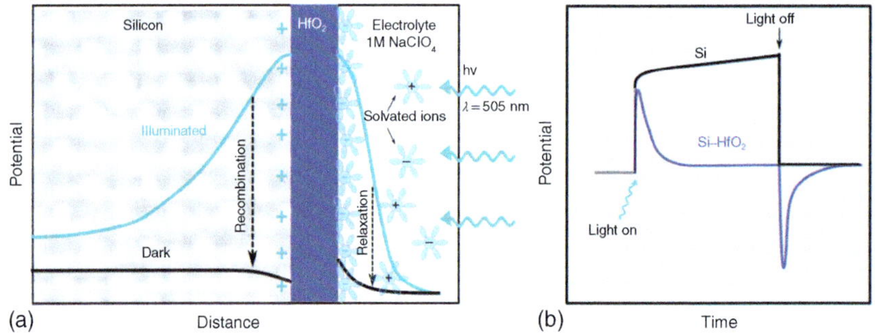

Figure 7.17 Schematic representation of charge distribution in p-Si with an oxide layer and within the electrolyte (image on the left) as well as the potential change (image on the right) when illumination is applied and terminated. When illuminated, photoelectrons are produced in p-Si, its surface is charged positively, and this charge is counterbalanced by electrons or non-specific adsorption of anions. When illumination is terminated, the photoelectrons recombine and the potential drops down as it is determined by the accumulated negative charge. With the time, the negative charge ceases and the electrode potential returns to an initial value in the dark. The potential–time diagram compares the curve for the Si-oxide electrode with that for the oxide-free electrode. Source: Figure authors: E. Juzeliūnas and A. Grigucevičienė.

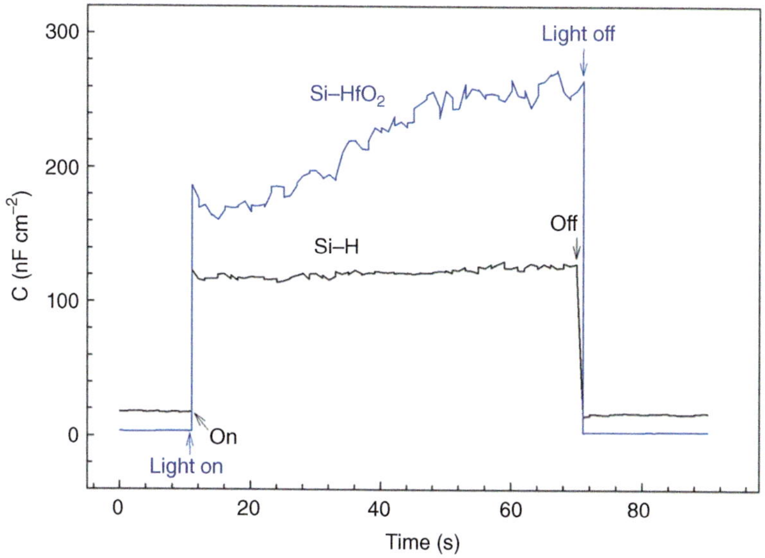

Figure 7.18 Capacitance variation when Si–H and Si–HfO$_2$ (10 nm) electrodes are illuminated by the light of wavelength $\lambda = 505$ nm and power $N = 50$ mW cm^{-2}. Source: Authors: P. Kalinauskas, L. Staišiūnas, and E. Juzeliūnas.

A distinctive difference is observed between E and C time variation when illumination is terminated: while the C drops down immediately (Figure 7.17), the E exhibits a prolonged relaxation (Figures 7.6 and 7.16, Table 7.1). The immediate drop of photocapacitance to zero is determined by the charge carrier recombination in the SCR, whereas the electrode potential depends on the charge at the interface metal oxide–solution, which ceases rather slowly.

The number of charge carriers in the dark Si electrode as well as the number of photoinduced electrons can be obtained from the Mott–Schottky plots (Figure 7.19). The electrode capacitance (C) was measured as a function of the electrode-negative polarization ($-\Delta E$) and plotted as C^{-2} vs. E, that is, in accordance with the relation (7.1). The data are demonstrated for acid (pH 3) and alkaline (pH 12) electrolytes. The electrode in the studied potential range is nearly ideally polarizable; no appreciable Faradaic process is observed under dark and illuminated conditions in this region.

The capacitance values of the dark electrode are on the order of nanofarads (nF cm^{-2}). Such low values indicate that the capacitances are attributable to the solid phase (e.g. SCR) rather than the solid–solution interface. A double layer, which is formed by water and electrolyte ions (Helmholtz layer), typically has the capacitance on the order of microfarads (μF cm^{-2}) (Bard and Faulkner 2001; Zhang 2004). As commonly known, C is inversely proportional to the distance between the positively and negatively charged zones. The SCR can extend up to 2000 Å, while the Helmholtz layer is only several angstroms wide (Bard and Faulkner 2001; Zhang 2004). The lowest C value will predict the total capacitance (in analogy with

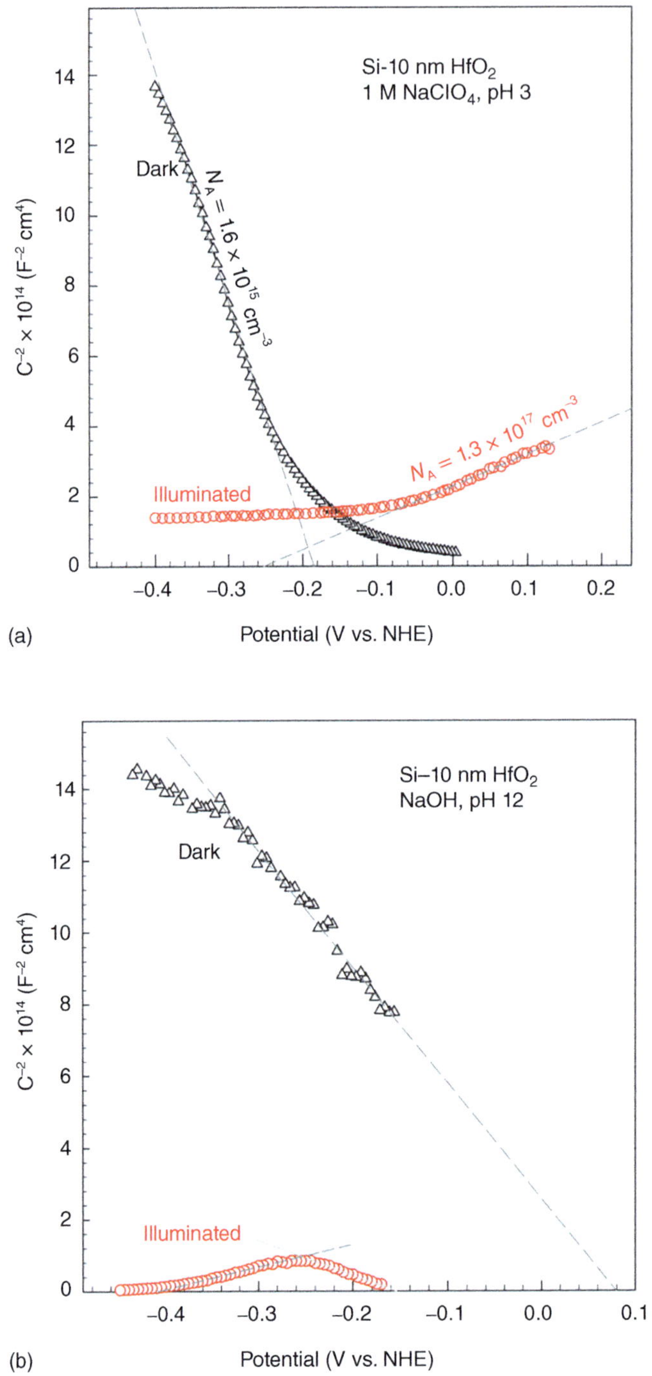

Figure 7.19 Mott–Schottky plots for p-Si with a 10 nm HfO_2 layer in the dark and under illumination ($\lambda = 505$ nm and $N = 50$ mW cm^{-2}) in (a) 1 M $NaClO_4$ at pH 3 and (b) NaOH at pH 12. The electrode potential has been stepped from the open circuit in a negative direction. Source: Reproduced from (Staišiūnas et al. 2022), which is an open-access article distributed under the terms of the Creative Commons Attribution 4.0 License (CC BY). http://creativecommons.org/licenses/by/4.0 (© The Authors, 2022).

the equivalent capacitance (C_{eq}) with actual capacitances (C_1 and C_2) connected in series: $1/C_{eq} = 1/C_1 + 1/C_2$).

The curve slope characterizes the doping level of the dark electrode or density of the photogenerated charges in the space–charge layer. Extrapolation of the C^{-2}–E curve to $C^{-2} = 0$ gives the flat band potential (E_{fb}), which indicates the potential, at which the energy band bending turns to zero. In an acid solution, a negative potential shift needs to be exerted for the dark electrode to achieve E_{fb}. This indicates that the dark silicon surface is positively charged. In an alkaline solution, there is a need to apply a positive potential to apply to obtain the energy flatness, and this indicates that the silicon surface is negatively charged. These data imply that the Si–HfO_2 electrode interfacial charge depends on the solution pH. The Si surface charge is determined by the specific adsorption of hydrogen or hydroxide ions. Adsorption of these ions determines the potential of the Helmholtz layer, which obeys the relation $\Delta E = B - 0.059\,\text{pH}$, that is, the potential varies by −59 mV per pH unit according to thermodynamic assumptions (Zhang 2004, p. 13). When the oxide film is present, the H^+ or OH^- ions can penetrate the HfO_2 film and charge the Si surface positively or negatively.

The capacitance of the dark electrodes (C_D) decreases with the negative potential shift ($-\Delta E$). Such feature is characteristic of a p-type semiconductor where holes (h^+) are major carriers. The number of the positive charges decreases with $-\Delta E$ and, consequently, C_D also decreases. The curve slope in Figure 7.19a gives the carrier number $N = 1.4 \times 10^{15}\,\text{cm}^{-3}$, which does not differ greatly from the value calculated for the sample bulk using the wafer conductivity as a criterion (Klaassen 1992) ($R = 10\,\Omega\,\text{cm}^{-1}$ and $N = 1.36 \times 10^{15}\,\text{cm}^{-3}$).

A distinctively different C vs. E trend is observed under illumination. The capacitance of the illuminated electrode (C^*) increases with the polarization to the level of several hundreds of nanofarads (e.g. 170–220 nF cm^{-2} in Figure 7.19a). A negative bias ($E \sim -0.25$ V) needs to be applied in the acid solution to attain the energy flatness; this indicates that the illuminated surface is charged negatively by photoelectrons. The C^* value increases with $-\Delta E$. The light generates minority carries – electrons, the number of these increases with $-\Delta E$ and, consequently, C^* also increases. The curve slope in Figure 7.19a indicates the number of charge carriers in the SCR of illuminated electrode to be $N \sim 1.16 \times 10^{17}\,\text{cm}^{-3}$. This number is significantly higher than that determined in the darkness. Note also that the crystalline Si contains $\sim 5 \times 10^{22}$ atoms per cm^3 (Bard and Faulkner 2001; Zhang 2004).

The shape of the E–C^* curve in an alkaline solution is more complicated: it shows an inversion with characteristic ascending and descending slopes (Figure 7.19b). The ascending part indicates that the Si surface is negatively charged, whereas the descending part implies a positive charge, as observed in the acid solution. The different capacitive behaviors in acid and alkaline solutions imply different structural properties of the interface at the solution side. As discussed above, in an alkaline solution, the illuminated Si surface will be charged negatively by adsorbed OH^- ions. Negative polarization will diminish the capacitance because the supplied electrons will reduce the number of positive charges. However, at the same time, electrolyte

Na^+ ions will be attracted to the interface because of the Coulombic force. The C^* will start to increase when Na^+ becomes dominant at the surface, as the curve in Figure 7.19b shows. It is worth mentioning that an interface layer model composed of the surface complexes ≡SiOH, ≡SiO–Na^+, and ≡SiO– has been proposed (Zhang 2004; Dove and Elston 1992).

The HfO_2 films can inhibit or accelerate the electron transfer to solution when compared to the coating-free silicon. Figures 7.20 and 7.21 characterize hydrogen reduction that is electron transfer through the hafnia film in acid (Figure 7.20) and alkaline (Figure 7.21) electrolytes. The effect of the hafnia film in both solutions is different. In an acid solution, the film enhances the reduction process when compared to that on the coating-free sample. In contrast, in an alkaline solution, an inhibition effect is observed. The acceleration effect implies surface protection from corrosion and prevention of the oxide layer formation, which has a higher passivating capacity than that of HfO_2. Si passivation may occur under acid or neutral conditions. It is also important that corrosion (oxide formation) could occur even at cathodic polarizations because of the chemical (not electrochemical) nature of the process (Hu et al. 2015). Silica is not stable in an alkaline environment; if formed, it dissolves exposing the electrode surface to corrosion. The hafnia film in such a case plays a protective role at the expense of electrode activity.

The discussed situation is rather typical: while mitigating corrosion, the oxide films reduce the surface activity and the electrode becomes electrochemically less

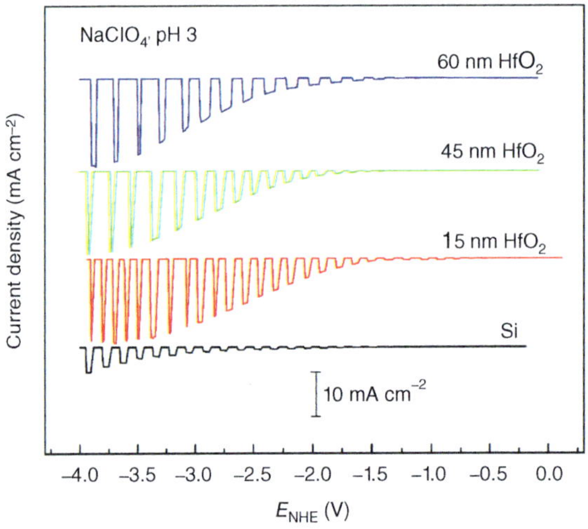

Figure 7.20 Cathodic photocurrents of hydrogen reduction in an acid electrolyte ($H^+ + 2e^- \rightarrow H_2$) over a wide range of cathodic voltages obtained for the p-Si–H electrode prepared by treatment in HF solution (marked as Si) and that coated with the HfO_2 layer with different thicknesses. Electrolyte: 1 M $NaClO_4$ (pH 3). An illumination of 505 nm wavelengths and an intensity of 50 mW cm^{-2} have been chopped at certain intervals. The potential sweep rate was 100 mV s^{-1}. Source: Reproduced from (Staišiūnas et al. 2022), which is an open-access article distributed under the terms of the Creative Commons Attribution 4.0 License (CC BY). http://creativecommons.org/licenses/by/4.0 (© The Authors, 2022).

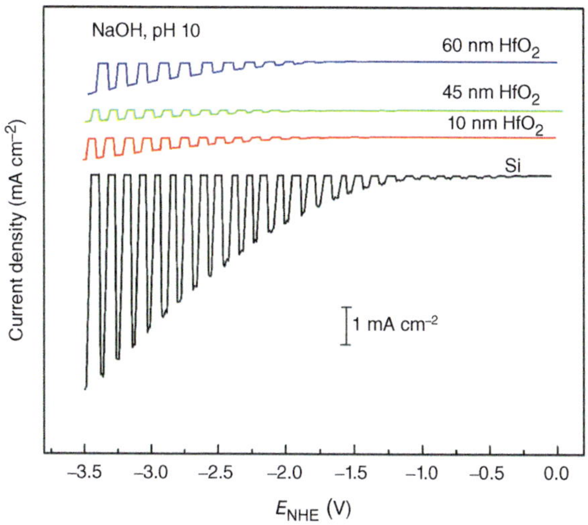

Figure 7.21 Cathodic photocurrents of hydrogen reduction in an alkaline electrolyte over a wide region of cathodic voltages obtained for the p-Si–H electrode prepared by treatment in HF solution (marked as Si) and that coated with the HfO_2 layer with different thicknesses. Electrolyte: NaOH solution at pH 12. The illumination of 505 nm wavelengths and an intensity of 50 mW cm^{-2} have been chopped at certain intervals. The potential sweep rate was 100 mV s^{-1}. Source: Reproduced from (Staišiūnas et al. 2022), which is an open-access article distributed under the terms of the Creative Commons Attribution 4.0 License (CC BY). http://creativecommons.org/licenses/by/4.0 (© The Authors, 2022).

active in certain potential regions. The dilemma is typically solved in favor of passivity because dissolution leads to continual destruction of the semiconductor and failure propagation throughout the active regions (Hu et al. 2015).

7.3.3 Si–Al_2O_3 Electrode

Alumina (Al_2O_3) ultrathin films can be applied for Si surface passivation and for PV applications (Repo et al. 2013; Hoex et al. 2008; Dingemans et al. 2010; Schmidt et al. 2009; Savin et al. 2015). This material exhibited high passivation ability, particularly to p-type emitters, which is due to low density of interface states and fixed negative charge present in the material (Repo et al. 2013). Effectiveness of such films has been demonstrated in the solar cells based on black silicon – a nanostructured material, which is capable of completely absorbing the light (Savin et al. 2015).

The Si sample with a Al_2O_3 layer shows the responses, whose shapes are analogous to those of the samples with silica and hafnia films (Figure 7.22). Again, the responses comprise the $\Delta E^{\uparrow}$ and $\Delta E^{\downarrow}$ spikes as well as the potential relaxations and the residual potential E_R. Thermal treatment of Al_2O_3 films on Si is typically used to improve the passivation capability of the film (Savin et al. 2015). The treatment did not change the basic characteristic of the photoresponses. Quantitatively, the photocurves are different: (i) the annealing shifts the entire curve by ~0.250 V positively, and this implies increased passivity of the electrode; (ii) the response $\Delta E^{\uparrow}$ of the thermally treated sample is by 25% higher when compared to the non-treated sample, and this indicates an improvement of photoresponsiveness of the thermally

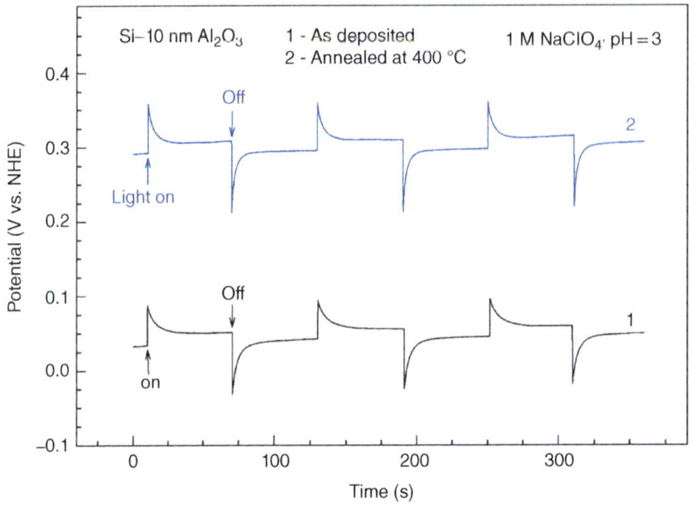

Figure 7.22 Photopotential variation at an open circuit of p-Si ALD coated with a 10 nm Al_2O_3 layer and annealed at 400 °C (one hour) in nitrogen. The electrolyte is 1 M $NaClO_4$ (pH 3), illumination ($\lambda = 505$ nm and $N = 48$ mW cm^{-2}) was chopped at one minute intervals. Source: Reproduced from (Kalinauskas et al. 2022), which is an open-access article distributed under the terms of the Creative Commons Attribution 4.0 License (CC BY). http://creativecommons.org/licenses/by/4.0 (© The Authors, 2022).

treated sample; (iii) in contrast to the 60 nm HfO_2 film discussed above, the 10 nm Al_2O_3 film is stable against thermal disintegration.

When summarizing the discussed PEC data, one could conclude that different oxide films (SiO_2, HfO_2, and Al_2O_3) exhibited the typicality of their photoresponses at an open circuit. The provided analysis, therefore, could be useful while studying other Si-oxide systems as well.

7.4 Quartz Crystal Nanogravimetry

The quartz crystal nanobalance (QCN) is a highly sensitive mass detector, which provides information with nanogram resolution about electrode mass change in situ and in real time. It was in 1959 when G. Sauerbrey published the dependence of quartz oscillation frequency on the change in surface mass (Sauerbray et al. 1959). Since then, the method has gained relevance in various fields of science and technology, in particular electrochemistry, materials science, and biosensors. The important fields of application also include corrosion of metals, anticorrosive stability of PVD alloys, adhesion of microorganisms, long-term effects in the atmosphere, and biosensors including the piezo-immuno ones. The principles of the QCN measurements as well as its application fields have been surveyed (Juzeliunas 2009; Hillman 2011). Dirri et al. have reviewed the QCN sensors used in space missions (Space Shuttle flights, NASA Space Transportation System, NASA STS) (Dirri et al. 2019). Among other materials, the outgassing characteristics of the Al alloy as well as surface contamination by Al particles 1–2 μm in size near scientific instruments (optics, detectors, telescope, and mirrors) were summarized.

The QCN development route falls into several stages: (i) metals under vacuum and gases (adsorption, oxidation, corrosion, thin films, and detection in gas chromatography); (ii) metals in solutions (electrochemistry of systems with rigidly attached mass); (iii) energy dissipating systems (viscoelastic films), and (iv) complex biochemical and biomimetic systems (protein, nucleic acid and living cell biosensors, and drug discovery). Thus, the QCN sensors make it possible to observe the bioelectrochemical events non-invasively in-situ and in real time in gaseous and liquid environments. The important QCN modifications include the nanobalance with foil and oxide electrodes as well as energy dissipation control.

The technical configuration of the QCN is shown in Figure 7.23. The QCN oscillator is designed from a quartz disc sliced from a single crystal, usually the AT plane, 35°10′ angle from the *z*-axis. The AT cut is preferred as it has minor temperature fluctuation in frequency at room temperatures. Both sides of the quartz are coated with metal electrodes, typically the inert ones such as Pt or Au. The quartz disc, being a piezoelectric material, deforms when the electric field is applied. The disc starts to oscillate with a characteristic fundamental frequency (f_0) when connected to the oscillator circuit with inductivity. The added or removed

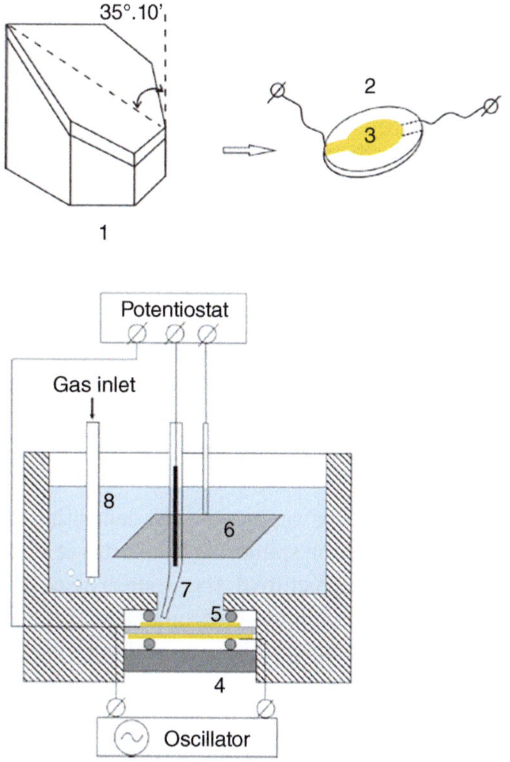

Figure 7.23 Principal configuration of the QCN: quartz crystal with AT section (1), quartz wafer (2) coated with metallic electrodes (3), oscillation equivalent circuit with a quartz resonator (Cg), inductivity (*L*), resistance (*R*), and conductivity (*C*). The electrochemical cell comprises a quartz resonator (2) mounted between two rubber rings (6, sections) and a jig (5), supporting (7) and reference electrodes (8), gas inlet (9), and solution. Source: Reproduced from (Juzeliūnas et al. 2009) with permission from the publisher ©Lietuvos Mokslų akademija (open-access article).

mass (Δm) relates to the change in oscillation frequency (Δf) according to the *Sauerbrey* equation (Sauerbrey 1959):

$$\Delta f = -2\Delta m n f_2^0 / \left(A\sqrt{\mu_Q \rho_Q}\right) \tag{7.3}$$

In this formula, n stands for the overtone number, A – surface area, μ_Q – shear modulus ($\mu_Q = 2.947 \times 10^{11}$ g cm^{-1} s^{-2}), and ρ_Q – density of quartz ($\rho_Q = 2.648$ g cm^{-3}). It should be noted that the above ratio is valid for relatively small mass change, and this should be rigidly bound to the quartz surface. The ratio becomes not accurate when the mass change exceeds c. 2% of the crystal mass.

Most of the nanobalances operate in the thickness shear mode of oscillation, i.e. the lateral surface motion is explored. The lateral amplitude of the oscillating quartz crystal is on the order of several nanometers. The linear dependence between the mass and the frequency change was verified by many researchers for rigid samples under vacuum, atmosphere, and liquids, and the results have been well surveyed in the literature (Ward 1995; Hillman 2003).

The QCN technique employs the quartz crystal oscillators with metallic electrodes, typically the inert ones such as gold and platinum. The material to be studied is covered on an inert electrode. Ruthenium dioxide (RuO_2) can be applied as a substitute of the metallic electrode (Leinartas et al. 2009). Two main types of ruthenium dioxide appears promising in various technical applications: the hydrous oxide $RuO_2 \times xH_2O$, which has mixed electron–proton conductivity (Swider-Lyons et al. 2005), and the anhydrous one, which exhibits a metal-like conductivity (Goodenough 1971). Apart from a high electric conductivity, both types also exhibit a high catalytic activity, supercapacitance properties, chemical inertness, and thermal stability. The properties of anhydrous oxide depend on formation conditions, for instance, the texture and resistivity of the RuO_2 thin film electrodes deposited by magnetron sputtering dependent on deposition power, oxygen pressure and O_2/Ar ratio, temperature, and substrate properties (Maiwa et al. 1994; Sakiyama et al. 1993; Butkutė et al. 1998). RuO_2 films have been grown epitaxially at 400–750 °C on an aluminum oxide substrate. Highly conductive crystalline RuO_2 has been epitaxially grown by pulsed laser deposition on $LaAlO_3$ substrates (Jia et al. 1995). Amorphous anhydrous RuO_2 films were deposited by the spray pyrolysis method (Gujar et al. 2007). An important property of the ruthenium oxide is extremely large pseudocapacitances. RuO_2 has shown to be probably the best electrode material in terms of a high specific capacitance, a wide potential window and a long shelf and cycle life.

The QCN sensor was prepared by depositing conductive RuO_2 on quartz disc using reactive magnetron sputtering (Leinartas et al. 2009). XRD measurements indicated the crystalline structure of the oxide. The oxide-based resonator operated stable in both gaseous and liquid environments. Water adsorption was demonstrated as an operational example, which was evaluated by exposing the ruthenium oxide to air at the humidity close to 100%. A rapid mass growth was observed during the first hour of exposure; later on, the increase in mass approached a limit. The total mass gain was estimated as high as $\Delta m = 7.5 \times 10^{-6}$ g cm^{-2}. This indicates that the thickness of the water layer was $d = 7.5 \times 10^{-8}$ cm provided the water does not penetrate the oxide structure, which is very unlikely. The surface roughness of the RuO_2 layer was determined on a micrometric scale by atomic force microscopy (AFM) to be as high as c. 1.4. An advantage of the sensor is the possibility to avoid the galvanic effects, which can occur when using the metallic electrodes. Thus, a stack of two oxides (SiO_2–RuO_2) can operate as a mass sensor.

It is worth mentioning some complications when applying the QCN in optical studies, e.g. PEC or spectroscopic measurements. A sudden increase in frequency was observed when laser illumination was applied to a corroding zinc or an inert gold electrode (Juzeliūnas et al. 1996). As the frequency shift did not depend on the electrode material, it has been attributed to the phenomena, which take place in the electrolyte. Contributions of thermal reduction of viscosity and density of electrolyte as well as warming of the quartz oscillator during illumination should be taken into consideration along with the electrode mass effects. The influence of illumination on zinc corrosion rate was determined as the difference between the frequency data in aerated and deoxygenated electrolytes. The QCN study did not show the indications of any significant photoinduced decomposition of the oxide layer, which developed on zinc during its corrosion in a naturally aerated 5% NaCl solution.

The mass changes of the Si–HfO_2 electrode determined by the QCN in acid and alkaline solutions are shown in Figure 7.24. The measurement reveals with nanogram resolution the electrode mass change after it comes in contact with the electrolyte. The QCN indicates the mass gain effect during the exposure.

The nanogravimetric data obtained for the Au–HfO_2 electrode indicate an increase in electrode mass during the exposure. With time, the mass becomes more or less constant for all the studied electrodes. Note that the mass change of the Au–HfO_2 electrode does not differ greatly from that of the Si–HfO_2 electrode. Because of the chemical inertness of the Au substrate, the mass gain is attributable to the events within the oxide film, most likely, the electrolyte absorption by the film. The mass of the film increased on an average up to $\Delta m \sim 1.0\ \mu g\ cm^{-2}$ during one hour of exposure. An increase in the mass of the film on the Si substrate was of similar value. There were no indications observed of oxide dissolution, which would be detected as a decrease in electrode mass. This implies effective electrode protection by the hafnia film both in acid and in alkaline media. As commonly known, silicon is stable in water and acidic solutions because of the formation of stable protective oxide. This oxide, however, is unstable in alkaline solutions. Chemical and electrochemical corrosion could contribute to the mass effect in alkaline solutions (Hu et al. 2015). The mass decrease of the electrode unprotected with hafnia coating indicates the corrosion process. With time, the dissolution ceases because of the surface passivation by corrosion products.

It is also important that the mass curve for the Si–HfO_2 electrode in an acid solution does not differ greatly from that in an alkaline solution. Thus, the amount of the absorbed electrolyte does not depend on solution acidity and the mass effect is similar, as the densities of both electrolytes do not differ considerably. The mass of the HfO_2 layer 70 nm in thickness is calculated to be $m = 68\ \mu g\ cm^{-2}$, assuming that its density is $d_{HfO_2} = 9.68\ g\ cm^{-3}$. The electrolyte intake can be approximated as

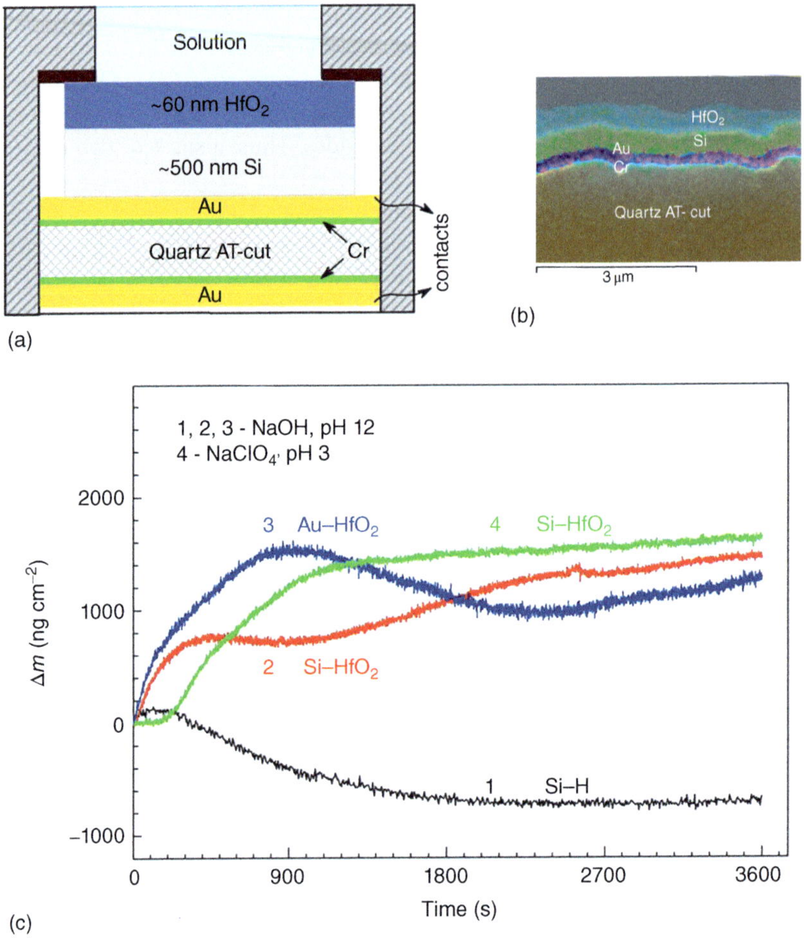

Figure 7.24 (a) Principal configuration of the QCN sensor for HfO_2 film studies on Au and Si substrates; (b) scanning electron microscopy (SEM) image of the sensor cross section composed of $Cr/Au/Si/HfO_2$ layers on quartz. The silicon layer was formed by rf magnetron sputtering and the HfO_2 layer was formed by ALD; (c) mass change determined for the samples with a 70 nm HfO_2 layer on Si (2, 4) and Au (3) substrates when exposed in NaOH (pH 12) (1–3) and 1 M $NaClO_4$ adjusted to pH 3 (4). Curve 1 shows the mass change of hafnia-free silicon pretreated in a HF solution. Source: Reproduced from (Staišiūnas et al. 2022), which is an open-access article distributed under the terms of the Creative Commons Attribution 4.0 License (CC BY). http://creativecommons.org/licenses/by/4.0 (© The Authors, 2022).

$\Delta m \sim 1.0\,\mu g\,cm^{-2}$; this corresponds to ~1.5% of the initial oxide mass. The volume of the accommodated solution will take approximately 15% of the initial oxide volume.

The QCN data show that the electrolyte penetrates the thin oxide film and transforms it to a hybrid electrode Si–HfO_2–electrolyte. The electrolyte plays here an essential role in interfacial charging.

The question also arises about perchlorate penetration into the oxide layer, first of all, because of the presence of non-uniformities and defect sites in the surface layer.

Raman spectroscopy implied that perchlorate ions do not enter appreciably the oxide layer during illumination; they are retained at the oxide/electrolyte boundary. This method is applied for effective detection and discrimination of oxyanions. Perchlorate possesses a relatively high Raman cross section (12.7×10^{-30} cm^2 sr^{-1} $molecule^{-1}$) (McCreery 2002) and can be determined by normal Raman as well as surface-enhanced Raman spectroscopy (SERS) approaches (Hao and Meng 2017). The detection limits range from micromolar to nanomolar levels (10^{-6}–10^{-9} M; 0.1 mg L^{-1}–0.1 μg L^{-1}). To that end, various SERS substrates mainly based on gold and silver are used: functionalized gold–silica (Au–SiO_2) (Wang et al. 2004), 2-dimethylaminoethanethiol (DMAE)-modified gold nanoparticles (Gu et al. 2009), silver, gold, silver–gold, and selective sorbents (Wang and Gu 2005; Nuntawong et al. 2013; Gu et al. 2004; Byram et al. 2018; Jubb et al. 2017). Also, the discrimination of perchlorate from salt mixtures has been demonstrated by Raman spectroscopy (Zapata and Garcia-Ruiz 2018). Our experiments, however, did not confirm the photoassisted adsorption of perchlorate ions in the hafnia layer on silicon. The p-type Si with 60 nm HfO_2 layer has been pre-exposed in a 1 M perchlorate electrolyte for five minutes, illuminated for five minutes by $\lambda = 505$ nm light, and withdrawn from the electrolyte. The Raman spectra of this sample were identical with those of the pristine counterpart, which was not treated in the above manner. Thus, one could conclude that the mass gain observed by QCN was because of the incorporation of water along with proton and hydroxyl. The protons could increase the state density at the Si/oxide interface, and such modification could form some areas of the interconnected network of defective regions (Topkar and Lal 1995). Penetration of water could create increased mobility paths for the protons (Topkar and Lal 1995).

The QCN data indicate high protective capability of the HfO_2 film: while bare silicon decreases its mass because of dissolution in an alkaline solution, no dissolution indications are observed for the Si–HfO_2 electrode. This electrode keeps the mass constant with no appreciable difference from the Au–HfO_2 electrode.

The QCN measurements in Figure 7.25 characterize the instability of the Si–SiO_2 interface, which leads to electrode mass change. Generally, a decrease in electrode mass implies dissolution of the oxide layer, whereas an increase in mass indicates electrolyte absorption by the oxide structure. It is assumed that the chemical interaction of the Si substrate with the electrolyte is negligible because of the protective action of the oxide film. This is very likely, especially in the early stages of exposure. Figure 7.25 shows that in a neutral electrolyte (pH 7), the electrode mass decreases at the beginning of exposure with the rate $\Delta m/\Delta t = 0.59$ ng s^{-1}. This indicates that silica dissolution is prevailing at the initial stages of exposure. After a certain time, the mass change stops indicating dissolution suppression. Furthermore, mass gain is observed, which implies that the electrolyte intake becomes dominant. The total increase in the electrode mass in a slightly acidic electrolyte (pH 5) does not deviate significantly from that in the neutral electrolyte. No dissolution indications are observed in the acid electrolyte (pH 3): the growth of electrode mass is observed from the very beginning. The increase in mass reaches a limit at $\Delta m \sim 500$ ng cm^{-2}; this mass makes approximately one-third of the silica mass of ($m = 1325$ ng cm^{-2}).

It is a common knowledge that silicon is stable in most aqueous solutions because of the formation of the stable oxide film. It is also known that SiO_2

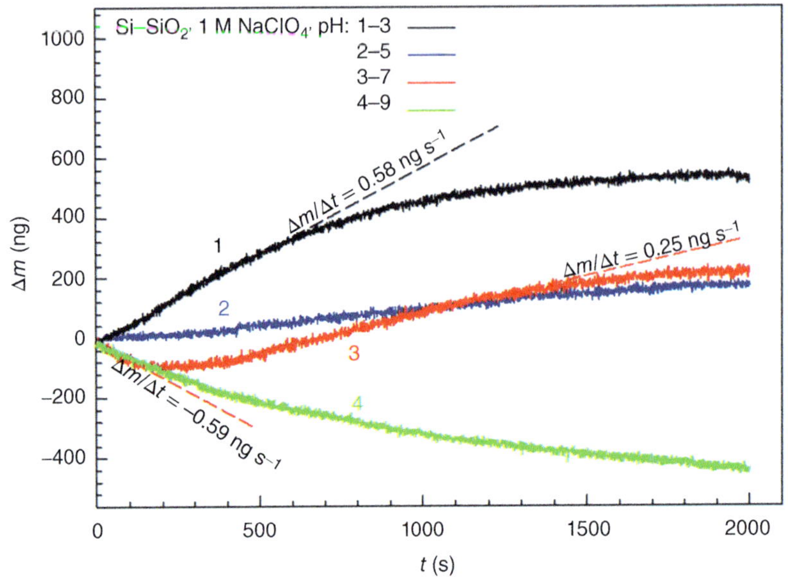

Figure 7.25 Mass change determined by the QCN of the magnetron-sputtered Si with the 10 nm SiO_2 layer during exposure to 1 M $NaClO_4$ adjusted to pH 3, 5, 7, and 9. Source: Reproduced from (Kalinauskas et al. 2022), which is an open-access article distributed under the terms of the Creative Commons Attribution 4.0 License (CC BY). http://creativecommons.org/licenses/by/4.0 (© The Authors, 2022).

tends to dissolve forming the soluble H_4SiO_4 when the dissolved silicon activity is low. One could assume that the initial decrease in electrode mass in a neutral electrolyte is due to SiO_2 dissolution at low activity of the dissolved silicon. With the exposure time, the silicon activity increases, the dissolution is suppressed, and, in turn, the mass effect is overwhelmed by the electrolyte intake into the oxide structure.

Silicon oxide has low solubility in water and acidic solutions, but it has high solubility in alkaline solutions. The mass decrease in an alkaline solution (pH 9) clearly indicates the dissolution process with a significant rate. Chemical and electrochemical corrosion could contribute to the mass effect in an alkaline solution. It has been shown that chemical rather than electrochemical mechanism dominates Si dissolution in alkaline solutions (Hu et al. 2015).

References

Bard, A. (1979). Photoelectochemistry and heterogeneous photo-catalysis at semiconductors. *J. Photochem.* 10: 59–75. https://doi.org/10.1016/0047-2670(79)80037-4.

Bard, A. and Faulkner, L.R. (2001). *Electrochemical Methods: Fundamentals and Applications*, 2e. New York/Chichester/Weinheim/Brisbane/Singapore/Toronto: Wiley. ISBN: 0-471-04372-9.

Bayerl, A. and Lanza, M. (2014). Hafnium-based thin oxides: versatile insulators for microelectronics. In: *Hafnium: Chemical Characteristics, Production and Applications* (ed. Y. HongYu), 113–135. Novascience.

Berdova, M. and Liu, X. (2016). Hardness, elastic modulus, and wear resistance of hafnium oxide-based films grown by atomic layer deposition. *J. Vacuum Sci. Technol. A.* 34: 051510. https://doi.org/10.1116/1.4961113.

Bersch, E., Rangan, S., Bartynski, R.A. et al. (2008). Band offsets of ultrathin high-κ oxide films with Si. *Phys. Rev. B* 78: 085114. https://doi.org/10.1103/PhysRevB.78.085114.

Butkutė, R., Lisauskas, V., and Vengalis, B. (1998). High-quality conductive In_2O_3 <Sn> and RuO_2 layers grown by magnetron sputtering. *Surf. Coat. Technol.* 100, 101: 305–308. https://doi.org/10.1016/s0257-8972(97)00638-5.

Byram, C., Moram, S., and Soma, V. (2018). Picosecond laser fabricated Ag, Au and Ag–Au nanoparticles for detecting ammonium perchlorate using a portable Raman spectrometer. *AIP Conf. Proc.* 1942: 050028. https://doi.org/10.1063/1.5028659.

Chaubey, S., Yao, Y., Makongo, J. et al. (2012). Microstructural and thermal investigations of HfO_2 nanoparticles. *RSC Adv.* 24: https://doi.org/10.1039/c2ra21003g.

Chen, Y.W., Liu, M., Kaneko, T., and McIntyre, P.C. (2010). Atomic layer deposited hafnium oxide gate dielectrics for charge-based biosensors. *Electrochem. Solid-State Lett.* 13: G29–G32. https://doi.org/10.1149/1.3280224.

Cho, S.K., Fan, F.-R.F., and Bard, A.J. (2012). Electrodeposition of crystalline and photoactive silicon directly from silicon dioxide nanoparticles in molten $CaCl_2$. *Angew. Chem. Int. Ed.* 51: 12740. https://doi.org/10.1002/anie.201206789.

Choi, M.J., Jung, J., Park, M. et al. (2014). Long-term durable silicon photocathode protected by a thin Al_2O_3/SiO_x layer for photoelectrochemical hydrogen evolution. *J. Mater. Chem. A* 2: 2928–2933. https://doi.org/10.1039/C3TA14443G.

Cui, J., Phang, S.P., Sio, H.C. et al. (2017a). Passivation of phosphorus diffused black multi-crystalline silicon by hafnium oxide. *Phys. Status Solidi RRL* 11: 1700296. https://doi.org/10.1002/pssr.201700296.

Cui, J., Wan, Y., Cui, Y. et al. (2017b). Highly effective electronic passivation of silicon surfaces by atomic layer deposited hafnium oxide. *Appl. Phys. Lett.* 110: 021602. https://doi.org/10.1063/1.4973988.

Dave, V., Mishra, P.K., and Chandra, R. (2018). Nanostructured hafnium oxide thin films for sensing carbon monoxide: an experimental investigation. *Mater. Today: Proc.* 5 (11): 23286–23292. https://doi.org/10.1016/j.matpr.2018.11.062.

Davis, K. and Tomozawa, M. (1995). Water diffusion into silica glass: structural changes in silica glass and their effect on water solubility and diffusivity. *J. Non-Cryst. Solids* 185: 203–220. https://doi.org/10.1016/0022-3093(95)00015-1.

Dingemans, G., Seguin, R., Engelhart, P. et al. (2010). Silicon surface passivation by ultrathin Al_2O_3 films synthesized by thermal and plasma atomic layer deposition. *Phys. Status Solidi RRL* 4: 10–12. https://doi.org/10.1002/pssr.200903334.

Dirri, F., Palomba, E., Longobardo, A. et al. (2019). A review of quartz crystal microbalances for space applications. *Sens. Actuators, A* 287: 48–75. https://doi.org/10.1016/j.sna.2018.12.035.

Dove, P.M. and Elston, S.F. (1992). Dissolution kinetics of quartz in sodium chloride solutions: analysis of existing data and a rate model for 25 °C. *Geochim. Cosmochim. Acta* 56: 4147–4156. https://doi.org/10.1016/0016-7037(92)90257-j.

Esposito, D.V., Levin, I., Moffat, T.P., and Talin, A.A. (2013). H_2 evolution at Si-based metal-insulator-semiconductor photoelectrodes enhanced by inversion channel charge collection and H spillover. *Nat. Mater.* 12: 562–568. https://doi.org/10.1038/nmat3626.

Fan, J., Liu, H., Kuang, Q. et al. (2012). Physical properties and electrical characteristics of H_2O-based and O_3-based HfO_2 films deposited by ALD. *Microelectron. Reliab.* 52: 1043–1049. https://doi.org/10.1016/j.microrel.2012.01.010.

Fohlerova, Z. and Mozalev, A. (2019). Anodic formation and biomedical properties of hafnium-oxide nanofilms. *J. Mater. Chem. B* 7: 2300–2310. https://doi.org/10.1039/C8TB03180K.

Gaskins, J.T., Hopkins, P., Merrill, D.R. et al. (2017). Review – investigation and review of the thermal, mechanical, electrical, optical and structural properties of atomic layer deposited high-κ dielectrics: beryllium oxide, aluminum oxide, hafnium oxide, and aluminum nitride. *ECS J. Solid State Sci. Technol.* 6: N189–N208. https://doi.org/10.1149/2.0091710jss.

Gerischer, H. (1975). Electrochemical photo and solar cells principles and some experiments. *J. Electroanal. Chem. Interfacial Electrochem.* 58: 263–274. https://doi.org/10.1016/S0022-0728(75)80359-7.

Goodenough, J.B. (1971). Metallic oxides. *Prog. Solid State Chem.* 5: 145–399. https://doi.org/10.1016/0079-6786(71)90018-5.

Gu, B., Tio, J., Wang, W. et al. (2004). Raman spectroscopic detection for perchlorate at low concentrations. *Appl. Spectrosc.* 58: 741–744. https://opg.optica.org/as/abstract.cfm?URI=as-58-6-741.

Gu, B., Ruan, C., and Wang, W. (2009). Perchlorate detection at nanomolar concentrations by surface-enhanced Raman scattering. *Appl. Spectrosc.* 63: 98–102. https://opg.optica.org/as/abstract.cfm?URI=as-63-1-98.

Gujar, T.P., Shinde, V.R., Lokhande, C.D. et al. (2007). Spray deposited amorphous RuO_2 for an effective use in electrochemical supercapacitor. *Electrochem. Commun.* 9: 504–510. https://doi.org/10.1016/j.elecom.2006.10.017.

Hao, J. and Meng, X. (2017). Recent advantages in SERS detection of perchlorate. *Front. Chem. Sci. Eng.* 11: 448–464. https://doi.org/10.1007/s11705-017-1611-9.

Hazarika, C. and Sharma, S. (2014). A mathematical model describing drift in SiO_2 gate pH ISFET's due to hydrogen ion diffusion. *Int. J. Appl. Engineer. Res.* 9: 21099–21113.

Hillman, A.R. (2003). The electrochemical quartz crystal microbalance. In: *Instrumentation and Electroanalytical Chemistry; Encyclopedia of Electrochemistry*, vol. 3 (ed. A.J. Bard, M. Stratmann and P.R. Unwin), 230–289. New York: Wiley.

Hillman, A.R. (2011). The EQCM: electrogravimetry with a light tough. *J. Solid State Electrochem.* 15: 1647–1660. https://doi.org/10.1007/s10008-011-1371-2.

Hoex, B., Schmidt, J., Pohl, P. et al. (2008). C-Si surface passivation by atomic layer deposited Al_2O_3. *J. Appl. Phys.* 104: 044903. https://doi.org/10.1063/1.2963707.

Hu, S., Lewis, N.S., Ager, J.W. et al. (2015). Thin-film materials for the protection of semiconducting photoelectrodes in solar-fuel generators. *J. Phys. Chem. C* 119: 24201–24228. https://doi.org/10.1021/acs.jpcc.5b05976.

Huang, X., Wang, J., Wang, J. et al. (2015). Preparation of graphene-hafnium oxide composite for selective enrichment and analysis of phosphopeptides. *RSC Adv.* 5: 89644–89651. https://doi.org/10.1039/C5RA17471F.

IUPAC (1997). *Compendium of Chemical Terminology*, 2end ed. (the "Gold Book"). Compiled by A.D. McNaught and A. Wilkinson. Oxford: Blackwell Scientific Publications. Online version (2019) created by S.J. Chalk. https://doi.org/10.1351/goldbook.

Jayaraman, V., Karthikeyan, S., Bharanidharan, G. et al. (2015). Insights into the binding of hafnium oxide nanoparticles to human serum albumin by various optical spectroscopy studies. *Trends Biomater. Artif. Organs* 29: 339–344.

Ji, L., McDaniel, M.D., Wang, S. et al. (2015). A silicon-based photocathode for water reduction with an epitaxial $SrTiO_3$ protection layer and a nanostructured catalyst. *Nat. Nanotechnol.* 10: 84–90. https://doi.org/10.1038/nnano.2014.277.

Jia, Q.X., Wu, X.D., Foltyn, S.R. et al. (1995). Heteroepitaxial growth of highly conductive metal oxide RuO_2 thin films by pulsed laser deposition. *Appl. Phys. Lett.* 67: 1677–1679. https://doi.org/10.1063/1.115054.

Jogiaas, T., Zabels, R., Tamm, A. et al. (2015). Mechanical properties of aluminum, zircomium, hafnium and tantalum oxides and their nanolaminates grown by atomic layer deposition. *Surf. Coat. Technol.* 282: 36–42. https://doi.org/10.1016/j.surfcoat.2015.10.008.

Jubb, A., Hatzinger, P., and Gu, B. (2017). Trace-level perchlorate analysis of impacted groundwater by elevated gold ellipse dimer nanoantenna surface-enhanced Raman scattering. *J. Raman Spectrosc.* 48: https://doi.org/10.1002/jrs.5070.

Juzeliunas, E. (2009). Quartz crystal microgravimetry: fifty years of application and new challenges. *Chemija* 20: 218–225.

Juzeliūnas, E., Kalinauskas, P., and Miečinskas, P. (1996). Electrochemical quartz crystal microgravimetry study of zinc electrode behaviour in a neutral solution by laser illumination. *J. Electrochem. Soc.* 143: 1525–1530.

Kalinauskas, P., Valsiūnas, I., Samulevičienė, M., and Juzeliūnas, E. (2001). Zinc photocorrosion in neutral solutions. *Corros. Sci.* 43: 2083–2092. https://doi.org/10.1016/S0010-938X(01)00013-0.

Kalinauskas, P., Staišiūnas, L., Grigucevičienė, A. et al. (2022). Photoelectrochemical and nanogravimetric study of electrolytic transformation of silicon-oxide interface. *J. Electrochem. Soc.* 169 (3): 036508. https://doi.org/10.3389/fchem.2022.859023.

Kim, H., McIntyre, P., and Saraswat, K. (2003). Effects of crystallization on the electrical properties of ultrathin HfO_2 dielectrics grown by atomic layer deposition. *Appl. Phys. Lett.* 82: 106. https://doi.org/10.1063/1.1533117.

Klaassen, D.B.M. (1992). A unified mobility model for device simulation – I. Model equations and concentration dependence. *Solid State Electronics* 35: 953–959. https://doi.org/10.1016/0038-1101(92)90325-7.

Kumar, S., Wang, Z., Huang, X. et al. (2017). Oxygen migration during resistance switching and failure of hafnium oxide memristors. *Appl. Phys. Lett.* 110: 103503. https://doi.org/10.1063/1.4974535.

Kushitashvili, Z., Bibilashvili, A., and Biyikli, N. (2017). Properties of hafnium oxide received by ultra violet stimulated plasma anodization. *IEEE Trans. Device Mater. Reliab.* 17: 667–671. https://doi.org/10.1109/TDMR.2017.2751078.

Leinartas, K., Miečinskas, P., and Juzeliūnas, E. (2009). Ruthenium dioxide quartz crystal nano-balance. *Sens. Actuators, B* 137: 762–767. https://doi.org/10.1016/j.snb.2008.12.003.

Li, M., Jin, Z., Zhang, W. et al. (2019). Comparison of chemical stability and corrosion resistance of group IV metal oxide films formed by thermal and plasma-enhanced atomic layer deposition. *Sci. Rep.* 9: 10438. https://doi.org/10.1038/s41598-019-47049-z.

Lohman, F. (1967). Fermi-niveau und flachbandpotential von molekülkristallen aromatischer kohlenwasserstoffe. *Z. Naturforsch.* 22a: 843–844.

Magiorella, L., Barouch, G., Devaux, C. et al. (2012). Nanoscale radiotherapy with hafnium oxide nanoparticles. *Future Oncol.* 8: 1167–1181. https://doi.org/10.2217/FON.12.96.

Maiwa, H., Ichinose, N., and Okazaki, K. (1994). Preparation and properties of Ru and RuO_2 thin film electrodes for ferroeletric thin films. *Jpn. J. Appl. Phys.* 33: 5223.

Mangla, O. and Gupta, V. (2016). Study of electrical properties of hafnium oxide thin film based metal-insulator-metal capacitors: pre and post metallic annealing. *J. Mater. Sci – Mater. Electron.* 27: 12527–12532. https://doi.org/10.1007/s10854-016-5783-8.

Marill, J., Anesary, N.M., Zhang, P. et al. (2014). Hafnium oxide nanoparticles: toward an in vitro predictive biological effect? *Radiat. Oncol.* 9: 150. https://doi.org/10.1186/1748-717X-9-150.

Marjani, S., Hosseini, S.E., and Faez, R. (2016). A silicon doped hafnium oxide ferroelectric p-n-p-n SOI tunneling field-effect transistor with steep subthreshold slope and high switching current ratio. *AIP Adv.* 6: 095010. https://doi.org/10.1063/1.4962969.

McCreery, R. (2002). Photometric standards for Raman spectroscopy. In: *Handbook of Vibrational Spectroscopy* (ed. J. Chalmas and P. Giffiths), 1–13. Chichester: Wiley.

Mendoza, J.G., Frutis, M.A.A., Flores, G.A. et al. (2010). Synthesis and characterization of hafnium oxide films for thermo and photoluminescence applications. *Appl. Radiat. Isot.* 68: 696–699. https://doi.org/10.1016/j.apradiso.2009.09.031.

Michaelis, A. (2008, 2008). Valve metal, Si and ceramic oxides as dielectric films for passive and active electronic devices. In: *Advances in Electrochemical Science and Engineering*, vol. 10 (ed. R.C. Alkire, D.M. Kolb, J. Lipkowski and P.N. Ross), 75. Wiley-VCH.

Müller, J., Polakowski, P., Müller, S. et al. (2016). High endurance strategies for hafnium oxide based ferroelectric field effect transistor. In: *In 2016 16th Non-volatile Memory Technology Symposium (NVMTS)*, Pittsburgh, 7781517. https://doi.org/10.1109/nvmts.2016.7781517.

Nozik, A.J. (1996). Physical chemistry of semiconductor-liquid interfaces. *J. Phys. Chem.* 100: 13061–13078. https://doi.org/10.1021/jp953720e.

Nuntawong, N., Eiamchai, P., Limwichean, S. et al. (2013). Trace detection of perchlorate in industrial-grade emulsion explosive with portable surface-enhanced Raman spectroscopy. *Forensic Sci. Int.* https://doi.org/10.1016/j.forsciint.2013.09.012.

Ortiz-Dosal, L.C., Eleazar, S., Kolosovas-Machuca, M. et al. (2017). Bioanalysis by immobilization of antibodies of hafnium (IV) oxide with 3-aminopropyltriethoxysilane. *Anal. Lett.* 50: 2937–2943. https://doi.org/10.1080/00032719.2017.1320666.

Oudot, E., Gros-Jean, M., Couroble, K. et al. (2018). Hydrogen passivation of silicon/silicon oxide interface by atomic layer deposited hafnium oxide and impact of silicon oxide underlayer. *J. Vac. Sci. Technol.* https://doi.org/10.1116/1.4999561.

Park, M., Lee, Y., Kim, H. et al. (2017). Surface and grain boundary energy as the key enabler of ferroelectricity in nanoscale hafnia-zirconia: a comparison of model and experiment. *Nanoscale* https://doi.org/10.1039/C7NR02121F.

Perram, J., Hunter, R., and Wright, H. (1974). The oxide-solution interface. *Aust. J. Chem.* 27: 461–475. https://doi.org/10.1071/CH9740461.

Plociennik, P., Zawadzka, A., Strzelecki, J. et al. (2014). Pulsed laser deposition (PLD) of hafnium oxide thin films. In: *2014 16th International Conference on Transparent Optical Networks (ICTON),* Graz, 1–3. https://doi.org/10.1109/ICTON.2014.6876620.

Polydorou, E., Botzakaki, M., Drivas, C. et al. (2018). Insights into passivation effect of atomic layer deposited hafnium oxide for efficiency and stability enhancement in organic solar cells. *J. Mat. Chem. C* 6: 8051–8059. https://doi.org/10.1039/c8tc02243g.

Pourbaix, M. (1974). *Atlas of Electrochemical Equilibria in Aqueous Solutions.* Houston: National Association of Corrosion Engineers.

Rahaman, S.Z., Lin, Y.D., Lee, H.Y. et al. (2017). The Rolle of Ti buffer layer thickness on the resistive switching properties of hafnium oxide-based resistive switching memories. *Langmuir* 33: 4654–4665. https://doi.org/10.1021/acs.langmuir.7b00479.

Ramzan, M., Rana, A.M., Ahmed, E. et al. (2015). Optical characterization of hafnium oxide thin films for heat mirrors. *Mater. Sci. Semicond. Process.* 32: 22–30. https://doi.org/10.1016/j.mssp.2014.12.079.

Repo, P., Haarahiltunen, A., Sainiemi, L. et al. (2013). Effective passivation of black silicon surfaces by atomic layer deposition. *IEEE J. Photovoltaics* 3: 90–94. https://doi.org/10.1109/JPHOTOV.2012.2210031.

Sakiyama, K., Onishi, S., Ishihara, K. et al. (1993). Deposition and properties of reactively sputtered ruthenium dioxide films. *J. Electrochem. Soc.* 140: 834–839.

Sauerbrey, G. (1959). Verwendung von Schwingquarzen zur Wägung dünner Schichten und zur Mikrowägung. *Z. Physik* 155: 206–222. https://doi.org/10.1007/BF01337937.

Savin, H., Repo, P., von Gastrow, G. et al. (2015). Black silicon solar cells with interdigitated back-contacts achieve 22.1% efficiency. *Nat. Nanotechnol.* 10: 624–629. https://doi.org/10.1038/nnano.2015.89.

Schmidt, J., Veith, B., and Brendel, R. (2009). Effective surface passivation of crystalline silicon using ultrathin Al_2O_3 films and Al_2O_3/SiN_x stacks. *Phys. Status Solidi RRL* 3: 287–289. https://doi.org/10.1002/pssr.200903272.

Seger, B., Pedersen, T., Laursen, A.B. et al. (2013). Using TiO_2 as a conductive protective layer for photocathodic H_2 evolution. *J. Am. Chem. Soc.* 135: 1057–1064. https://doi.org/10.1021/ja309523t.

Shang, J., Xue, W., Ji, Z. et al. (2017). Highly flexible resistive switching memory based on amorphous-nanocrystalline hafnium oxide films. *Nanoscale* 9: 7037–7046. https://doi.org/10.1039/C6NR08687J.

Sharma, A., Longo, V., Verheijen, M. et al. (2017). Atomic layer deposition of HfO_2 using $HfCp(NMe_2)_3$ and O_2 plasma. *J. Vac. Sci. Technol., A* https://doi.org/10.1116/1.4972210.

Sherck, N.J. and Won, Y.Y. (2017). Technical note: a simulation study on the feasibility of radiotherapy dose enhancement with calcium tungstate and hafnium oxide nano- and microparticles. *Z. Med. Phys.* 44: 6583–6588. https://doi.org/10.1002/mp.12588.

Singh, R., Vandana, Panigrahi, J., and Singh, P.K. (2016). Plasma assisted atomic layer deposited hafnium oxide films for silicon surface passivation. *RCS Adv.* 6: 97720–99727. https://doi.org/10.1039/c6ra19442g.

Staišiūnas, L., Leinartas, K., Juzeliūnas, E. et al. (2020). Anticorrosion performance of hafnium oxide ultrathin films on AZ31 magnesium alloy. *Surf. Coat. Technol.* 397: 126046. https://doi.org/10.1016/j.surfcoat.2020.126046.

Staišiūnas, L., Kalinauskas, P., Juzeliūnas, E. et al. (2022). Silicon passivation by ultrathin hafnium oxide layer for photoelectrochemical applications. *Front. Chem.* 10: 859023. https://doi.org/10.3389/fchem.2022.859023.

Sun, K., Madsen, K., Andersen, P. et al. (2012). Metal on metal oxide nanowire Co-catalyzed Si photocathode for solar water splitting. *Nanotechnology* 23: 194013. https://doi.org/10.1088/0957-4484/23/19/194013.

Swider-Lyons, K., Love, C.T., and Rolison, D.R. (2005). Selective vapor deposition of hydrous RuO_2 thin films. *J. Electrochem. Soc.* 152: C158–C162. https://doi.org/10.1149/1.1859713.

Tan, G.L., Lemon, M.F., Jones, D.J., and French, R.H. (2005). Optical properties and London dispersion interaction of amorphous and crystalline SiO_2 determined by vacuum ultraviolet spectroscopy and spectroscopic ellipsometry. *Phys. Rev. B* 72: 205117. https://doi.org/10.1103/Phys.RevB.72.205117.

Topkar, A. and Lal, R. (1995). Ionic penetration into reoxidized nitride oxides in electrolyte-oxide-semiconductor structures. *Thin Solis Films.* 259: 259–263. https://doi.org/10.1016/0040-6090(94)06437-7.

Verelli, E. and Tsoukalas, D. (2011). Optimization of hafnium oxide for use in nanoparticle memories. *Microelectron. Eng.* 88: 1189–1193. https://doi.org/10.1016/j.mee.2011.03.074.

Wang, W. and Gu, B. (2005). New surface-enhanced Raman spectroscopy substrates via self-assembly of silver nanoparticles for perchlorate detection in water. *Appl. Spectrosc.* 59: 1509–1515. https://doi.org/10.1366/000370205775142458.

Wang, C., Yang, S., Yuan, Y. et al. (2017). Corrosion behavior of hafnium in anhydrous isopropanol and acetonitrile solutions containing bromide ions. *Trans. Nonferous Metals Soc. China* 27: 1896–1906. https://doi.org/10.1016/S1003-6326(17)60214-0.

Ward, M.D. (1995). Principles and application of the electrochemical quartz crystal microbalance. In: *Physical Electrochemistry: Principles, Methods and Applications* (ed. I. Rubinstein), 293. CRC Press.

Wolf, S. and Tauber, R.N. (2000). *Silicon Processing for the VLSI ERA*, vol. 1. Sunset Beach, CA: Process Technology, Lattice Press.

Ye, H., Park, H.S., Akhavan, V. et al. (2011). Photoelectrochemical characterization of $CuInSe_2$ and $Cu(In_{1-x}Ga_x)Se_2$ thin films for solar cells. *J. Phys. Chem. C* 115: 234–240. https://doi.org/10.1021/jp108170g.

Zapata, F. and Garcia-Ruiz, C. (2018). The discrimination of 72 nitrate, chlorate and perchlorate salts using IR and Raman spectroscopy. *Spectrochim. Acta A Mol. Biomol. Spectrosc.* 189: 535–542. https://doi.org/10.1016/j.saa.2017.08.058.

Zhang, X.G. (2004). *Electrochemistry of Silicon and Its Oxide*. New York/Boston/Dordrecht/London/Moscow: Kluwer Academic Publishers. ISBN: 0-306-46541-8.

Zhang, Y.H., Campbell, S.A., and Karthikeyan, S. (2017a). Preparation of hollow out-of-plane hafnium oxide microneedle arrays for transdermal drug delivery. *Microsyst. Technol.* 23: 5887–5892. https://doi.org/10.1007/s00542-017-3354-4.

Zhang, L., Zhang, J., Jiao, H. et al. (2017b). Thickness-dependent surface morphology and crystallization of HfO_2 coatings prepared by ion-assisted deposition. *Thin Solid Films* 642: 359–363. https://doi.org/10.1016/j.tsf.2017.10.010.

Zhang, X., Hsu, C., Lien, S. et al. (2019). Temperature-dependent HfO_2/Si interface, structural evolution and its mechanism. *Nanoscale Res. Lett.* 14: 83. https://doi.org/10.1186/s11671-019-2915-0.

Zou, X., Ji, L., Ge, J. et al. (2019). Electrodeposition of crystalline silicon films from silicon dioxide for low-cost photovoltaic applications. *Nat. Commun.* 10: 5772. https://doi.org/10.1038/s41467-019-13065-w.

8

Electro-Deoxidation of Solid Compounds in Molten Salts

Traditional production and purification of silicon requires large amount of energy and emits greenhouse gases. A fundamental challenge is to extract metals from oxides (ores) producing friendly oxygen instead of CO_2, which is generated in traditional production of metals (e.g. carbothermic reduction). Zero carbon footprint is assumed when using electrons as absolutely clean reducing agents generated from sustainable sources, such as solar or wind. Electrochemistry can turn this scientists' dream into reality.

An opportunity provides molten oxide electrolysis (MOE), in which metals could be extracted from a feedstock in a liquid state. This route is technically simpler when compared to traditional extraction metallurgy. The primary MOE advantages include reductions in energy consumption and mitigation of greenhouse gas emission. Also, an important advantage is the opportunity to achieve high purity product while avoiding its additional purification, which is costly (Birat et al. 2009; Allanore et al. 2013; Fray 2013). However, MOE typically includes carbon anodes, which are consumable during electrolysis and emit carbon oxides due to the interaction of the anode material and the oxide ions in the melt:

$$C + xO^{2-} \rightarrow CO_x \tag{8.1}$$

A challenging goal is creation of the electrode materials, which remains stable in extreme environment at high temperatures, do not corrode anodically, and operate without greenhouse-gas emissions. A new anode material based on Cr–Fe alloys has been proposed for iron extraction from molten oxide (Allanore et al. 2013). Extraction of iron and oxygen has been achieved at $T = 1565\,°C$ with limited changes in dimensions of the $Cr_{90}Fe_{10}$ anode. Apart from the practical importance, the study challenged high-temperature oxidation theory also by demonstrating formation of a novel alloy microstructure. The study by Allanore et al. (2013) brings the opportunity of electrification of iron production closer to reality (Fray 2013).

Another sustainable route of metals extraction is the electrochemical deoxidation of solid oxides in molten salts. The method, named as FFC Cambridge, has been reported in the year 2000 (Chen et al. 2000). The authors succeeded to electrochemically reducing titanium from titanium dioxide in molten $CaCl_2$. Oxygen is ionized, dissolved in electrolyte, and discharged at the anode leaving pure titanium at the cathode. There is no need in this process to dissolve a mineral in an electrolyte and

Silicon: Electrochemistry, Production, Purification and Applications, First Edition. Eimutis Juzeliūnas.

to deposit metal on the cathode. Simplicity and rapidity of the process as well as applicability to a wide range of metal oxides attracted global interest. The solid-to-solid conversion allows avoiding the dendritic growth, which is typical for traditional electrowinning of refractory metals. Also, oxidation–reduction shuttle of intermediate cation between the cathode and anode is prevented (Wang et al. 2008).

The FFC process opened cheaper and environment-friendly route for electrochemical production of metals and alloys – over 30 metals or semimetals have been extracted from solid compounds. To mention a few: Ti, Cr, Mn, Fe, Co, Ni, Cu, Sn, Al, Si, Zr, Nb, Mo, Hf, Ta, and W. The most important compositions (alloys) formed by FFC include: Fe–Ti, Ni–Ti, Tb–Ni, Nd–Co, Co–Cr, La–Ni, Ce–Ni–Cu, Zr–Cr–Ni, Ca–Si, and Si–C–SiC. Lanthanides and actinides have also been extracted: Ce, Pr, Nd, Sm, Gd, Tb, Dy, U, Np, Pu, and Am. The extracted metals and alloys have been surveyed in the following review articles: (Wang et al. 2008, Xiao and Wang 2014, Juzeliūnas and Fray 2020).

Basically, the FFC method relies on the concept that a substance X may be removed from the metal compound (Me1X) by electrolysis in a melt of other metal compound (Me2Y) under such conditions that the reaction of Me1 is preferential at the electrode rather than that of Me2. The reaction proceeds via a direct one-step deoxidation, as in case of NiO reduction to Ni (Qiu et al. 2005), or stepwise, as in the case of titanium reduction TiO_2–Ti_3O_5–Ti_2O_3–TiO–Ti (Dring et al. 2005; Schwandt et al. 2010; Alexander et al. 2011). Figure 8.1 shows the microstructure evolution when extracting Ti from TiO_2.

The TiO_2 in the form of discs with 25–30% open porosity were polarized cathodically for the time periods ranging from 0.5 to 120 hours in molten $CaCl_2$ at 900 °C (Schwandt et al. 2010). Figure 8.2 shows the cathodic current variation during the first stages of electrolysis. Products of the electro-deoxidation are identified at certain time intervals in the figure. The current at the initial phase, which extends over one hour, was concluded to be too large to sustain the oxide reduction alone. Calcium incorporation into the cathode, which occurs in parallel to the deoxidation process, has been discussed.

The following reactions, which occur during the initial stages of electrolysis (~1 hour), have been considered:

$$5TiO_2 + Ca^{2+} + 2e^- \rightarrow Ti_4O_7 + CaTiO_3 \qquad (E^\circ = 1.797\,V) \tag{8.2}$$

$$4Ti_4O_7 + Ca^{2+} + 2e^- \rightarrow 5Ti_3O_5 + CaTiO_3 \qquad (E^\circ = 1.668\,V) \tag{8.3}$$

$$3Ti_3O_5 + Ca^{2+} + 2e^- \rightarrow 4Ti_2O_3 + CaTiO_3 \qquad (E^\circ = 1.564\,V) \tag{8.4}$$

The conclusion about these reactions was based on such factors as the elemental and phase composition, the mass of quenched samples, and the charge amount, which passes through the electrode during the electrolysis. These data illustrate that the reduction reaction $TiO_2 \rightarrow Ti$ is a rather complicated process; it comprises a variety of Ti-suboxides.

Molten $CaCl_2$ has predominantly been used as electrolyte in the FFC processes. Major physical–chemical advantages of molten $CaCl_2$ include low vapor pressure, high electric conductivity, and ability to dissolve the reaction products, such as

Figure 8.1 Microstructure evolution during TiO_2 deoxidation: (a) – TiO_2; (b) – Ti_2O_3 – $CaTiO_3$; (c) $CaTi_2O_4$; (d) $CaTi_2O_4/TiO$; (e) $Ti[O]_x$. Cross-marks in (b) – (d) identify the locations where the elemental composition was determined by the SEM. The electro-deoxidation of porous TiO_2 disc (8 g) was performed in molten $CaCl_2$ electrolyte using a graphite anode. The applied voltage was 2.5 V in respect to graphite anode. Source: Schwandt et al. (2010), Trans Tech Publications Ltd.

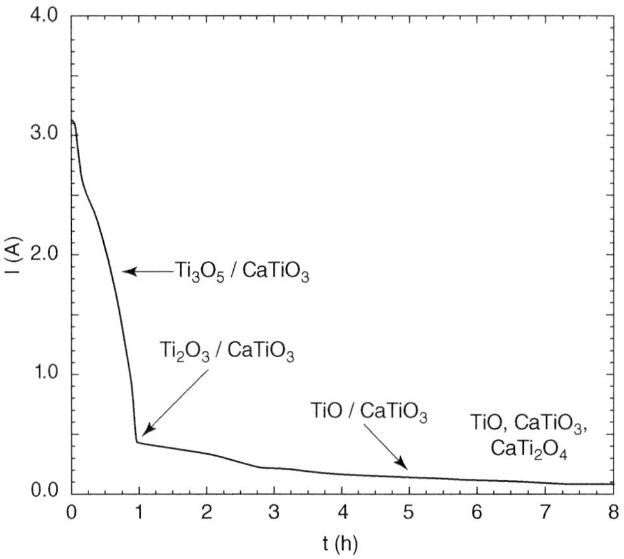

Figure 8.2 Current variation during electro-deoxidation of TiO_2 in molten $CaCl_2$ at the voltage of 2.5 V against a graphite anode. Source: Schwandt et al. (2010)/Trans Tech Publications Ltd./CC BY-4.0.

oxygen ions, silicates, or carbon dioxide. From an economical point of view, the salt is not expensive, and it is widely available in the market for large-scale applications. Also, calcium chloride is nontoxic and well soluble in water, which makes it possible to easily remove the remaining solidified electrolyte from product by simple washing at room temperature. To reduce operation temperatures various eutectics are also used as electrolytes. Thus, mixtures of $CaCl_2$–KCl–LiCl, $CaCl_2$–CaO, or $CaCl_2$–NaCl have been used for electrochemical reduction of silicon. Juzeliūnas and Fray have surveyed the electrolyte compositions (Juzeliūnas and Fray 2020).

Molten $CaCl_2$ has a wide electrochemical window, which makes it possible to reduce metals or semimetals from oxides without co-deposition of calcium metal. The oxide decomposition voltage (E_d, V) relates directly to the Gibbs-free energy (ΔG, J mol^{-1}):

$$\Delta G = -nFE_d \tag{8.5}$$

where n stands for the number of electrons in electrochemical reaction, and F is the Faraday constant. The E_d means the voltage between a positive and negative electrode in the electrochemical cell, which is needed to decompose an oxide. Electrochemists, however, are in the habit of operating with the electrode potential (E), which is measured as a voltage between the cathode (working electrode) and a reference electrode, whose potential is precisely known under standard conditions. In aqueous electrolytes, typical reference electrodes are silver/silver chloride (Ag/AgCl), calomel (Hg/Hg_2Cl_2) or hydrogen (H^+/H) standard electrodes. In molten salts, application of reference electrodes is more complicated due to operation at high temperatures. Pure metals, such as Pt or Mo, are frequently used as pseudo-reference electrodes in molten salts and the potential of such electrodes is calibrated with respect to that of the electrolyte decomposition and the equilibrium potential of the redox pair Ca^{2+}/Ca ($E^{\circ}_{Ca^{2+}/Ca}$).

Wang et al. summarized the ΔG and E_d values of many metal oxides (Wang et al. 2008). The energy values referred to below were taken from this publication. The reaction, which determines the electrochemical stability of the molten $CaCl_2$ electrolyte

$$CaCl_2 \rightarrow Ca + Cl_2 \tag{8.6}$$

is characterized by the parameters $\Delta G = 667.4\,kJ\,mol^{-1}$ and $E_d = 3.459$ V. This means the electrochemical window of the electrolyte is as wide as 3.459 V between a cathode and anode. The E_d values of most metal oxides are less when compared with that of $CaCl_2$. For our discussions, the most relevant is the characteristics of the decomposition of silica

$$SiO_2 \rightarrow Si + O_2 \tag{8.7}$$

The reaction is characterized by the parameters $\Delta G = 700.3\,kJ\,mol^{-1}$ and $E_d = 1.815$ V. Thus, the reaction implies that silica can be electrochemically decomposed at an energy less than that needed for the decomposition of calcium chloride. In other words, at certain voltage, the electrolyte will remain electrochemically stable, while silicon can be reduced from its oxidized state.

Another important reactions are decomposition of carbon oxides

$$CO_2 \rightarrow C + O_2 \tag{8.8}$$

$$2CO \rightarrow 2C + O_2 \tag{8.9}$$

The dioxide is decomposed at $E_d = 1.026$ V and the monoxide is decomposed at $E_d = 1.117$ V. Such conversion is important in several aspects. Carbon anodes, which are typically used in molten salt electrolysis, can react with the oxygen released from an oxide (e.g. 8.7), producing CO_x. A detrimental CO_x decomposition can occur along with the desired deoxidation reaction and this may lead to contamination of the product by carbon. This could be an issue when seeking to produce high purity material, such as solar-grade silicon. On the other hand, carbon reduction provides an opportunity for synthesis of important compositions. Zou et al. proposed a facile electrosynthesis of silicon carbide nanowires from silica/carbon precursors in molten salt (Zou et al. 2017). Juzeliūnas et al. synthesized in a molten salt the photoactive carbon–carbide structure on silicon (Zou et al. 2017; Juzeliūnas et al. 2018).

Decomposition in molten salts of lightweight metals oxides (MgO and Al_2O_3) are of great interest because electrodeposition of Mg and Al from aqueous electrolytes is problematic due to hydrolysis and high electronegativity of these metals. The E_d value of MgO is 2.283 V and that of Al_2O_3 is 2.252 V. Both values are below that of $CaCl_2$ decomposition.

The oxides of refractory metals make another important oxide class. Deposition of refractory metals is important because of their specific properties, such as heat, corrosion, and abrasion resistance. However, deposition of refractory metals from aqueous electrolytes is challenging. Theoretically, there are good perspectives to the extraction of refractory metals from oxides, as their E_d values are below that of $CaCl_2$ (in V): HfO_2 – 2.36; ZrO_2 – 2.274; TiO_2 – 1.897; V_2O_3 – 1.473; Nb_2O_3 – 1.441; V_2O_5 – 1.098; WO_3 – 0.939; Cr_2O_3 – 1.43. The E_d values of many other metal oxides are also less than that of $CaCl_2$ (in V): Fe_2O_3 – 0.892; SnO_2 – 0.877; MnO_2 – 0.800; NiO – 0.692; Co_3O_4 – 0.596; Cu_2O – 0.431.

Lanthanides and actinides can be electrochemically extracted from their oxides in molten $CaCl_2$ (E_d in V): Tb_2O_3 – 2.651; Dy_2O_3 – 2.623; Nd_2O_3 – 2.553; Pr_2O_3 – 2.549; La_2O_3 – 2.525; UO_2 – 2.285; CeO_2 – 2.194.

Thus, the decomposition voltages of many oxides are below the voltage of decomposition of $CaCl_2$, that is, the molten salt electrolyte. This means that the elements could be electrochemically deoxidized in molten $CaCl_2$ with no reduction of calcium metal. However, it is worth mentioning that in certain cases electrodeposition of calcium metal is desired, as in the case of formation of photoresponsive porous silicon through its etching by a liquid calcium silicide (Ca_xSi_y) (Juzeliūnas et al. 2016). The reduced calcium interacts with the silicon producing liquid silicide, which dissolves and leaves behind on the substrate a porous structure. The reduced calcium could also act as a reducing agent of the oxidized species present in the electrolyte, as in the case of electrochemical synthesis of photoactive carbon–carbide structure from CO_2 on Si substrate (Juzeliūnas et al. 2018).

Technical aspects of the FFC process have been widely discussed (Chen et al. 2004; Wang et al. 2008; Fray 2008; Schwandt et al. 2010; Fray 2012). Typically, a metal oxide (Me_xO_y) powder is used as a precursor, although solid precursors, such as quartz, are also used. The powder is pressed into a pellet to which a metallic current collector (e.g. Mo, Ni wire, or foil) is attached. The pellet-collector system is placed into molten salt electrolyte. The electrochemical cell is also supplied by carbon or inert anode (counter electrode). The voltage between the cathode and the anode is supplied, at which metal oxide is decomposed:

$$MeO + 2e^- \rightarrow Me + O^{2-} \tag{8.10}$$

The cell voltage is selected so that the above reaction occurs, avoiding deposition of the electrolyte cation:

$$Ca^{2+} + 2e^- \rightarrow Ca \tag{8.11}$$

The oxygen released in (8.10) is transferred to the anode where it is oxidized to O_2 on an inert anode (e.g. $CaTi_xRu_{(1-x)}O_3$ (Fray 2012)) or to CO_x on a carbon anode (8.1). As already discussed, there is agreat difficulty in creating effective and chemically inert anodes to reduce the process carbon footprint.

It is important that the Me products have porous or powder-like structures. These could be used as the functional materials in relevant applications such as printing technologies and production of composite materials. Alternatively, the powders could also further be treated by the methods of conventional metallurgy.

The FFC method of electrochemical reduction may also be extended to compounds other than oxides. An example is electro-desulfidation of molybdenum sulfide MoS_2 in molten chloride (Li et al. 2007). The electrolysis resulted in reduction of the metal on the cathode and oxidation of S^{2-} to elemental sulfur on the anode. Thus, the method yielded simultaneous extraction of two elements – the metal and the sulfur.

The cathodic reduction processes in the MOE and the FFC methods are associated with oxygen gas development on the anode, provided it is made of an inert material. This suggests an exciting application field that is oxygen production in space exploration (Sanderson 2009; Fray 2013; Schwandt ét al. 2012). Lunar or Mars soils are composed of fine fractions of the regolith – unconsolidated, loose deposits covering the planet surface. Regolith is formed due to disintegration of the rocks by meteoric impacts and bombardment by solar and interstellar atomic particles so that, over time, the surface is ground to finer particle size forming the surface dust with micrometer sized particles. Common lunar minerals are composed of iron, magnesium, calcium, aluminum, silicon, titanium, and oxygen. Lunar rocks have been found rich in anorhite ($CaAl_2Si_2O_8$); some rocks have high content of titanium in the form of ilmenite ($FeTiO_3$) (Crawford 2015). The lunar minerals in a molten form can be processed electrochemically. The regolith can be decomposed to obtain oxygen needed to sustain life or to be used as a fuel-oxygen mixture for rockets. Opportunity to produce oxygen on the moon provides significant cost savings of the space expedition. The rocket's weight could be substantially reduced, as there is no need to carry the liquid oxygen necessary for return journey, the oxygen would be produced on the moon.

In the FFC reactor, there is no need to melt the rock; the oxygen production could be achieved at lower temperature than the melting point of regolith minerals (around 900 °C, or less). In the MOE, much higher temperatures are required in melting the minerals. On the other hand, the advantage of MOE is that the regolith is directly converted to electrolyte; thus, there is no need of regolith consolidation. The fine sandy nature of the Moon rock makes it difficult to prepare a solid-state precursor. However, an opportunity provides reduction of a mineral suspension in the molten salt electrolyte, such as $CaCl_2$.

References

Alexander, D.T.L., Schwandt, C., and Fray, D.J. (2011). The electro-deoxidation of dense titanium dioxide precursor in molten calcium chloride giving a new reaction pathway. *Electrochim. Acta* 56: 3286–3295. https://doi.org/10.1016/j.electacta.2011.01.027.

Allanore, A., Yin, L., and Sadoway, D.R. (2013). A new anode material for oxygen evolution in molten oxide electrolysis. *Nature* 497: 353–356. https://doi.org/10.1038/nature12134.

Birat, J.-P., Lorrain, J.-P., and de Lassat, Y. (2009). The "CO_2 tool": CO_2 emissions and energy consumption of existing and breakthrough steelmaking routes. *Rev. Métall.* 106: 325–336. https://doi.org/10.1051/metal/2009060.

Chen, G.Z., Fray, D.J., and Farthing, T.W. (2000). Direct electrochemical reduction of titanium dioxide to titanium in molten calcium chloride. *Nature* 407: 361–364. https://doi.org/10.1038/35030069.

Chen, G.Z., Gordo, E., and Fray, D.J. (2004). Direct electrolytic preparation of chromium powder. *Metall. Mater. Trans. B* 35: 223–233. https://doi.org/10.1007/s11663-004-0024-6.

Crawford, I.A. (2015). Lunar resources: a review. *Prog. Phys. Geog.* 39: 137–167. https://doi.org/10.1177/0309133314567585.

Dring, K., Dashwood, R., and Inman, D.J. (2005). Predominance diagrams for electrochemical reduction of titanium oxides in molten $CaCl_2$. *J. Electrochem. Soc.* 152: D184–D190. https://doi.org/10.1149/1.2034552.

Fray, D.J. (2008). Novel methods for the production of titanium. *Int. Mater. Rev.* 53: 317–325. https://doi.org/10.1179/174328008X324594.

Fray, D.J. (2012). Exploring novel uses of molten salts. *ECS Trans.* https://doi.org/10.1149/05011.0003ecst.

Fray, D.J. (2013). Iron production electrified. *Nature* 497: 324–325. https://doi.org/10.1038/nature12102.

Juzeliūnas, E. and Fray, D.J. (2020). Silicon electrochemistry in molten salts. *Chem. Rev.* 120: 1690–1709. https://doi.org/10.1021/acs.chemrev.9b00428.

Juzeliūnas, E., Coxon, P.R., Fray, D.J. et al. (2016). Photoelectrochemical activity of macroporous silicon formed in molten calcium chloride. *Chemija* 27: 11–16.

Juzeliūnas, E., Fray, D.J., Kalinauskas, P. et al. (2018). Electrochemical synthesis of photoactive carbon-carbide structure on silicon in molten salt. *Electrochem. Commun.* 90: 6–10. https://doi.org/10.1016/j.elecom.2018.03.003.

Li, G.M., Wang, D.H., Jin, X.B., and Chen, G.Z. (2007). Electrolysis of solid MoS_2 in molten $CaCl_2$ for Mo extraction without CO_2 emission. *Electrochem. Commun.* 9: 1951–1957. https://doi.org/10.1016/j.elecom.2007.05.007.

Qiu, G.H., Ma, M., Wang, D.H. et al. (2005). Metallic cavity electrodes for investigation of powders: electrochemical reduction of NiO and powders in molten $CaCl_2$. *J. Electrochem. Soc.* 152: E328–E336. https://doi.org/10.1149/1.2012567.

Sanderson, K. (2009). How to breathe on the moon. *Nature* https://doi.org/10.1038/news.2009.803.

Schwandt, C., Doughty, G.R., and Fray, D.J. (2010). The FFC-Cambridge process for titanium metal winning. *Key Eng. Mater.* 436: 13–25. https://doi.org/10.4028/www.scientific.net/KEM.436.13.

Schwandt, C., Hamilton, J.A., Fray, D.J., and Crawford, I.A. (2012). The production of oxygen and metal from lunar regolith. *Planet. Space Sci.* 74: 49–56. https://doi.org/10.1016/j.pss.2012.06.011.

Wang, D., Jin, X., and Chen, G.Z. (2008). Solid state reactions: an electrochemical approach in molten salts. *Annu. Rep. Prog. Chem. Sect. C* 104: 189–234. https://doi.org/10.1039/B703904M.

Xiao, W. and Wang, D. (2014). The electrochemical reduction process of solid compounds in high temperature molten salts. *Chem. Soc. Rev.* 43: 3215–3228. https://doi.org/10.1039/c3cs60327.

Zou, X., Ji, L., Lu, X., and Zhou, Z. (2017). Facile electrosynthesis of silicon carbide nanowires from silica/carbon precursors in molten salt. *Sci. Rep.* 7, art. no. 9978. https://doi.org/10.1038/s41598-017-10587-5.

9

Voltammetry and Basic Reactions of Silicon Electrode in Molten $CaCl_2$

Pure silicon (Si) at room temperature is an insulating material; its low conductivity limits electrodeposition and prevents formation of a compact layer. Relevant example is the Si deposition from ionic liquids (ILs) – the room temperature molten salts. Only very thin nonuniform Si layers were deposited from ILs due to the insulating properties of the deposits at room temperatures (Endres et al. 2008; Katayama et al. 2001; Zein El Abedin et al. 2004; Tsuyuki et al. 2018). Silicon conductivity, however, rises with temperature owing to its narrow energy gap between the valence and conduction bands. Thus, when the electrolyte temperature is increased, silicon transforms from an insulating material to an active electrode.

Voltammetry studies provide important information about silicon electrochemical reactions and their mechanisms in molten salts. To this end, a three-electrode system is typically used, which comprises the working, the reference, and the counter electrodes. Figure 9.1 shows schematically the typical electrochemical installation used in molten salt experiments. A resistant crucible made of metal oxides, such as Al_2O_3–ZrO_2, is employed as a body of the electrochemical cell. The crucible is filled with the salt and placed in a reactor, which is closed by a lid equipped with gas inlet and outlet as well as electrical feedthroughs for electrodes. The cell is put into a vertical furnace; the temperature is raised according to certain protocol, and the salt is melted. Majority of the silicon reduction studies has been performed in molten calcium chloride ($CaCl_2$) (Table 9.1). Advantages of this electrolyte include low vapor pressure, wide electrochemical window, high conductivity, and ability to dissolve the oxygen ions O^{2-} electrochemically released from an oxide. The salt is nontoxic, cheap, and readily available in the market.

Table 9.1 summarizes the major experimental conditions of silica deoxidation experiments applied by numerous authors: electrolyte and its temperature, Si feedstock (precursor) as well as the materials used to prepare the contacting, the counter, and the reference electrodes. The data period covers the last two decades, starting from 2003 when the first SiO_2 electro-deoxidation was reported (Nohira et al. 2003). The authors mainly sought to demonstrate the principal opportunity of silica deoxidation and produce high purity silicon (Jin et al. 2004; Yasuda et al. 2005a,b,c; Xiao et al. 2010; Pistorius and Fray 2006; Ergül et al. 2011; Akpinar et al. 2017; Yang et al. 2017). Considerable efforts have been made to form silicon nanowires (NWs) (Yang et al. 2009; Nishimura et al. 2011; Fang et al. 2015; Zou

Silicon: Electrochemistry, Production, Purification and Applications, First Edition. Eimutis Juzeliūnas.

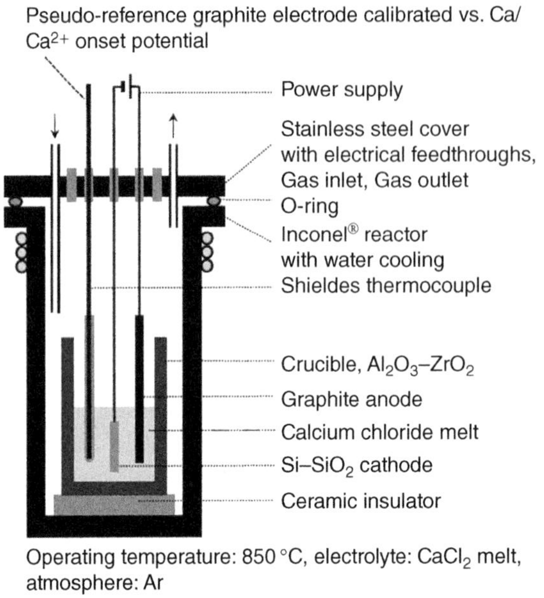

Figure 9.1 Schematic representation of typical unit for electrochemical metal extraction from oxides in molten salt electrolyte. Body of the electrochemical cell is made of highly resistant oxides, such as alumina, zirconia, or silica (quartz). The cell is placed in a hermetic reactor made of Ni–Cr based superalloy (Inconel®), to which an inert gas (argon) is supplied. The reactor is placed in a vertical furnace, in which the salt is melted. Electrodes are introduced to the cell and connected via temperature resistant leads to power supply (e.g. potentiostat). Source: Schwandt et al. (2010)/Trans Tech Publications Ltd./CC BY-4.0.

et al. 2017a). Another important field was the formation of photo-responsive silicon (Cho et al. 2012a,b; Juzeliūnas et al. 2010, 2018; Xiao et al. 2013). An emerging field is the formation of a p–n junction for the manufacturing of low-cost photovoltaic devices (Zou et al. 2017a).

Table 9.1 refers to a variety of experimental conditions, which have been applied by numerous groups. The working electrode for the $SiO_2 \rightarrow Si$ conversion has been used in various forms, such as monolithic quartz, SiO_2 powders pressed into pellets, or granules confined in a container. A thin SiO_2 layer thermally formed on Si substrate to obtain surface nano-micro architectures has also been studied.

Most of the studies were conducted in molten $CaCl_2$ electrolyte at temperatures of 850–900 °C. Yasuda et al. lowered the operating temperature to 500 °C; this increases the processing energy efficiency (Yasuda et al. 2005c). To this end, the authors used the electrolyte composition $CaCl_2$–KCl–LiCl instead of conventionally used single $CaCl_2$. Ergül et al. conducted the reduction at 750 °C using binary $CaCl_2$–NaCl electrolyte (Ergül et al. 2011).

Table 9.1 lists a variety of materials, which have been used for preparation of contacting electrodes: molybdenum (Mo), tungsten (W), nickel (Ni), silicon (n-Si, p-Si), stainless steel, platinum (Pt), zinc (Zn), silver (Ag), and graphite. The contacting electrode is attached to the SiO_2 sample in order to induce the electro-deoxidation.

Table 9.1 Experimental conditions of Si electrochemical reduction in the $CaCl_2$-based molten electrolytes using various SiO_2 precursors.

Electrolyte	Si precursor	Contacting electrode, substrate	Counter electrode	Reference electrode	Temperature (°C)	Process specification, reference
$CaCl_2$	Quartz plate	Mo	Graphite	Pt – Ca^{2+}/Ca Ag/AgCl in $CaCl_2$ Mo – Ca^{2+}/Ca	850	SiO_2 deoxidation (Nohira et al. 2003; Yasuda et al. 2005a,b,c)
$CaCl_2$	Quartz, SiO_2 pellet	W, Ni	Graphite	Pt, Ag/AgCl in NaCl–KCl	850	SiO_2 deoxidation (Jin et al. 2004; Xiao et al. 2010)
$CaCl_2$	Quartz plate	p-Si (100)	Glassy carbon	Ag/AgCl in $CaCl_2$ Mo–Ca/Ca^{2+}	850	SoG-Si production (Yasuda et al. 2013)
$CaCl_2$–KCl–LiCl	Quartz plate	Mo	Graphite	Ag/AgCl in LiCl–KCl–$CaCl_2$	500	SiO_2 deoxidation (Yasuda et al. 2005c)
$CaCl_2$–CaO	SiO_2 powder pellet	Ni	Graphite	Graphite – Ca/CaO	900	SiO_2 deoxidation (Pistorius and Fray 2006)
$CaCl_2$	Quartz plate, SiO_2 powder pellet	Mo	Not given	Pt	900	Si–NWs (Yang et al. 2009)
$CaCl_2$–NaCl	SiO_2 powder pellet	Stainless steel, Ni	Graphite	—	750–850	SiO_2 deoxidation (Ergül et al. 2011)
$CaCl_2$	SiO_2 powder pellet	Mo	Vitreous carbon	Ag/AgCl in $CaCl_2$	850	Si–NWs (Nishimura et al. 2011)
$CaCl_2$–CaO	Quartz, SiO_2 pellet	Ni	Graphite	Ag/AgCl in NaCl–KCl	850–900	Si dissolution- deposition (Xiao et al. 2012)
$CaCl_2$	SiO_2–NiO powders in pellet	Mo	Graphite	—	900	Si–NWs (Fang et al. 2015)
$CaCl_2$	SiO_2 granules	n-Si (100)	Glassy carbon	Ag/AgCl in $CaCl_2$	850	SiO_2 deoxidation (Yang et al. 2016)

(Continued)

Table 9.1 (Continued)

Electrolyte	Si precursor	Contacting electrode, substrate	Counter electrode	Reference electrode	Temperature (°C)	Process specification, reference
$CaCl_2$	Quartz plate, tube, SiO_2 powder	Si, Pt, Ni	Glassy carbon	Ag/AgCl in $CaCl_2$	850	SoG-Si production (Akpinar et al. 2017)
$CaCl_2$	SiO_2 granules, Si powder and granules	Graphite	Glassy carbon	Ag/AgCl in $CaCl_2$	850	SiO_2 deoxidation (Yang et al. 2016)
$CaCl_2$	SiO_2 granules	Mo, liquid Zn and Zn–Si alloy	Graphite	Ag/Ag^+ Mo – Ca/Ca^{2+}	850	SoG-Si production (Yang et al. 2017)
$CaCl_2$	SiO_2 NPs (5–15 nm in electrolyte)	Mo, Ag, n-Si (100) substrates	Glassy carbon, graphite	Mo – Ca/Ca^{2+}, graphite	850	Si layer formation (Cho et al. 2012b)
$CaCl_2$	SiO_2 NPs (5–15 nm in electrolyte)	Ag substrate	Glassy carbon, graphite	Mo – Ca/Ca^{2+}	850	Deposition of photoactive Si (Cho et al. 2012a)
$CaCl_2$	SiO_2 layer (0.3–2.0 μm) on Si	Mo	Graphite	Graphite	850–900	Si structuring, photoresponsive layer (Juzeliūnas et al. 2010, 2011, 2012, 2016, 2018)
$CaCl_2$–CaO	$CaSiO_3$, SiO_2 powder	Graphite, p-Si	Graphite	Graphite	850	p–n junction of Si films (Zou et al. 2017b)
$CaCl_2$	SiO_2 pellet	Mo	Graphite	Ag/AgCl	900	Photoresponsive, nanostructured Si (Xiao et al. 2013)
$CaCl_2$	SiO_2 and C powder, pellet	Ni	Graphite	Pt	900	Si–C NWs (Zou et al. 2017a)
$CaCl_2$	Silica glass tube	W	Vitreous carbon, Ag	Ag/Ag^+ Mo – Ca/Ca^{2+}	850	SiO_2 reduction, *in situ* synchrotron X-ray diffraction study (Katasho et al. 2020)

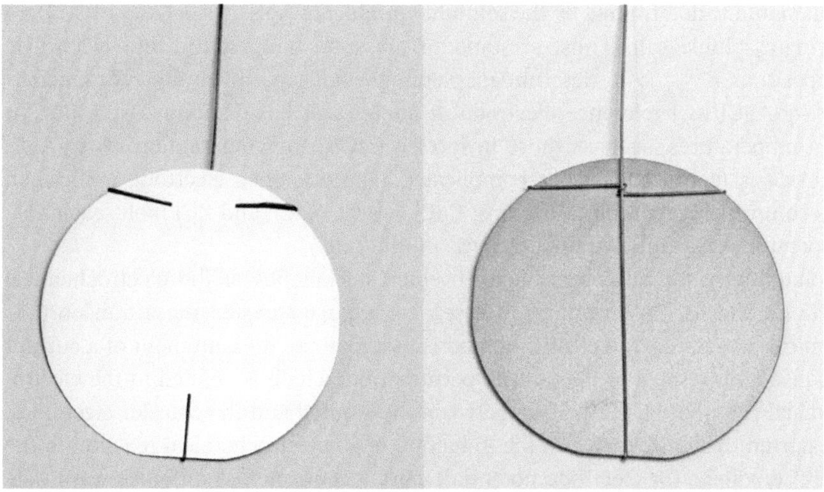

Figure 9.2 The Si–Mo electrodes prepared from p-Si wafer of 22 cm^2 with 2 μm SiO_2 thermal coating in a Mo wire (Ø 0.5 mm) frame, which has a function of a contacting electrode at which silica conversion to silicon is initiated. The front side (left) is polished and the back side (right) is unpolished.

The process is induced locally and then spreads over the silica surface. The produced silicon, being at high temperature conductive, takes the role of a contacting electrode, so that 3D process evolution is possible.

Figure 9.2 shows the design of a working electrode, which comprises the Si wafer coated with SiO_2 layer and the Mo wireframe as a contacting electrode. The electrochemical deoxidation is initiated at the Mo/SiO_2 contact points. The Mo frame also plays the role of a current collector from the entire area where the isolating silica was converted to conductive silicon.

Table 9.1 also shows that graphite is a material of choice by many authors for the preparation of the counter electrode. From an environmental point of view, the application of such an electrode is detrimental due to CO_x emission by the anodic reaction. The oxygen O^{2-} released from SiO_2 interacts with the anode producing CO or CO_2 gas. Therefore, search for alternative materials for the counter electrode is a challenge. The electrodes should be chemically inert, conductive, stable at high temperatures, and resistant to electrochemical disintegration.

Graphite or platinum is frequently used as a reference electrode, that is, the electrode with respect to which potential of the working electrode is measured (Table 9.1). Such electrode is referred to as a quasi-reference one, as its equilibrium potential is not precisely determined. In $CaCl_2$ electrolyte, it is rational to compare the potential of the quasi-reference electrode with the equilibrium potential of the calcium pair $E^\circ_{Ca/Ca^{2+}}$. Calcium metal is electrochemically deposited $Ca^{2+} + 2e^- \rightarrow Ca$ on a foreign metal substrate for such control.

In aqueous electrolytes, it is a common practice to use a silver/silver chloride (Ag/AgCl) in KCl solution as a reference electrode. The equilibrium potential of this electrode ($E^\circ_{Ag/AgCl}$) depends on Ag^+ ions concentration in the electrolyte, which is fundamentally determined by the solubility product of AgCl (SP = [Ag+]/[Cl^-]) – a sparingly soluble salt. Thus, at standard pressure, temperature and electrolyte composition $E^\circ_{Ag/AgCl}$ is determined precisely and experimentally reproducible. The Ag/AgCl/KCl reference electrode in molten salt has distinctive specifics. At high temperatures, such as those in molten $CaCl_2$, the construction of Ag/AgCl electrode is technically rather complicated. The reference electrode for molten salts comprises certain proportions of $CaCl_2$, AgCl, NaCl, and KCl melt sealed in a temperature-resistant container (quartz, mullite, etc.).

Voltammetry measurements help to elucidate kinetics of the electrochemical reactions, that is, the limiting processes, e.g. charge transfer, mass transport, or chemical processes. One of this method is based on the measurement of a current response with respect to the voltage perturbation, which is applied to the electrochemical cell, that is, the voltage between the studied and the counter electrodes. The current response is due to the induction of an electrochemical reaction in the cell. The voltage (or electrode potential) can be controlled in different ways. The potentiostatic control refers to a constant potential step(s) and the current flows as a function of time. The potentiodynamic method refers to an electrode potential change at a constant rate ($v = dE/dt$). In many cases, the electrode potential is repeatedly swept forward and backward; this method is referred to as cyclic voltammetry (CV). Readers seeking information on the principles of voltammetry are referred to the book written by Eliaz and Gileadi, which reviews the most important techniques of physical electrochemistry and their applications in materials science, engineering, nanoscience, nanotechnology, and various industrial fields (Eliaz and Gileadi 2019).

Figure 9.3 displays the CV curves obtained for Si electrode in molten $CaCl_2$. The inset photographs show the Si electrode with the attached Mo-wire frame as a contacting electrode, which doubles as a current collector. For experiments of this nature, both the silicon and the contacting electrode are in contact with the molten salt electrolyte; this means that the measured current is the sum of the contributions from both electrodes. The contribution of the current collector has to be subtracted from the overall response in order to study the behavior of the silicon solely. Figure 9.3 displays the curve obtained for the single Mo frame, that is, without Si sample, for comparison. Note that Mo has been used as a contacting electrode in many studies (Table 9.1). It is worth noting that the curves for Mo electrode did not depend on potential sweep rate: the curves were almost coincident at the sweep rate of 5 and 500 $mV\,s^{-1}$. This clearly indicates the kinetic process without transport limitations in the electrolyte.

A fundamental characteristic of any electrochemical system is the electrochemical window, which predicts the feasibility zone of electrochemical reactions. No electrochemical reaction, whose redox potential is beyond the window, can be performed. In molten $CaCl_2$, the window is limited by the reactions of calcium reduction and chlorine evolution:

$$Ca^{2+} + 2e^- \rightarrow Ca \tag{9.1}$$

$$2Cl^- \rightarrow Cl_2 + 2e^- \tag{9.2}$$

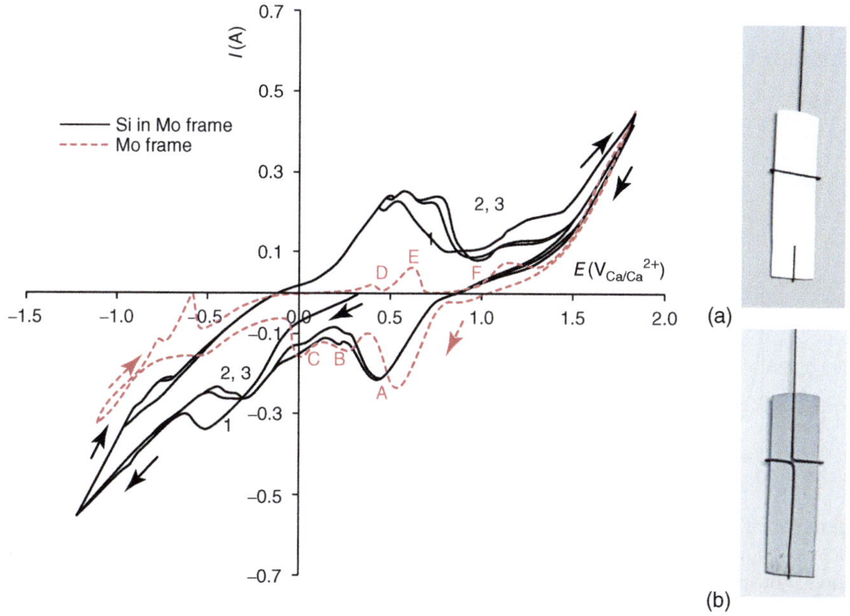

Figure 9.3 Cyclic voltammetry curves obtained for Mo (red) and p-Si/Mo (dark) electrodes in molten $CaCl_2$ at $T = 850\,°C$ and potential sweep rate $v = 100\,mV\,s^{-1}$. Potentials are given in respect of the equilibrium potential of the calcium electrode Ca/Ca^{2+} ($E^{\circ}_{Ca/Ca^{2+}}$). The first cycle was started at $E_{ocp} = 0.33\,V$ for Si–Mo and $E_{ocp} = 0.85\,V$ for Mo electrode. Three cycles are displayed for Si–Mo electrode, which show a closed voltammetric loop for 2nd and 3rd cycles, for which the starting and the final points are coincident. The 3rd cycle is shown for Mo electrode as well, which also exhibits a closed loop. The samples were prepared from p-Si (100) wafer ($S = 1.3 \times 5.0 = 6.50\,cm^2$). Mo wire (Ø 0.5 mm) frame has been used as a contacting electrode and a current collector, as shown in the photographs (a) (front side, polished) and (b) (back side, unpolished). Source: E. Juzeliūnas (Author).

Molybdenum has several oxidation states ranging from Mo^0 to Mo^{6+} and the most common states are Mo^{4+} and Mo^{6+} (Braithwaite and Haber 1994). The overall CV curve for Mo is quite complicated. Three characteristic peaks of cathodic reduction (A–B–C) were observed during the negative potential sweep from the anodic limiting zone. The corresponding anodic peaks appear during the positive sweep (D–E–F). These peaks imply a transition through several Mo oxidation states during the potential sweep, most probably, the redox transitions Mo–Mo^{2+}–Mo^{4+}–Mo^{6+}. The exponential current growth on Mo electrode at $E < -0.8\,V$ and the corresponding oxidation peak within −0.8 to −0.5 V indicate excessive calcium reduction on this electrode below $E \sim -0.8\,V_{Ca/Ca^{2+}}$. It is important to note that Ca does not form an alloy with Mo (Sakanaka et al. 2015; Massalski et al. 1990).

Distinctive behavior demonstrates the Si electrode. An initial rectilinear current increase starting from $E = 0.33\,V$ (Figure 9.3) is attributable to a residual current, which is always observed in high-temperature molten salts. Such current is due to electronic conductivity of molten electrolyte as well as presence of contaminants. Appreciable reduction of Ca on Si appears below the calcium redox potential

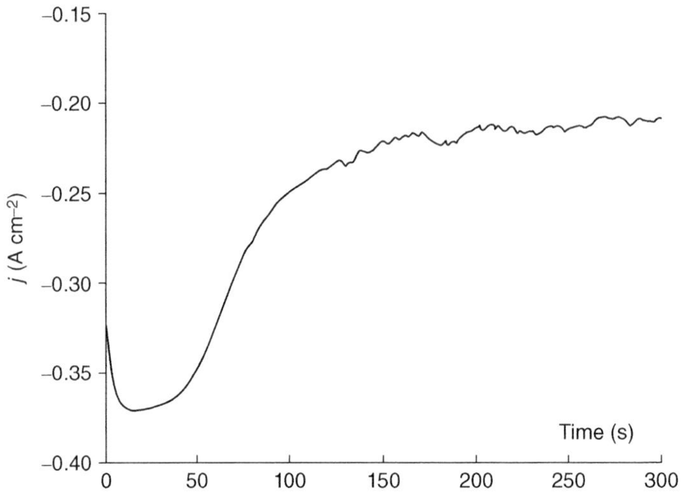

Figure 9.4 Current density variation during potentiostatic polarization of the electrode whose configuration is shown in Figure 9.2 at $E = -0.4\ V_{Ca/Ca^{2+}}$ in molten $CaCl_2$, $T = 850\,°C$. Source: Author E. Juzeliūnas.

($E < \sim 0.0\ V_{Ca/Ca^{2+}}$), that is, in accordance with the thermodynamic predictions. Further polarization leads to some electrode passivity as the peak at $E \sim -0.4\ V_{Ca/Ca^{2+}}$ indicates.

Potentiostatic polarization to the peak region shows the passivation tendency as well (Figure 9.4). The current increases immediately after polarization to $j \sim 0.32\ A\,cm^{-2}$, reaching slowly a maximum at $j \sim 0.37\ A\,cm^{-2}$ and decreases to some limiting value $j \sim 0.2\ A\,cm^{-2}$.

The passivity behavior is explained by the formation of calcium silicide, which reduces conductivity of the electrode. X-ray diffraction (XRD) analysis in Figure 9.5 identifies the silicides, which developed in the peak region at $E = -0.4\ V_{Ca/Ca^{2+}}$. The XRD data identifies Ca_2Si, $CaSi_2$, and $Ca_{14}Si_{19}$. Deposition of silicides at $E < E^{\circ}_{Ca/Ca^{2+}}$ on Si was demonstrated by Sakanaka et al. (2015). These phenomena will be extensively discussed in Chapter 16.

It is important to note that calcium also can be reduced at the potential more positive than the potential of the redox couple Ca/Ca^{2+} ($E > E^{\circ}_{Ca/Ca^{2+}}$) (Xiao and Wang 2014; Xiao et al. 2010; Nohira et al. 2003; Gao et al. 2005; Yasuda et al. 2005a,b). Such phenomenon when metal (M) is reduced on foreign substrate at more positive potentials than its redox potential $E^{\circ}_{Me/Me^{z+}}$ is referred to as underpotential deposition (UPD). Generally, it is assumed in UPD that only up to one monolayer of metal on foreign metal substrate (S) is deposited and the deposition occurs when the interaction between M and S is stronger than that between S and S (Zangari 2015). Anomalous alloy deposition has been explained by the UPD concept when more negative alloy component is preferentially deposited. A relevant example is an electrochemical deposition technically important Zn–Co alloy, which is widely used for corrosion protection of steel (Stankevičiūtė et al. 1998). The alloy contains minor

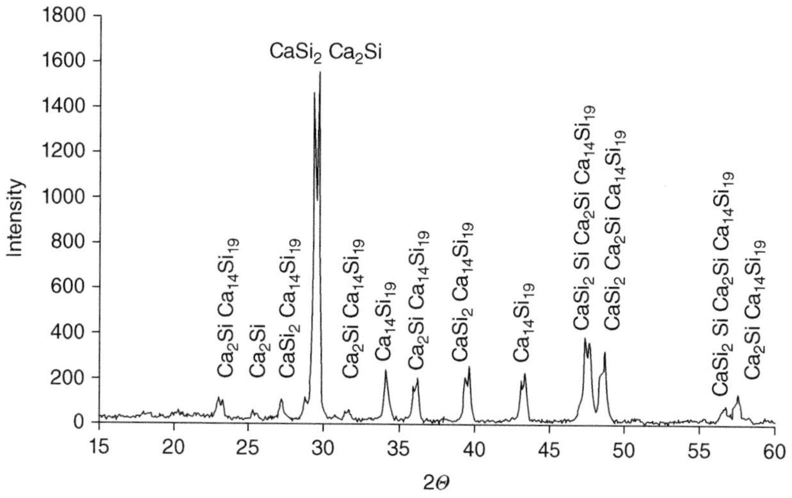

Figure 9.5 XRD pattern, which indicate development of calcium silicide on p-Si polarized for five minutes at $E = -0.4\ V_{Ca/Ca^{2+}}$ in molten $CaCl_2$ electrolyte at $T = 850\,°C$. Electrode configuration is shown in Figure 9.3.

amounts of Co, typically, less than 1%. Deposition of the less noble component Zn ($E^\circ_{Zn/Zn^{2+}} = -0.76$ V vs. NHE) in such alloy is preferential compared to the nobler Co ($E^\circ_{Co/Co^{2+}} = -0.28$ V vs. NHE). According to the UPD hypothesis, the deposition proceeds continuously on an alloy surface that is different from the parent metals and, thus, deposition of less noble components could be preferential (Nicol and Philip 1976). Similarly, one could expect deposition of Ca at $E > E^\circ_{Ca/Ca^{2+}}$ on Si leading to Ca–Si formation

$$y\text{Si} + x\text{Ca}^{2+} + 2xe^- \rightarrow \text{Ca}_x\text{Si}_y \tag{9.3}$$

At high temperature, penetration of deposited Ca into Si matrix is possible, which results in permanent retention of Si–Ca composition, enabling continuity of the Ca UPD process. Yasuda et al. concluded that the formation of CaSi and $CaSi_2$ at 0.35 V vs. $E^\circ_{Ca/Ca^{2+}}$ occurs (Yasuda et al. 2005a).

The above data imply the opportunity for electrochemical synthesis of calcium silicide – an important semiconductor and superconductor as well as a prospective material for hydrogen storage. Also, silicide formation leads to macro-porosity of the silicon surface, which is attractive in applications where high surface-to-volume rate is important (Juzeliūnas et al. 2016).

The final sections of the cathodic CV curves in Figure 9.3 identify the reaction of calcium reduction, which limits window of the electrochemical system. The current-potential characteristic of Ca reduction–oxidation should follow classical Butler–Volmer equation providing that the electrochemical oxidation–reduction rate is determined solely by electron transfer (Bard and Faulkner 2001):

$$j = j_0 \left[e^{-\frac{\alpha nF}{RT}(E-E^\circ)} - e^{\frac{(1-\alpha)nF}{RT}(E-E^\circ)} \right] \tag{9.4}$$

where j stands for current density (A cm^{-2}), j_0 – exchange current density, α – charge transfer coefficient, which characterizes symmetry of energy barrier of electron transfer Ca^{2+} – Ca, n – number of electrons, F – Faraday constant, R – universal gas constant, T – absolute temperature, and E (V) – electrode potential in respect to the equilibrium potential $E°$. Deviation from the equilibrium $\eta = E - E^0$ is the charge transfer overpotential. For large negative η, the relationship (9.4) transforms into:

$$\eta = E - E° = \frac{RT}{\alpha nF}\ln j_0 - \frac{RT}{\alpha nF}\ln j \tag{9.5}$$

This relationship is known as Tafel equation, which predicts rectilinear dependence between the overpotential (electrode polarization) and the current logarithm. Kinetics of such a process is characterized by the Tafel constant, which is measured as a slope of the exponential voltammetric curve $E - \log j$

$$b = \frac{dE}{d\log j} = \frac{2.3RT}{\alpha nF} \tag{9.6}$$

Theoretical curve slope at $T = 1198$ K (900 °C) equals to $b = 0.241$ V assuming symmetry of the energy barrier ($\alpha = 0.5$). The constant b is true at sufficiently high deviations from $E°$ that is at sufficiently high overpotentials where the contribution of opposite reaction is negligible. This condition, however, is not true close to the equilibrium potential where contribution of opposite reaction is significant: the measured cathodic current is reduced by the contribution of anodic current and vice versa. The Tafel constants at low overpotentials could be estimated from the voltammetric data represented by Allen-Hickling coordinates (Bard and Faulkner 2001) (Figure 9.6):

$$\log Y = \log\left(\frac{j}{1 - e^{-\frac{2F|\eta|}{RT}}}\right) \tag{9.7}$$

The $\eta - \log j$ transformation makes it possible to subtract the contribution of an opposing reaction, that is the cathodic component in anodic region and the anodic component in cathodic region. Thus, it is possible to determine the Tafel constants close to equilibrium that is at minor deviations from $E°_{Ca/Ca^{2+}}$.

Figure 9.6 shows the voltammetric curves obtained for Si wafer in Mo frame represented by Allen-Hickling coordinates. The measurement commenced in a negative region ($E = -1.5\ V_{Ca/Ca^{2+}}$) where excessive reduction of Ca occurs. The Tafel constant in this far cathodic region is $b_c = 1.4$ V, which is even greater than the values determined using single Mo and graphite electrodes.

At low overpotentials ($\eta = \pm 50$ mV, Figure 9.6), the Tafel constants for both cathodic and anodic processes are equal $b_c = b_a = 0.24$ V (dotted lines) and correspond well to the theoretical constant $b = 0.241$ V. According to the classical model, this is indicative of a kinetic process, which is limited by a reversible two-electron transfer with the energy barrier being symmetrical ($\alpha = 0.5$). At higher polarizations, the deviation from the 0.24 V is observed, which is most likely due to a change in composition of the electrolyte layer close to the electrode. Ca reduction leads to enrichment of the surface layer by Cl^- anions, and this creates an additional

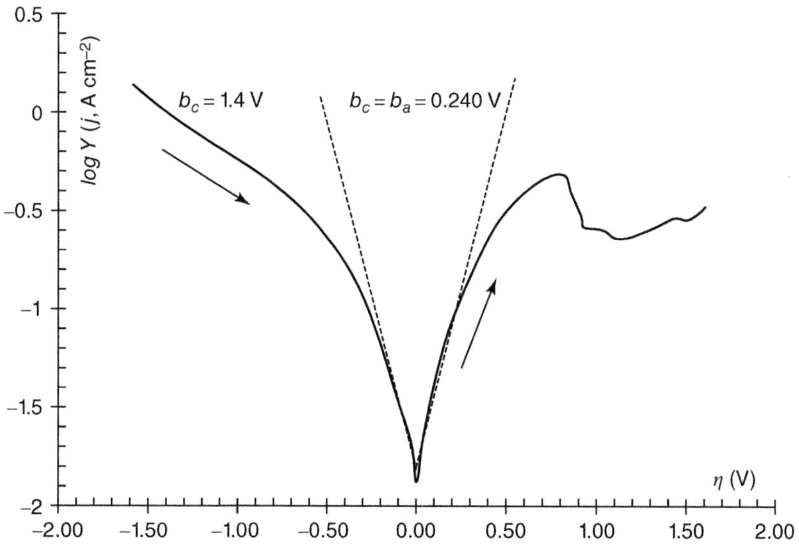

Figure 9.6 Overpotential (η,V) vs. current (j, A cm^{-2}) represented by Allen-Hickling coordinates $\eta - \log Y$ (9.7) obtained for Si wafer ($S = 23$ cm^2) in Mo wire frame ($S = 2$ cm^2 Ø 0.5 mm, Figure 9.2) in molten CaCl$_2$ at 900 °C. The potential sweep was started from $E = -1.8\ V_{Ca^{2+}/Ca}$ in positive direction at the rate $v = 100$ mV s^{-1}. Dotted lines with the slope 0.240 V dec^{-1} represent the Tafel slopes when the reverse component, e.g. cathodic current in anodic region or vice versa, is subtracted.

barrier to the charge transfer. Analogously, in anodic area, the surface layer should be enriched by Ca^{2+} ions.

A wide bell-shaped anodic curve ranging from 0.0 to 1.0 V in Figure 9.3, as well as the large maximum in Figure 9.6 indicates the oxidation of the cathodically reduced calcium $Ca \rightarrow Ca^{2+} + 2e^-$. Another possible process is silicon oxidation $Si \rightarrow Si^{z+} + ze^-$. One could see from the final part of the anodic curve in Figure 9.3 that the ending points for Mo and Mo–Si electrodes actually coincide. The current in this region is due to chlorine evolution (9.2). The coincidence of the currents for Mo and Si–Mo electrodes indicate that there is no contribution from silicon – the process occurs exclusively on molybdenum. This means that silicon becomes passive during the oxidation $Si \rightarrow Si^{4+}$ process due to formation of sparingly soluble layer. This phenomenon is of practical significance in electrochemical Si purification processes. Thus, Kongstein et al. have shown that Si anode becomes passive during electrorefining process due to SiO_2 formation (Kongstein et al. 2007). A liquid Si–Cu anode was proposed to avoid the passivation phenomena (Cai et al. 2012). On the other hand, this region provides an opportunity to form macro-porous structures on silicon. This will be discussed in more detail in Chapter 15.

To summarize the voltammetric studies, one can conclude that: (i) silica layer, which is inevitably present on silicon under ambient conditions, is dissolved by CaO, which is inevitably present in molten CaCl$_2$; this creates an opportunity for surface structuring by silicon redeposition; (ii) calcium is reduced on silicon at the potentials greater than the equilibrium potential $E^\circ_{Ca/Ca^{2+}}$; this leads to silicide formation and opens a root for electrochemical synthesis of semiconductors; (iii) silicide forms also at $E < E^\circ_{Ca/Ca^{2+}}$, which provides the possibility of surface macro-structuring; (iv) silicon oxidizes in a wide anodic range; this provides possibility of surface structuring by electrochemical re-deposition; (v) silicon is passive in far anodic region; this gives an opportunity for macro-porous surface structuring. In subsequent paragraphs, these aspects will be discussed in detail.

References

Akpinar, B., Erdogan, M., Akduman, B., and Karakaya, I. (2017). The use of silicon wafer barriers in the electrochemical reduction of solid silica to form silicon in molten salts. *ECS Trans.* 77: 2011–2016. https://doi.org/10.1149/07711.2011ecst.

Bard, A.J. and Faulkner, L.R. (2001). *Electrochemical Methods. Fundamentals and Applications*, 100–104. Wiley.

Braithwaite, E.R. and Haber, J. (ed.) (1994). *Molybdenum: An Outline of Its Chemistry and Uses*, Studies in Inorganic Chemistry, vol. 19, 478. Elsevier. ISBN: 0-444-88198-0.

Cai, J., Luo, X., Haarberg, G.M. et al. (2012). Electrorefining of metallurgical grade silicon in molten $CaCl_2$ based salts. *J. Electrochem. Soc.* 159: D155–D158. https://doi.org/10.1149/2.041203jes.

Cho, S.K., Fan, F.R.F., and Bard, A.J. (2012a). Electrodeposition of crystalline and photoactive silicon directly from silicon dioxide nanoparticles in molten $CaCl_2$. *Angew. Chem. Int. Ed.* 51: 12740–12744. https://doi.org/10.1002/ange.201206789.

Cho, S.K., Fan, F.R.F., and Bard, A.J. (2012b). Formation of a silicon layer by electroreduction of SiO_2 nanoparticles in $CaCl_2$ molten salt. *Electrochim. Acta* 65: 57–63. https://doi.org/10.1016/j.electacta.2012.01.008.

Eliaz, N. and Gileadi, E. (2019). *Physical Electrochemistry. Fundamentals, Techniques, and Applications*, 2e. Weinheim: Wiley-VCH Verlag GmbH & Co. https://doi.org/10.1002/ange.201104618.

Endres, F., Abbot, A.P., and MacFarlane, D.R. (2008). *Electrodeposition from Ionic Liquids*. Wiley-VCH https://doi.org/10.1002/9783527622917.

Ergül, E., Karakaya, I., and Erdogan, M. (2011). Electrochemical decomposition of SiO_2 pellets to form silicon in molten salts. *J. Alloys Compd.* 509: 899–903. https://doi.org/10.1016/j.jallcom.2010.09.126.

Fang, S., Wang, H., Yang, J. et al. (2015). Electrochemical preparation of silicon nanowires from porous NiO/SiO_2 blocks in molten $CaCl_2$. *Mater. Lett.* 160: 1–4. https://doi.org/10.1016/j.matlet.2015.07.082.

Gao, P., Jin, X.B., Wang, D.H. et al. (2005). A quartz sealed Ag/AgCl reference electrode for $CaCl_2$ based molten salts. *J. Electroanal. Chem.* 579: 321–328. https://doi.org/10.1016/j.jelechem.2005.03.004.

Jin, X., Gao, P., Wang, D. et al. (2004). Electrochemical preparation of silicon and its alloys from solid oxides in molten calcium chloride. *Angew. Chem. Int. Ed.* 43: 733–736. https://doi.org/10.1002/anie.200352786.

Juzeliūnas, E., Cox, A., and Fray, D.J. (2010). Silicon surface texturing by electro-deoxidation of a thin silica layer in molten salt. *Electrochem. Commun.* 12: 1270–1274. https://doi.org/10.1016/j.elecom.2010.06.034.

Juzeliūnas, E., Cox, A., and Fray, D.J. (2011). Silicon surface texturing by electro-deoxydation of a thin silica layer. In: *TMS2011, 140th Annual Meeting and Exhibition, February 27-March 3, 2011. San Diego, California, USA*, 709–716.

Juzeliūnas, E., Cox, A., and Fray, D.J. (2012). Electro-deoxidation of thin silica layer in molten salt – Globular structures with effective light absorbance. *Electrochim. Acta* 68: 123–127. https://doi.org/10.1016/j.electacta.2012.02.054.

Juzeliūnas, E., Coxon, P.R., Fray, D.J. et al. (2016). Photoelectrochemical activity of macroporous silicon formed in molten calcium chloride. *Chemija* 27: 11–16.

Juzeliūnas, E., Fray, D.J., Kalinauskas, P. et al. (2018). Electrochemical synthesis of photoactive carbon-carbide structure on silicon in molten salt. *Electrochem. Commun.* 90: 6–10. https://doi.org/10.1016/j.elecom.2018.03.003.

Katasho, Y., Norikawa, Y., Yamamoto, T. et al. (2020). *In situ* synchrotron X-ray diffraction study of the electrochemical reduction of SiO_2 in molten $CaCl_2$. *Electrochem. Commun.* 115: 106740. https://doi.org/10.1016/j.elecom.2020.106740.

Katayama, Y., Yokomizo, M., Miura, T., and Kishi, T. (2001). Preparation of novel fluorosilicate salt for electrodeposition of silicon at low temperature. *Electrochemistry* 69 (11): 834–836. https://doi.org/10.5796/electrochemistry.69.834.

Kongstein, O.E., Wollan, C., Sultana, S., and Haarberg, G.M. (2007). Electrorefining of silicon in molten calcium chloride. *ECS Trans.* 3: 357–361. https://doi.org/10.1149/1.2798679.

Massalski, T.B., Okamoto, H., Subramanian, P.R., and Kacprzak, L. (ed.) (1990). *Binary Alloy Phase Diagrams*, vol. 2, 953. ASM International.

Nicol, M.J. and Philip, H.I.J. (1976). Underpotential deposition and its relation to the anomalous deposition of metals in alloys. *Electroanal. Chem.* 70: 233. https://doi.org/10.1016/S0022-0728(76)80109-X.

Nishimura, Y., Nohira, T., Kobayashi, K., and Hagiwara, R. (2011). Formation of Si nanowires by direct electrolytic reduction of porous SiO_2 pellets in molten $CaCl_2$. *J. Electrochem. Soc.* 158: E55–E59. https://doi.org/10.1149/1.3575635.

Nohira, T., Yasuda, K., and Ito, Y. (2003). Pinpoint and bulk electrochemical reduction of insulating silicon dioxide to silicon. *Nat. Mater.* 2: 397–401. https://doi.org/10.1038/nmat900.

Pistorius, P.C. and Fray, D.J. (2006). Formation of silicon by electro-deoxidation, and implications for titanium metal production. *J. S. Afr. Inst. Min. Metall.* 106: 31–41. https://hdl.handle.net/10520/AJA0038223X_3168.

Sakanaka, Y., Goto, T., and Hachiya, K. (2015). Electrochemical formation of Ca–Si in molten $CaCl_2$–KCl. *J. Electrochem. Soc.* 162: D186–D191. https://doi.org/10.1149/2.1021504jes.

Schwandt, C., Doughty, G.R., and Fray, D.J. (2010). The FFC-Cambridge process for titanium metal winning. *Key Eng. Mater.* 436: 13–25. https://doi.org/10.4028/www.scientific.net/KEM.436.13.

Stankevičiūtė, A., Leinartas, K., Bikulčius, G. et al. (1998). Deposition of Zn–Co by constant and pulse current. *J. Appl. Electrochem.* 28: 89–95. https://doi.org/10.1023/A:1003206019963.

Tsuyuki, Y., Takai, H., Fukunaka, Y., and Homma, T. (2018). Formation of Si thin films by electrodeposition in ionic liquids for solar cell applications. *Jpn. J. Appl. Phys.* 57 (8S3): 08RB11. https://doi.org/10.7567/JJAP.57.08RB11.

Xiao, W. and Wang, D. (2014). The electrochemical reduction process of solid compounds in high temperature molten salts. *Chem. Soc. Rev.* 43: 3215–3228. https://doi.org/10.1039/c3cs60327j.

Xiao, W., Jin, X.B., Deng, Y. et al. (2010). Rationalisation and optimization of solid state electro-reduction of SiO_2 to Si in molten $CaCl_2$ in accordance with dynamic three-phase interlines based voltammetry. *J. Electroanal. Chem.* 639: 130–140. https://doi.org/10.1016/j.jelechem.2009.12.001.

Xiao, W., Wang, X., Yin, H. et al. (2012). Verification and implications of the dissolution-electrodeposition process during the electro-reduction of solid silica in molten $CaCl_2$. *RSC Adv.* 2: 7588–7593. https://doi.org/10.1039/C2RA20698F.

Xiao, W., Jin, X.B., and Chen, G.Z. (2013). Up-scalable and controllable electrolytic production of photo-responsive nanostructured silicon. *J. Mater. Chem. A* 1 (35): 10243–10250. https://doi.org/10.1039/C3TA11823A.

Yang, J., Lu, S., Kan, S. et al. (2009). Electrochemical preparation of silicon nanowires from nanometer silicon in molten calcium chloride. *Chem. Commum.* 22: 3273–3275. https://doi.org/10.1039/b902029b.

Yang, X., Yasuda, K., Nohira, T. et al. (2016). The role of granule size on the kinetics of electrochemical reduction of SiO_2 granules in molten $CaCl_2$. *Metall. Mater. Trans. B* 47: 788–797. https://doi.org/10.1007/s11663-015-0456-1.

Yang, X., Ji, L., Zou, X. et al. (2017). Toward cost-effective manufacturing of silicon solar cells: electrodeposition of high-quality Si films in a $CaCl_2$-based molten salt. *Angew. Chem. Int. Ed.* 56: 15078–15082. https://doi.org/10.1002/anie.201707635.

Yasuda, K., Nohira, T., and Ito, Y. (2005a). Effect of electrolysis potential on reduction of solid silicon dioxide in molten $CaCl_2$. *J. Phys. Chem. Solids* 66: 443–447. https://doi.org/10.1016/jpcs.2004.06.037.

Yasuda, K., Nohira, T., Amezawa, K. et al. (2005b). Mechanism of direct electrolytic reduction of solid SiO_2 to Si in molten $CaCl_2$. *J. Electrochem. Soc.* 152: D69–D74. https://doi.org/10.1149/1.1864453.

Yasuda, K., Nohira, T., Ogata, Y.H., and Ito, Y. (2005c). Direct electrolytic reduction of solid silicon dioxide in molten LiCl–KCl–$CaCl_2$ at 773 K. *J. Electrochem. Soc.* 152: D208–D212. https://doi.org/10.1149/1.2042910.

Yasuda, K., Nohira, T., Kobayashi, K. et al. (2013). Improving purity and process volume during direct electrolytic reduction of solid SiO_2 in molten $CaCl_2$ for the production of solar-grade silicon. *Energy Technol.* 1 (4): 245–252. https://doi.org/10.1002/ente.201300001.

Zangari, G. (2015). Electrodeposition of alloys and compounds in the era of microelectronics and energy conversion technology. *Coatings* 5: 195–218. https://doi.org/10.3390/coatings5020195.

Zein El Abedin, S., Borissenko, N., and Endres, F. (2004). Electrodeposition of nanoscale silicon in a room temperature ionic liquid. *Electrochem. Commun.* 6: 510–514. https://doi.org/10.1016/j.elecom.2004.03.013.

Zou, X., Ji, L., Lu, X., and Zhou, Z. (2017a). Facile electrosynthesis of silicon carbide nanowires from silica/carbon precursors in molten salt. *Sci. Rep.* 7: 1–9. art. no. 9978. https://doi.org/10.1038/s41598-017-10587-5.

Zou, X., Ji, L., Yang, X. et al. (2017b). Electrochemical formation of a p–n junction of thin film silicon deposited in molten salt. *J. Am. Chem. Soc.* 139: 16060–16063. https://doi.org/10.1021/jacs.7b09090.

10

$Si-SiO_2$ Electrode in Molten $CaCl_2$

Silicon electrode coated with a thin silica layer ($Si–SiO_2$) is an important precursor to electrochemically obtaining the surface nano-micro-architectures. To this end, either an electrochemical conversion $SiO_2 \rightarrow Si$, or vice versa process – reduction of dissolved silica – can be performed (Juzeliūnas et al. 2010, 2011, 2012, 2016, 2018). Such surface processing requires understanding the interaction between the $Si–SiO_2$ system and the molten salt electrolyte.

Potentiometry measurements reveal informatively an interaction of a SiO_2 layer on Si with the molten $CaCl_2$ electrolyte. This method measures the electrode potential with respect to that of a reference electrode, while no current is passed through the electrodes. Figure 10.1 shows such measurement, that is, variation of the open circuit potential (E_{ocp}) of p-Si thermally coated with 2 μm SiO_2 layer. Molybdenum contacting electrode has been attached to silica surface, whose configuration is shown in Figure 9.3. Variation of the E_{ocp} within half an hour or less is displayed. The data for single Mo as well as for that attached to bare Si are given for comparison. One can see that the initial potential value of the $Si–SiO_2–Mo$ system is greater than that of Si–Mo or single Mo. This implies some positive charging of the SiO_2 surface, probably, due to silica dissociation. With exposure time, the potential values of the $Si–SiO_2$ sample scatter randomly and disorderly approaching the curve of bare Si. After a certain period, the curves for $Si–SiO_2$ and bare Si become the same. This indicates the removal of the silica layer during the exposure.

It is important to note that some CaO is inevitably present in molten $CaCl_2$ electrolyte due to the highly hygroscopic nature of this salt. Hydrolysis of $CaCl_2$ leads to CaO formation. To avoid such a process, anhydrous $CaCl_2$ salt is kept under vacuum at elevated temperatures. The heating pattern typically involves a stepwise increase in the temperature from ambient to ~200 °C and holding up to 24 hours. However, even in prudently pretreated molten electrolyte, presence of CaO has always been observed in practice (Xiao and Wang 2014). The CaO can interact with SiO_2-producing silicates (SiO_3^{2-}, SiO_4^{4-}, $Si_2O_7^{6-}$), which dissolve in the melt (Xiao et al. 2010, 2012, 2013, Xiao and Wang 2014; Cho et al. 2012; Zou et al. 2017)

$$SiO_2 + CaO \rightarrow CaSiO_3 \tag{10.1}$$

$$CaSiO^3 \rightarrow Ca^{2+} + SiO_3^{2-} \tag{10.2}$$

Silicon: Electrochemistry, Production, Purification and Applications, First Edition. Eimutis Juzeliūnas.

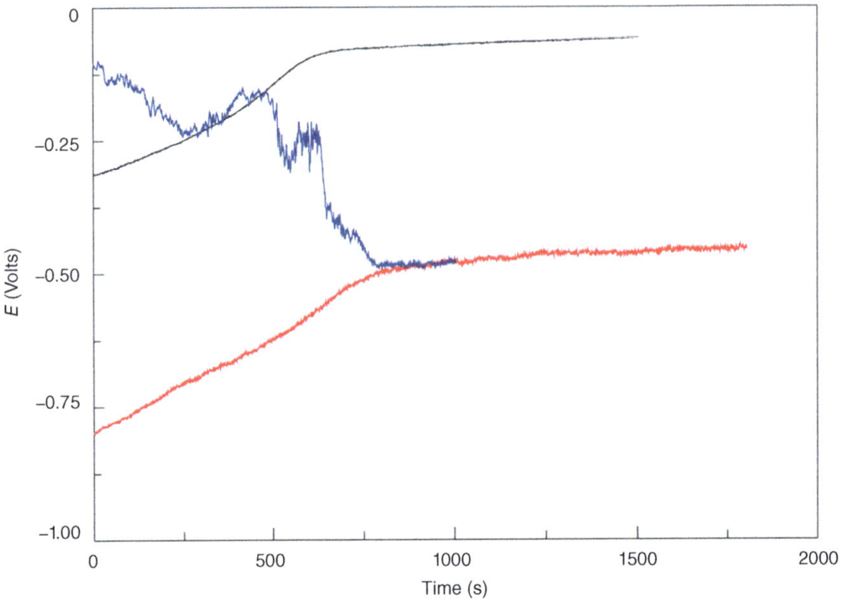

Figure 10.1 Open circuit potential (E_{ocp}) vs. time for Mo frame (black line), Si in Mo frame (red), and Si with 2 μm SiO_2 layer in Mo frame (blue) in molten $CaCl_2$ at 850 °C. The electrode configuration is given in Figure 9.3. Potential has been measured against graphite quasi-reference graphite electrode. The SiO_2 layer was thermally coated on p-Si wafer. Source: Reproduced from Juzeliūnas et al. (2012) with permission from of Elsevier.

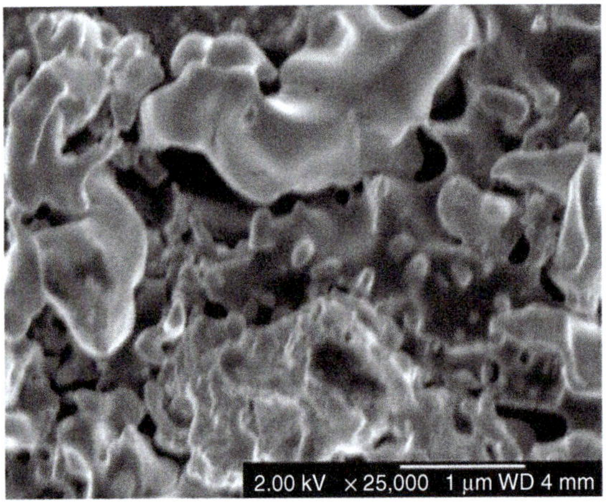

Figure 10.2 SEM image of p-Si thermally coated with 2 μm SiO_2 layer exposed for 20 minutes in molten $CaCl_2$ at 850 °C. The initial sample surface was polished to a mirror shiny appearance. The image demonstrates the corrosion features of an initially mirror shiny surface. Source: E. Juzeliūnas (Author).

The Gibbs free energy change of reaction (10.1) is $\Delta G = -139.76\ \text{kJ mol}^{-1}$ at 1123 K (Xiao and Wang 2014).

The SEM image in Figure 10.2 provides additional evidence for silica instability in molten $CaCl_2$. The image shows characteristic corrosion features determined on a micrometer scale after exposure of the Si wafer, which has been coated with 2 μm SiO_2 layer and polished to a mirror-shiny appearance. Yang et al. have observed transformation of nano-porous SiO_2 pellet into agglomerated network structure during 30 minutes exposure in molten $CaCl_2$ at 900 °C (Yang et al. 2009). In this experiment, the starting SiO_2 was in the form of an amorphous powder with nanometer-scale porosity. The behavior of this material in molten salt was different from the behavior of bulk amorphous SiO_2. The authors expected that the softening and melting temperature of nanometer-sized silica particles is lower when compared with bulk silica, this is especially due to the fluxing action of molten $CaCl_2$ (Yang et al. 2009). It was suggested that silica exists at the SiO_2-electrolyte interface in a colloid or liquid state.

An electrolyte composed of $CaCl_2$ and CaO melt was used for quartz or silica dissolution and subsequent manufacturing or refining of silicon. Kongstein et al. dissolved metal-grade silicon in $CaCl_2$–NaCl–CaO melt and re-deposited highly pure Si on the cathode (Kongstein et al. 2007).

Xiao et al. studied the silica-electrolyte interactions by measuring a change in quartz mass (Xiao et al. 2012). They found that the quartz rod reduced its mass after immersion in molten $CaCl_2$. The mass loss was enhanced by more than one order of magnitude when adding to the electrolyte 2 mol% of CaO. The dissolution was verified by electrolysis experiment using a silica pallet wrapped in nickel foam, which had a function of contacting electrode. Silicon was deposited on the outer surface of the nickel, which had no contact with the silica feedstock. This implied silica dissolution in the electrolyte and its electrodeposition on the area outside the silica precursor.

Thus, one can conclude from the data referred to above that both silicon and silica are not stable in molten $CaCl_2$ electrolyte – the materials could soften, melt or dissolve, and their structures undergo morphological transformations. This is particularly important in the experiments when using SiO_2 nanoparticles as the precursors (Cho et al. 2012). Dissolution of SiO_2 nano-powder in molten $CaCl_2$ (10.1) and (10.2) and subsequent Si electrodeposition

$$SiO_3{}^{2-} + 4e^- \rightarrow Si + 3O_2{}^- \tag{10.3}$$

is more likely than the claimed direct de-oxidation of the nanoparticles on the cathode

$$SiO_2 + 4e^- \rightarrow Si + O_2{}^- \tag{10.4}$$

An important point follows from the above discussion: a native (ambient) oxide, which is always present on silicon, should be removed in molten $CaCl_2$. This means that silicon in the electrolyte is in a state without a passivation layer, that is, the electrode is electrochemically active.

References

Cho, S.K., Fan, F.R.F., and Bard, A.J. (2012). Formation of a silicon layer by electroreduction of SiO_2 nanoparticles in $CaCl_2$ molten salt. *Electrochim. Acta* 65: 57–63. https://doi.org/10.1016/j.electacta.2012.01.008.

Juzeliūnas, E., Cox, A., and Fray, D.J. (2010). Silicon surface texturing by electro-deoxidation of a thin silica layer in molten salt. *Electrochem. Commun.* 12: 1270–1274. https://doi.org/10.1016/j.elecom.2010.06.034.
Juzeliūnas, E., Cox, A., and Fray, D.J. (2011). Silicon surface texturing by electro-deoxydation of a thin silica layer. In: *TMS2011, 140th Annual Meeting and Exhibition, February 27-March 3, 2011. San Diego, California USA*, 709–716.
Juzeliūnas, E., Cox, A., and Fray, D.J. (2012). Electro-deoxidation of thin silica layer in molten salt – Globular structures with effective light absorbance. *Electrochim. Acta* 68: 123–127. https://doi.org/10.1016/j.electacta.2012.02.054.
Juzeliūnas, E., Coxon, P.R., Fray, D.J. et al. (2016). Photoelectrochemical activity of macroporous silicon formed in molten calcium chloride. *Chemija* 27 (1): 11–16.
Juzeliūnas, E., Fray, D.J., Kalinauskas, P. et al. (2018). Electrochemical synthesis of photoactive carbon-carbide structure on silicon in molten salt. *Electrochem. Commun.* 90: 6–10. https://doi.org/10.1016/j.elecom.2018.03.003.
Kongstein, O.E., Wollan, C., Sultana, S., and Haarberg, G.M. (2007). Electrorefining of silicon in molten calcium chloride. *ECS Trans.* 3: 357–361. https://doi.org/10.1149/1.2798679.
Xiao, W. and Wang, D. (2014). The electrochemical reduction process of solid compounds in high temperature molten salts. *Chem. Soc. Rev.* 43: 3215–3228. https://doi.org/10.1039/c3cs60327j.
Xiao, W., Jin, X.B., Deng, Y. et al. (2010). Rationalisation and optimization of solid state electro-reduction of SiO_2 to Si in molten $CaCl_2$ in accordance with dynamic three-phase interlines based voltammetry. *J. Electroanal. Chem.* 639: 130–140. https://doi.org/10.1016/j.jelechem.2009.12.001.
Xiao, W., Wang, X., Yin, H. et al. (2012). Verification and implications of the dissolution-electrodeposition process during the electro-reduction of solid silica in molten $CaCl_2$. *RSC Adv.* 2: 7588–7593. https://doi.org/10.1039/C2RA20698F.
Xiao, W., Jin, X.B., and Chen, G.Z. (2013). Up-scalable and controllable electrolytic production of photo-responsive nanostructured silicon. *J. Mater. Chem. A* 1 (35): 10243–10250. https://doi.org/10.1039/C3TA11823A.
Yang, J., Lu, S., Kan, S. et al. (2009). Electrochemical preparation of silicon nanowires from nanometer silica in molten calcium chloride. *Chem. Commun.* 3273–3275. https://doi.org/10.1039/b902029b.
Zou, X., Ji, L., Lu, X., and Zhou, Z. (2017). Facile electrosynthesis of silicon carbide nanowires from silica/carbon precursors in molten salt. *Sci. Rep.* 7 (1): 1–9, art. no. 9978. https://doi.org/10.1038/s41598-017-10587-5.

11

Formation of Silicon Oxide Layer

Native oxide (silica, SiO_2) film is an inevitable component of silicon. A very thin layer forms spontaneously under ambient conditions whose thickness varies from several angstroms to several nanometers depending on the surface preparation, air humidity, and composition as well as exposure duration (Zhang 2004). Despite the thinness of the layer, it plays a significant role in Si electrode kinetics as well as its electronic properties.

Passivation of silicon by thin silica film is a well-known approach, which is applied in order to increase efficiency of photovoltaic (PV) devices. The films reduce recombination losses, improve carrier transport and enhance stability of the devices, especially for those, which are based on Si nano–micro architectures. Thin SiO_2 layer significantly improves electric characteristics of Si-based devices due to reduction in number of dangling bounds up to 5 orders of magnitude (Michaelis 2008; Hazarika and Sharma 2014). A passivation stack composed of silica-hafnia (SiO_2–HfO_2) reduced the density of interface traps down to $\sim 1.0 \times 10^{11}$ cm^{-2} eV^{-1} level and the negative charge density of the film down to $\sim -1.0 \times 10^{11}$ cm^{-2} level (Oudot et al. 2018).

Another important aspect of silica film application on silicon is the substrate protection from an aggressive environment as well as improving its electrochemical stability (Hu et al. 2015). The SiO_x layers have been proposed for protection of p-Si photocathodes for sunlight-driven water splitting (Esposito et al. 2013; Choi et al. 2014).

The passivation term in semiconductor physics (or PV) and electrochemistry addresses different phenomena. In physics, electronic passivation means the reduction of the interface state density (D_{if}) (unsaturated bonds) on the semiconductor surface, e.g. by the chemical reduction with passivating dielectrics (Cui et al. 2017a,b; Singh et al. 2016). Another important type is the field-effect passivation, which relates to electrostatic shielding of the charge carriers from the interface by the internal electric field, e.g. the films having negative (Al_2O_3, SiO_2+HfO_2) or positive (SiN_x) charge (Cui et al. 2017b; Singh et al. 2016; Oudot et al. 2018).

In (photo)electrochemistry, an additional component – electrolyte – plays an essential role. In particular, specific adsorption of H^+ or OH^- ions is important. The adsorbed ions along with photo-generated surface charges can produce charged interface and create a field effect by analogy with the effect under atmospheric

Silicon: Electrochemistry, Production, Purification and Applications, First Edition. Eimutis Juzeliūnas.

conditions. Electrochemical passivation refers primarily to an inhibition of the charge transfer from (to) semiconductor to (from) electroactive species in electrolyte. The oxide, in such a case, plays the role of insulator, which inhibits the charge exchange between the solid and liquid conductors. An oxide layer on a photocathode protects the substrate from an aggressive environment and electrochemical disintegration. The layer also inhibits electron exchange between the semiconductor and the active species in the electrolyte (Sun et al. 2012; Seger et al. 2013; Esposito et al. 2013; Choi et al. 2014; Hu et al. 2015; Ji et al. 2015). Another kind of passivation is spontaneous formation of protective film from corrosion products (Hu et al. 2015). It is desirable for a photo-electrode that the passive layer be able to reduce recombination velocity in the semiconductor, protect the Si-electrolyte junction against corrosion, and at the same time should not inhibit substantially the electron transfer to electrolyte. However, in most cases, the oxide films while mitigating Si corrosion, simultaneously reduce surface activity: the electrode becomes electrochemically inactive in certain potential regions. The dilemma is typically solved in favor of passivity because dissolution leads to continual destruction of the semiconductor and failure propagation throughout active regions.

It is important to note that photoelectrochemical studies provide information, which correlates with that of the solid-state cells (Cho et al. 2012; Ye et al. 2011). Such correlation lays the basis for electrochemical testing of Si surface photo-responsiveness, so that there is no prior need to design a solar cell, which is a time, cost, and labor-consuming process. Also, the photopotential responses in electrolytes can characterize efficiency of electronic passivation of silicon surface. Silicon with SiO_2 layer was used as a reference to compare the passivation performance of other oxides, such as hafnia (HfO_2) or alumina (Al_2O_3) (Staišiūnas et al. 2022; Kalinauskas et al. 2022).

Unlike in gaseous environments, the thin film oxides are not stable in electrolytes due to the absorption of water and electrolyte ions. Distortion of oxide surfaces by penetration of water and ions is a phenomenon of general nature (Perram et al. 1974). Water incorporation into oxide could reduce its density, refraction index, or acoustic velocity (Davis and Tomozawa 1995). Proton and water penetration into oxide is of importance to understand instability of dielectrics during operation of ion-sensitive field-effect transistors (ISFETs). The presence of water or hydrogen-related species (–H, –OH) in ISFETs could enhance interface state generation and charge trapping (Topkar and Lal 1995; Hazarika and Sharma 2014).

Electrolyte absorption by oxide films on silicon has been studied by the quartz crystal nanobalance (QCN) – a highly sensitive mass detector, which provides information with nanogram resolution (ng cm^{-2}) *in situ* and in real time (Staišiūnas et al. 2022; Kalinauskas et al. 2022). Ultrathin film of silicon 10 nm in thickness was formed on Si and Au substrates by radio frequency magnetron sputtering (RFMS). The QCN measurements characterized the instability of the Si–SiO_2 interface, which was detected as electrode mass change. Dissolution of silica film has been detected to cause a decrease in electrode mass, whereas, mass gain indicated electrolyte absorption by the oxide. In neutral electrolyte (pH 7), the mass of the Si–SiO_2 electrode decreased at the beginning of exposure at the rate $\Delta m/\Delta t = 0.59\,ng\,s^{-1}$ (Kalinauskas et al. 2022). This indicated silica dissolution at initial stages of exposure. After a certain period, the mass change stopped indicating the dissolution suppression. Further mass gain implied the electrolyte intake. No dissolution indications have been observed in an acid electrolyte (pH 3); only mass growth was observed during the exposure to some limiting mass of $\Delta m \sim 500\,ng\,cm^{-2}$, and this made approximately one-third of the entire mass of the silica film ($m = 1325\,ng\,cm^{-2}$). It is a common knowledge that SiO_2 has low solubility in water or acidic solutions but it is soluble in alkaline solutions. The QCN experiments demonstrated effectiveness of this nanogravimetric approach to study dynamics of the SiO_2 degradation in various media.

Well-defined silica layer could be formed on silicon by thermal oxidation. Silicon is in this respect not unique: thermal oxidation is applicable to various materials. The process is usually performed at temperatures between 800 and 1200 °C in a vertical or horizontal furnace, operating under dry or wet oxidation conditions. In wet process, water vapor is applied, which oxidizes silicon and produces hydrogen:

$$Si + 2H_2O \rightarrow SiO_2 + 2H_2 \tag{11.1}$$

In a dry process, pure oxygen is used:

$$Si + O_2 \rightarrow SiO_2 \tag{11.2}$$

Dry thermal process is employed to produce a thinner oxide layer (on the order of nanometers), whereas wet process is more suitable for the production of thick layers (on the order of micrometers). Some hydrochloric acid could also be added to the oxidizing ambient to remove metal contaminants from the oxide. Silicon is consumed and ambient oxygen is accommodated within the substrate so that its volume expansion occurs. It is assumed that the thicknesses of the thermally produced silica are about two times more than the thickness of the silicon.

Generally, Deal-Grove model predicts the rate of oxide growth, which was proposed by the authors in 1965 (Deal and Grove 1965). The model takes into consideration the reaction phenomena at outer and inner boundaries of the oxide layer including the diffusion process within the oxide. According to this model, the oxide growth time (t) and its thickness (X_o) is expressed by the following equation:

$$t = \frac{X_o^2}{B} + \frac{X_o}{B/A} \tag{11.3}$$

The parameters A and B relate to the physico-chemical constants of the oxidation reactions. The ratio (11.3) is true to bare Si surface. A correction factor τ is necessary to introduce when the oxide is grown on the sample with already pre-existing oxide layer:

$$X_o(t) = A/2\left[\sqrt{1 + \frac{4B}{A^2}(t + \tau)} - 1\right] \tag{11.4}$$

The Deal-Grove model describes well oxidation of single-crystal Si under most conditions. The above relationships have been proven to be in excellent agreement with the experimental data obtained for both oxygen and water oxidants over the temperature range of 700–1300 °C, the partial pressure 0.1–1.0 atm, and the oxide thickness 30–2000 nm. It is, however, important to note that growth of very thin oxides (<~25 nm) deviates from the above model: more rapid growth occurs in silicon nanostructures, such as nanowires (Liu et al. 2016). Properties of thermal Si oxidation are well surveyed (Massoud 1997; Massoud and Plummer 1985; Jaeger 2002; Zhang 2004). The oxidation thickness calculators, which bases on the above model, are widely available. The calculators take into consideration comprehensive oxidation factors, such as initial thickness of the oxide, temperature, crystal orientation, environment (wet or dry), partial pressure, dopant concentration, and type, as well as oxidation time.

Thermally formed oxide layer has an amorphous structure. The weight density of the amorphous product is slightly less than that of crystalline SiO_2 (2.2 and 2.65 g cm^{-3}, respectively). The thermal oxide has excellent electric resistivity on the order of 10^{19} Ω cm^{-1}. Tan and coworkers studied the optical properties of amorphous and crystalline SiO_2 by vacuum ultraviolet spectroscopy and spectroscopic ellipsometry (Tan et al. 2005). The band gap of amorphous SiO_2 film was determined as high as $E_{g,SiO_2} = 7.64$ eV. At high temperatures, such as those in molten $CaCl_2$ electrolyte, the oxide is also highly resistant.

Srivastava et al. studied the electrical conductivity of SiO_2 thermally grown on p-Si (boron-doped) and n-Si (phosphorous-doped) (Srivastava et al. 1985). DC and AC electrical conductivities were measured in the wide temperature range from 25 to 1100 °C. The authors found that below 450 °C, the conductivity did not depend on temperature. The conductivity is increased at higher temperatures. Conduction in SiO_2 was concluded to be both ionic and electronic in the temperature range of 500–960 °C. The ionic conduction was due to the transport of the oxygen ions, and this conductivity type was predominant. Also, the AC conductivity of the oxide on n-Si was lower than the conductivity grown on p-Si. According to this study, the conductivity of thermal SiO_2 at 850 °C is in the range 10^{-10} to 10^{-11} Ω^{-1} cm^{-1}. These data are relevant for studies of Si–SiO_2 electrode in molten $CaCl_2$.

The silica films are good diffusion masks for common Si wafer dopants; they have good selectivity with respect to Si substrate etching with hydrofluoric acid.

Apart from thermal oxidation, there have been other methods developed for SiO_2 layer formation on silicon, such as sol–gel or physical vapor deposition (PVD). Advantage of the sol–gel process includes possibility to prepare multi-element oxides at relatively low temperatures. The method is attractive due to technical convenience: there is no need for expensive specific equipment. Typically, dip coating or spin coating is used. The method has been applied to produce SiO_2 antireflection coatings on Si basis (Skolik et al. 2019; Jannat et al. 2016). Sol–gel approach has also been used to incorporate Si-nanoparticles into various dielectrics (SiO_2, ZrO_2, and TiO_2) (Jettanasen 2015) or into sol–gel matrix, which exhibited good photoluminescence stability (Khan et al. 2013). A great challenge is the production of a compact silicon layer by a sol–gel route in organic solvents. This would prevent silicon oxidation. However, shrinkage of the gel upon drying limits the opportunities to obtain compact layer.

Of great interest is the formation of silicon with silica layer on foreign substrates, such as aluminum, steel, copper, and glass. Silicon films with efficient light absorbance on metal or glass substrates give possibility of combining structural materials with silicon for facing panels in buildings, which could be used to harness solar energy. Such films protected by an oxide film could also effectively protect the metal structures from corrosion. Note, however, that SiO_2 coatings are sensitive to alkaline medium.

The Si and SiO_2 layers on metal substrates could be prepared by the methods of PVD, such as plasma spray or RFMS. Figure 11.1a shows an example of the Si film deposited on stainless steel by RFMS. P-type (boron doped) silicon wafer has been

(a)

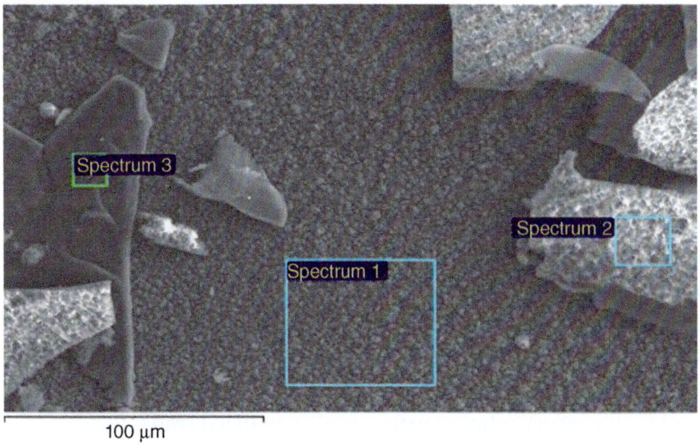

Figure 11.1 (a) SEM micrograph of the cross-section of the Si layer deposited by RFMS on stainless steel (AISI 304) and SiO_2 film; the Si layer thickness is ~11.1 μm. Pt thin film was also sputtered on top of the sample. The cross-section was prepared by the focused ion beam of Ga^+ ions. (b) Si fragments after four hours exposure of the Si film at 900 °C. The temperature has been increased from ambient to 900 °C at the rate of 5 °C min^{-1}. Identified are the characteristic surface areas as Spectrum, which were analyzed by EDX. Spectrum 1 (in at %): Si – 11.15, O – 52.07, Cr – 11.92, Mn – 1.43, Fe – 11.79, Ni – 3.39. Spectrum 2: Si – 45.8, O – 43.01, Cr – 0.51, Mn – 0.22, Fe – 0.35, Ni – 2.76. Spectrum 3: Si – 74.47, O – 16.62, Cr – 0.11, Mn – 0.04, Fe – 0.24, Ni – 0.11. (c) SEM image maps the elements for the cross section at the area Spectrum 3: yellow – Si, red – O, and aqua/cyan – Fe. Source: E. Juzeliūnas(Author).

(c)

Figure 11.1 (*Continued*)

used as a sputtering target, which made it possible to deposit rather thick silicon films. The SEM micrograph shows the cross-section of the Si deposited by the RFMS on a stainless steel AISI 304 substrate. Rather thick Si layer of c. 11 μm thickness has been deposited. The cross-section was prepared by the focused ion beam of\break Ga^+ ions.

Figure 11.1b shows the Si fragments after four hours exposure of the Si film at 900 °C. The temperature has been raised from ambient to 900 °C at the rate of 5 °C min^{-1}. The silicon film is disintegrated during thermal treatment into fragments, whose size is hundreds of micrometers. Identified in the figure are the characteristic surface areas, marked as Spectrum 1, 2, and 3, which were analyzed by EDX. Figure 11.1b,c provides the SEM and EDX analysis data of three characteristic locations on the heated surface. The locations Spectrum 3 and Spectrum 2 relate to fragments of the disintegrated silica layer; they are composed mainly of silicon and oxygen. At the rest of the surface (Spectrum 1), oxygen dominates, while content of silicon is significantly less, around 10 at.%. Elements of the stainless steel basement (Fe, Cr, Ni) are identified in the Spectrum 1 region; in other locations, content of these elements is much lower. In-depth analysis of the cross-section in the area Spectrum 3 clearly identifies Si (yellow), whose surface is oxidized (red). Basement iron is also identified (aqua/cyan), whose surface is also oxidized (red).

Thermal disintegration of SiO_2 or Si films on foreign substrate is due to different expansion coefficients of the film and the substrate. Thermal expansion of fused silica or those deposited by means of plasma-induced chemical vapor deposition (PCVD) is extremely small ($\alpha_{SiO_2} \sim 0.5 \times 10^{-6}$ m°C^{-1}) (Bachmann et al. 1988). Common metals, alloys and glasses exhibit much greater expansion ($\times 10^{-6}$ m °C^{-1}): $\alpha_{glasses} = 5.9$–9.0; $\alpha_{steels} = 11$–16; $\alpha_{Al} = 21$–24; $\alpha_{Cu} = 16.0$–16.7. Due to the different expansions, silica films on 304 stainless steel substrate delaminated in molten

$CaCl_2$ at 850 °C. Similar behavior has been observed for Si on copper substrate. In contrast, plasma-sprayed titanium dioxide on ferritic steels was successfully deoxidized in the molten salt at 850 °C, as the thermal expansions of the steel and the titanium differ less than 20% within the relevant temperature range (Schwandt and Fray 2007).

References

Bachmann, P.K., Wiechert, D.U., and Meeuwsen, T.P.M. (1988). Thermal expansion coefficients of doped and undoped silica prepared by means of PCVD. *J. Mater. Sci.* 23: 2584–2588. https://doi.org/10.1007/BF01111918.

Cho, S., Fan, F., and Bard, A. (2012). Electrodeposition of crystalline and photoactive silicon directly from silicon dioxide nanoparticles in molten $CaCl_2$. *Angew. Chem. Int Ed.* 124: 12912–12916. https://doi.org/10.1002/anie.201206789.

Choi, M.J., Jung, J., Park, M. et al. (2014). Long-term durable silicon photocathode protected by a thin Al_2O_3/SiO_x layer for photoelectrochemical hydrogen evolution. *J. Mater. Chem. A* 2: 2928–2933. https://doi.org/10.1039/c3ta14443g.

Cui, J., Phang, S., Sio, H. et al. (2017a). Passivation of phosphorus diffused black multi-crystalline silicon by hafnium oxide. *Phys. Status Solidi RRL* 11: 1700296. https://doi.org/10.1002/pssr.201700296.

Cui, J., Wan, Y., Cui, Y. et al. (2017b). Highly effective electronic passivation of silicon surfaces by atomic layer deposited hafnium oxide. *Appl. Phys. Lett.* 110: 021602. https://doi.org/10.1063/1.4973988.

Davis, K. and Tomozawa, M. (1995). Water diffusion into silica glass: structural changes in silica glass and their effect on water solubility and diffusivity. *J. Non-Cryst. Solids* 185: 203–220. https://doi.org/10.1016/0022-3093(95)00015-1.

Deal, B.E. and Grove, A.S. (1965). General relationship for the thermal oxidation of silicon. *J. Appl. Phys.* 36: 3770. https://doi.org/10.1063/1.1713945.

Esposito, D.V., Levin, I., Moffat, T.P., and Talin, A.A. (2013). H_2 evolution at Si-based metal-insulator-semiconductor photoelectrodes enhanced by inversion channel charge collection and H spillover. *Nat. Mater.* 12: 562–568. https://doi.org/10.1038/nmat3626.

Hazarika, C. and Sharma, S. (2014). A mathematical model describing drift in SiO_2 gate pH ISFET's due to hydrogen ion diffusion. *Int. J. Appl. Eng. Res.* 9: 21099–21113.

Hu, S., Nathan, S.L., Ager, J.W. et al. (2015). Thin-film materials for the protection of semiconducting photoelectrodes in solar-fuel generators. *J. Phys. Chem.* 119: 24201–24228. https://doi.org/10.1021/acs.jpcc.5b05976.

Jaeger, R.C. (2002). *Thermal Oxidation of Silicon. Introduction to Microelectronics Fabrication*, 2e. Upper Saddle River: Prentice Hall. ISBN: 0-201-44494-1.

Jannat, A., Lee, W., Akhtar, M.S. et al. (2016). Low cost sol–gel derived SiC–SiO_2 nanocomposites as anti reflection layer for enhanced performance of crystalline silicon cells. *Appl. Surf. Sci.* 369: 545–551. https://doi.org/10.1016/j.apsusc.2016.02.098.

Jettanasen, J. (2015). Study of silicon nanoparticles in dielectric oxides obtained by sol–gel route. *J. Experim. Nanosci.* 10: 1319–1326. https://doi.org/10.1080/17458080.2015.1007096.

Ji, L., McDaniel, M.D., Wang, S. et al. (2015). A silicon-based photocathode for water reduction with an epitaxial $SrTiO_3$ protection layer and a nanostructured catalyst. *Nat. Nanotechnol.* 10: 84–90. https://doi.org/10.1038/nnano.2014.277.

Kalinauskas, P., Staišiūnas, L., Grigucevičienė, A. et al. (2022). Photoelectrochemical and nanogravimetric study of electrolytic transformation of silicon-oxide interface. *J. Electrochem. Soc.* 169: 036508. https://doi.org/10.1149/1945-7111/ac5938.

Khan, M.N., Al Dwayyan, A.S., and Al Hoshan, M. (2013). Morphology and optical properties of a porous silicon-doped sol–gel host. *Electron. Mater. Lett.* 9: 697–703. https://doi.org/10.1007/s13391-013-2241-0.

Liu, M., Peng, J. et al. (2016). Two-dimensional modeling of the self-limiting oxidation in silicon and tungsten nanowires. *Theor. Appl. Mech. Lett.* 6: 195–199. https://doi.org/10.1016/j.taml.2016.08.002.

Massoud, H.Z. (1997). Thermal oxidation of silicon in the ultrathin regime. *Solid State Electron.* 41: 929–934. https://doi.org/10.1016/S0038-1101(97)00001-4.

Massoud, H.Z. and Plummer, J.D. (1985). Thermal oxidation of silicon in dry oxygen: accurate determination of the kinetics rate constants. *J. Electrochem. Soc.* 132: 2693–2700. https://doi.org/10.1149/1.2113649.

Michaelis, A. (2008). Valve metal, Si and ceramic oxides as dielectric films for passive and active electronic devices. In: *Advances in Electrochemical Science and Engineering*, vol. 10 (ed. R. Alkire, D. Kolb, J. Lipkowski and P. Ros), 75. Wiley-VCH https://doi.org/10.1002/9783527625307.

Oudot, E., Gross-Jean, M., Courouble, K. et al. (2018). Hydrogen passivation of silicon/silicon oxide interface by atomic layer deposited hafnium oxide and impact of silicon oxide underlayer. *J. Vac. Sci. Technol., A* 36: 01A116. https://doi.org/10.1116/1.4999561.

Perram, J., Hunter, R., and Wright, H. (1974). The oxide-solution interface. *Aust. J. Chem.* 27: 461–475. https://doi.org/10.1071/CH9740461.

Schwandt, C. and Fray, D.J. (2007). Preparation of titanium coatings on steel through electro reduction of plasma sprayed and screen printed titanium dioxide deposits. *Ironmak. Steelmak.* 34: 225–230. https://doi.org/10.1179/174328107X167968.

Seger, B., Pedersen, T., Laursen, A.B. et al. (2013). Using TiO_2 as a conductive protective layer for photocathodic H_2 evolution. *J. Am. Chem. Soc.* 135: 1057–1064. https://doi.org/10.1021/ja309523t.

Singh, R., Vandana, Panigrahi, J., and Singh, P.K. (2016). Plasma assisted atomic layer deposited hafnium oxide films for silicon surface passivation. *RCS Adv.* 97720–97727. https://doi.org/10.1039/C6RA19442G.

Skolik, M., Domanowska, A., Karasinski, P. et al. (2019). Double layer sol–gel derived antireflective coatings on silicon – design, optical and auger electron spectroscopy characterization. *Mater. Lett.* 251: 210–213. https://doi.org/10.1016/j.matlet.2019.05.071.

Srivastava, J.K., Prasad, M., and Wagner, J.B. Jr., (1985). Electrical conductivity of silicon dioxide thermally grown on silicon. *J. Electrochem. Soc.* 132: 955–963. https://doi.org/10.1149/1.2113993.

Staišiūnas, L., Kalinauskas, P., Juzeliūnas, E. et al. (2022). Silicon passivation by ultrathin hafnium oxide layer for photoelectrochemical applications. *Front. Chem.* 10: 859023. https://doi.org/10.3389/fchem.2022.859023.

Sun, K., Madsen, K., Andersen, P. et al. (2012). Metal on metal oxide nanowire co-catalyzed Si photocathode for solar water splitting. *Nanotechnology* 23: 194013. https://doi.org/10.1088/0957-4484/23/19/194013.

Tan, G.L., Lemon, M.F., Jones, D.J., and French, R.H. (2005). Optical properties and London dispersion interaction of amorphous and crystalline SiO_2 determined by vacuum ultraviolet spectroscopy and spectroscopic ellipsometry. *Phys. Rev. B* 72: 205117. https://doi.org/10.1103/PhysRevB.72.205117.

Topkar, A. and Lal, R. (1995). Ionic penetration into reoxidized nitride oxides in electrolyte-oxide-semiconductor structures. *Thin Solid Films* 259: 259–263. https://doi.org/10.1016/0040-6090(94)06437-7.

Ye, H., Park, H., Akhavan, V. et al. (2011). Photoelectrochemical characterization of $CuInSe_2$ and $Cu(In_{1-x}Ga_x)Se_2$ thin films for solar cells. *J. Phys. Chem. C* 115: 234–240. https://doi.org/10.1021/jp108170g.

Zhang, X.G. (2004). *Silicon and Its Oxide*, 58–64. Kluwer Academic Publishers.

12

In Situ Studies of $SiO_2 \rightarrow Si$ Conversion – Synchrotron X-ray Diffraction

The electrochemical silica–silicon conversion in a molten salt electrolyte is a dynamic process. The transformation reaction

$$SiO_2(s) + 4e^- \rightarrow Si(s) + 2O^{2-} \tag{12.1}$$

is associated with such complex processes as oxygen diffusion in the electrolyte and within the silicon pores, formation of highly inhomogeneous surface with different areas of conductivity and activity, propagation of the reaction over the surface and in-depth penetration, precipitation of sparingly soluble compounds, which passivated the electrode surface, etc. It is, therefore, a great challenge to obtain process information *in situ* and in real time.

This electrochemically driven conversion propagates into an insulator (SiO_2) as a three-phase interline process (3PI) (Xiao et al. 2006). It is initiated at three-phase contact point, that is, a conducting contact electrode, isolating SiO_2, and the molten salt electrolyte. Furthermore, the reduced Si takes the role of contacting electrode and the process is governed by the propagation of the Si–SiO_2-electrolyte 3PI. This is a dynamic process, which rapidly spreads over silica surface and penetrates much slowly into silica depth. The SiO_2 reduction is mainly determined by the diffusion of the released oxygen ions O^{2-} (12.1) (Xiao et al. 2006). The group led by G. Z. Chen developed a model, which describes the process penetration into insulators such as SiO_2 or AgCl (Xiao et al. 2006, 2007). The initial conditions of the model assume the potentiostatic electrolysis of a cylindrical precursor. The reduction current density (j, $A\,cm^{-2}$), which flows through the cylindrical cross-section is expressed by the equation

$$j = \frac{a}{\sqrt{1 + 2abt/q}} \tag{12.2}$$

where a ($mA\,cm^{-2}$) and b (cm^{-1}) are constants for a given potential (E, V) and t (s) is the electrolysis time. The reduction depth (L, cm) normal to the cross-section is determined by the equation

$$L = \sqrt{\frac{1}{b^2} + \frac{2at}{bq}} - \frac{1}{b} \tag{12.3}$$

where q ($C\,cm^{-3}$) is the charge needed to reduce a unit volume of the oxide. The authors assumed that after prolonged electrolysis, the total electrode

Silicon: Electrochemistry, Production, Purification and Applications, First Edition. Eimutis Juzeliūnas.

polarization (η, V) is predicted mainly by the ohmic (η^*_{ohm}) and concentration (η^*_C) polarizations within the reduced porous Si layer. In such a case, the expressions for j and L transform into following equations:

$$j = \frac{\eta^*_{ohm}}{\sqrt{\rho^2 L^2_{min} + 2\rho\eta^*_{ohm}(t - t_{min})/q}} \tag{12.4}$$

$$L = \sqrt{\frac{2\eta^*_{ohm}}{q\rho}(t - t_{min}) + L^2_{min}} \tag{12.5}$$

where t_{min} stands for the minimum reduction time, after which the Eqs. (12.4) and (12.5) becomes valid. The penetration depth L_{min} means the depth reached at t_{min}. The resistance ρ (Ohm cm) refers to the sum of the resistances of the generated porous Si layer and the ionic resistivity of the solution confined by the pores. Note also that the parameter t_{min} is much less in comparison with the electrolysis time needed to reduce the entire oxide. This model was verified for reduction of silver from AgCl in aqueous electrolyte, and furthermore, it has been applied to silicon reduction in quartz in molten $CaCl_2$ (Xiao et al. 2006). Another important finding of the authors is the cathodic passivation of the electrode at certain potentials. When sweeping the electrode potential in negative direction, the current decreased linearly, however, in a certain region, the voltammetric curve showed a minimum. This behavior was attributed to cathodic passivation, which is due to the increase in concentration of O^{2-} and precipitation of CaO, which inhibits the cathodic process. Further polarization leads to CaO reduction and formation of calcium silicide (Ca–Si alloy) on the produced Si surface.

The provided observations show that highly inhomogeneous and porous interface develops during the silica–silicon conversion. The process is characterized by permanently altering conditions of silicon crystallization. Very little is known about the electric properties of the interface, for instance, charging of Helmholtz layer or formation of diffusion layer. Identification of the intermediate products developed during $Si^{4+} \rightarrow Si$ reduction would help to better understand the factors relevant to structural tuning. This is, however, not a trivial task.

Conventional methods such as dc-voltammetry or electrochemical impedance spectroscopy (EIS) are widely used to study the dynamics of electrochemical reaction in molten salt electrolytes. Substantial difficulties appear when seeking to perform *in situ* analytical studies. High-temperature specifics limit considerably the opportunities. Experimental installations require materials, which are stable at high temperatures such as super-alloys, machinable glass ceramics, quartz, zirconia, or alumina. Highly stable conductive materials have also been used in the preparation of contacting and counter electrodes. In typical configurations, the reaction zone is difficult to observe visually or transmit to it an optical signal for analytical purposes. An opportunity for optically transparency suggests the reactor is made of quartz (Cho et al. 2012; Zou et al. 2017, 2019). However, it was found that such a cell suffers from erosion and corrosion in molten $CaCl_2$ due to the solubility of the produced O^{2-} ions (Dobberstein 2007). Xiao et al. also found that the ionized oxygen can react with solid silica forming calcium silicates (Xiao et al. 2012). An electrolyte composed of $CaCl_2$ and CaO melt was used for quartz dissolution and subsequent manufacturing or refining of silicon (Olsen 2002). The CaO is also added to electrolyte in order to induce continuous ionization of SiO_2 precursor to form silicate ions, which can further be reduced to silicon (Zou et al. 2017). Thus, quartz is not stable in molten $CaCl_2$; it dissolves in such electrolyte due to spontaneously present or purposefully introduced CaO as well as electrochemically produced O^{2-} ions during SiO_2–Si conversion.

In view of aforesaid, it is not surprising that studies of silicon conversion in molten salts have predominantly been done using *ex situ* methods such as SEM, STEM, XRD, or XPS. Such investigations, however, relate to the cooled and washed samples. Cooling could lead to structural changes within the sample and washing leads to the removal of some soluble reaction products as well as the formation of additional phases. Also, in certain cases, the surfaces could change due to exposure under vacuum in an analytical chamber.

An opportunity of *in situ* experimentation provides the synchrotron X-ray diffraction analysis. The technique has been applied to directly observe the phase transformations of electrochemical SiO_2 conversion to Si in molten $CaCl_2$ at 850 °C (Katasho et al. 2020). The approach has first been applied for *in situ* phase identification of Ti reduction from TiO_2 and $NiTiO_3$ in molten $CaCl_2$ (Bhagat et al. 2010; Jackson et al. 2010). An analogous study was also performed for electrochemical deoxidizing of uranium dioxide (UO_2) in molten LiCl–KCl eutectic (Brown et al. 2015). The technique takes advantage of high penetration of synchrotron radiation, which is applicable to the samples heated in a liquid environment as well as tunable wavelength and beam polarization.

Figure 12.1 shows the experimental configuration applied by Katasho and coworkers to study *in situ* the SiO_2 electrochemical reduction by synchrotron X-ray diffraction (Katasho et al. 2020). The experiments were performed in molten $CaCl_2$ electrolyte at 850 °C. The electrochemical cell was equipped with a vitreous carbon crucible, which along with a silver plate played also the role of counter electrode. The working electrode was prepared by inserting a tungsten rod into an amorphous SiO_2 (silica glass). The Ag^+/Ag electrode prepared from Ag wire in $CaCl_2$ melt with 0.5 mol% AgCl was used as a reference electrode. The salts were placed in a mullite tube.

The study enabled the direct detection of the intermediate products, which form during SiO_2–Si conversion. Thus, Ca_2SiO_4 phase was detected inside the SiO_2/Si electrode during the reduction. The lifetime of this phase was estimated to be as long as 25 minutes under applied conditions. It is important that hypothesis of Si reduction from silicates has been raised by other authors when discussing the mechanism based on the voltammetry or *ex situ* XPS data. Due to the presence of CaO, SiO_2 dissolves producing silicates (SiO_3^{2-}, SiO_4^{4-}, $Si_2O_7^{6}$), which can further be reduced electrochemically (Yasuda et al. 2007; Xiao and Wang 2014; Juzeliūnas et al. 2012; Zou et al. 2017). The *in situ* data obtained by synchrotron X-ray diffraction confirmed that the hypothesis was realistic.

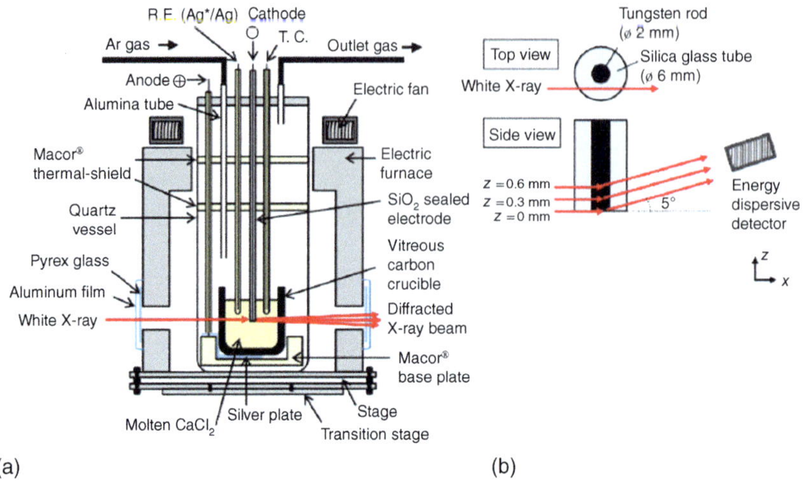

Figure 12.1 Experimental configuration of *in situ* study of SiO_2 electrochemical reduction by synchrotron X-ray diffraction in molten $CaCl_2$. Figure (a) shows the electrochemical cell composed of the vitreous carbon crucible, which played also the role of counter electrode, the W-SiO_2 precursor and the Ag^+/Ag reference electrode prepared from Ag wire in $CaCl_2$ melt with 0.5 mol% AgCl in a mullite tube. The cell is placed in a closed quartz reactor equipped with Ar^+ gas supply and the thermo-couples for temperature control. Vertical furnace is equipped with four windows for white X-rays supply. Figure (b) represents schematics of the tungsten contact electrode sealed in SiO_2 glass and used as a precursor for $SiO_2 \rightarrow Si$ conversion. The beam path is also identified. Source: Reproduced from Katasho et al. (2020)/with permission of Elsevier.

References

Bhagat, R., Dye, D., Raghunathan, S.I. et al. (2010). In situ synchrotron diffraction of the electrochemical reduction pathway of TiO_2. *Acta Mater.* 58: 5057–5062. https://doi.org/10.1016/j.actamat.2010.05.041.

Brown, L.D., Abdulaziz, R., Jervis, R. et al. (2015). Following the electroreduction of uranium dioxide to uranium in LiCl–KCl eutectic in situ using synchrotron radiation. *J. Nucl. Mater.* 464: 256–262. https://doi.org/10.1016/j.jnucmat.2015.04.037.

Cho, S.K., Fan, F.R.F., and Bard, A.J. (2012). Formation of a silicon layer by electroreduction of SiO_2 nanoparticles in $CaCl_2$ molten salt. *Electrochim. Acta* 65: 57–63. https://doi.org/10.1016/j.electacta.2012.01.008.

Dobberstein, U.H. (2007). Method for the purification of a semiconductor material by application of an oxidation-reduction reaction. Patent: WO 2007/126309 A2, 08.11.2007.

Jackson, B.K., Dye, D., Inman, D. et al. (2010). Characterization of the FFC Cambridge process for NiTi production using in situ X-ray synchrotron diffraction. *J. Electrochem. Soc.* 157: E57–E63. https://doi.org/10.1149/1.3299369.

Juzeliūnas, E., Cox, A., and Fray, D.J. (2012). Electro-deoxidation of thin silica layer in molten salt – Globular structures with effective light absorbance. *Electrochim. Acta* 68: 123–127. https://doi.org/10.1016/j.electacta.2012.02.054.

Katasho, Y., Norikawa, Y., Yamamoto, T. et al. (2020). *In situ* synchrotron X-ray diffraction study of the electrochemical reduction of SiO_2 in molten $CaCl_2$. *Electrochem. Commun.* 115: 106740. https://doi.org/10.1016/j.elecom.2020.106740.

Olsen, E. (2002). Electrolyte and method for manufacturing and/or refining of silicon. Patent WO/2002/099166, Publication date: 12.12.2002.

Xiao, W. and Wang, D. (2014). The electrochemical reduction process of solid compounds in high temperature molten salts. *Chem. Soc. Rev.* 43: 3215–3228. https://doi.org/10.1039/c3cs60327.

Xiao, W., Jin, X., Deng, Y. et al. (2006). Electrochemically driven three-phase interlines into insulator compounds: electroreduction of solid SiO_2 in molten $CaCl_2$. *Chem. Phys. Chem.* 7: 1750–1758. https://doi.org/10.1002/cphc.200600149.

Xiao, W., Jin, X., Deng, Y. et al. (2007). Three-phase interlines electrochemically driven into insulator compounds: a penetration model and its verification by electroreduction of solid AgCl. *Chem. Eur. J.* 13 (2): 604–612. https://doi.org/10.1002/chem.200600172.

Xiao, W., Wang, X., Yin, H. et al. (2012). Verification and implications of the dissolution-electrodeposition process during the electroreduction of solid silica in molten $CaCl_2$. *RSC Adv.* 2: 7588–7593. https://doi.org/10.1039/C2RA20698F.

Yasuda, K., Nohira, T., Hagiwara, R., and Ogata, Y.H. (2007). Diagrammatic representation of direct electrolytic reduction of SiO_2 in molten $CaCl_2$. *J. Electrochem. Soc.* 154: E95–E101. https://doi.org/10.1149/1.2736641.

Zou, X., Ji, L., Yang, X. et al. (2017). Electrochemical formation of a p–n junction on thin film silicon deposited in molten salt. *J. Am. Chem. Soc.* 139: 16060–16063. https://doi.org/10.1021/jacs.7b09090.

Zou, X., Ji, L., Ge, J. et al. (2019). Electrodeposition of crystalline silicon films from silicon dioxide for low-cost photovoltaic applications. *Nat. Commun.* 10 (1): 5772. https://doi.org/10.1038/s41467-019-13065-w.

13

Molten Oxide Electrochemistry at Ultra-High Temperatures

Silicon dioxide is an attractive feedstock for silicon production by electrodeposition. This mineral is inexpensive and readily available in rather pure form. In previous chapters, we discussed perspectives of silicon production by direct electrochemical reduction from SiO_2, e.g. the FFC Cambridge process. The process is based on the electrolysis of a solid silica precursor in a molten salt electrolyte. Temperature lowering of the process is important in terms of cost efficiency as well as environmental footprint. To reduce the temperature, various molten salt eutectics are applied instead of single $CaCl_2$.

Another opportunity provides operation at ultra-high temperatures, at which SiO_2 is in a liquid state. Liquid silicon is electrochemically obtained from molten silica. Such process requires much higher temperatures when compared with those applied in the reduction of solid precursor. Silica melting point is $T_m = 1710\,°C$ and that of silicon is $T_m = 1410\,°C$. We will discuss further the principles of silicon extraction from SiO_2 at temperatures higher than the melting point of silicon. Such processes are referred to as molten oxide electrolysis (MOE) (Allanore et al. 2013) or ultra-high temperature electrochemistry (UTE) (Wang et al. 2022).

According to the thermal classification provided by Wang et al., the electrochemical engineering of metals' extraction falls into three categories: the processes at low temperature ($<100\,°C$), those at high temperature ($100–1000\,°C$), and those at ultra-high temperature ($>1000\,°C$) (Wang et al. 2022). The first type refers to metal electrodeposition from aqueous solutions. The second type means the processes in high-temperature molten salts, e.g. molten $CaCl_2$ at temperatures of about $850\,°C$. The UTE deals with molten oxides, whose melting point is higher than $1000\,°C$, such as SiO_2, Al_2O_3, TiO_2, FeO, and MgO.

The above classification is indicative; it approximates the temperatures and the methods. For instance, at room temperatures, the electrodeposition of metals can be performed not only from aqueous solutions but also from ionic liquids (the ionic liquids are also referred to as room temperature molten salts). Also, the molten salt electrochemistry can encompass a wider temperature range. For instance, as it will be further discussed, in molten fluoride electrolytes the extraction processes operate at temperatures well above $1000\,°C$.

Silicon: Electrochemistry, Production, Purification and Applications, First Edition. Eimutis Juzeliūnas.

When compared with traditional extraction metallurgy, major advantages of the MOE are technical simplicity, reductions in energy consumption, and opportunity to achieve high purity products while avoiding costly purification.

De Mattei et al. reported in 1981, that silicon extraction from silica in BaO–SiO_2–BaF_2 molten electrolyte at the temperatures above the silicon melting point, that is, in excess of 1410 °C, so that the silicon is electrodeposited in a liquid form (De Mattei et al. 1981a,b). While barium oxide (BaO) was proposed as a preferred component, other alkaline earth oxides, such as CaO or SrO, or alkali metal oxides, such as Na_2O, Li_2O, or K_2O, can be used as well. It was assumed that SiO_2 reacts with BaO and forms barium silicate ($BaSiO_3$), from which silicon is electrochemically reduced. Another component of the electrolyte was a fluoride, preferably of the same metal as the oxide, for instance, BaF_2. Fluoride reduces the melting point of the electrolyte as well as its viscosity, which facilitates electrodeposition processes. The starting materials (SiO_2, $BaCO_3$, and BaF_2) were heated to about 1450 °C. At this temperature, $BaCO_3$ decomposes to BaO and CO_2, and the starting mixture is melted. Graphite anode and cathode were used and silicon was reduced in the form of droplets, which are removed from the bath and solidified. The current efficiency of 15% was attained. The silicon yield rate was rather low (about $0.03\,g\,h^{-1}$). However, rather pure silicon of approximately 99.98% purity, was extracted.

Elwell also reported in 1981 the silicon electrowinning from silica dissolved in alkali metal fluoride/alkaline earth fluoride eutectics (e.g. SiO_2, NaF, CaF_2) at 1150 °C (Elwell 1981). Graphite cathode was used, and silicon was obtained in the form of small particles dispersed in the solidified solvent. A subsequent thermal process was needed to isolate the silicon.

The electrodeposition of silicon from solutions of SiO_2 in NaF/CaF_2 and related fluoride eutectics is reported for the first time. The silicon is in the form of small particles dispersed in the solidified solvent, and a subsequent thermal process is required to isolate it.

Fluorides reduced significantly in the SiO_2 dissolution temperature, and this is important in terms of processing cost efficiency as well as environment friendliness. The phase diagrams of eutectic composition of SiO_2–KF showed a single eutectic type with the eutectic composition at 50 mol% of KF and temperature at 720 °C (Suzuki et al. 2019). Obviously, the electrolyte temperature can be appropriately reduced below that of single SiO_2. The SiO_2 melting point is even lower with LiF or LiF + NaF additives. Thus, the melting point of LiF–KF system is 492 °C at concentration ratio 51 : 49 mol%, respectively. The analogous value for LiF–NaF–KF at 46.5 : 11.5 : 42.0 mol% is 457 °C. The Raman spectra of these systems with SiO_2 additive did not differ significantly from the spectra determined for KF–SiO_2 system (Suzuki et al. 2019). The authors expect that O^{2-} ions would cleave the Si—O—Si bonds in silica, forming tetrahedron species $(SiO_3F)^{3-}$ Li_2O was added to the electrolyte as an oxygen source, and the oxide promoted the formation of $(SiO_3F)^{3-}$. Very thin layers of silicon, which are only a few micrometers in thickness, were deposited in Li_2O-free electrolyte, whereas the layers over 100 μm were deposited when adding 5.0 mol% of Li_2O. Current efficiency was also enhanced by Li_2O from ~13% to ~50%.

Molten oxides are divided into alkaline (CaO, MgO, Na_2O) or acidic (SiO_2, TiO_2, Al_2O_3) (Wang et al. 2022). Molten SiO_2 forms a characteristic network structure with O^{2-} ions. The oxygen ions can assume various positions in the structure: bridging, non-bridging, and free oxygen. There are several theories concerning the structure of molten oxides: molecular, ionic, or polymer theory. According to the polymers theory, silicon and oxygen in the melt form a polymer from a tetrahedral silicate SiO_4^{4-}.

In the UTE processes, the molten oxides are used simultaneously as an electrolyte and as a feedstock. Thus, the concentration of the reactant in the UTE is much higher than the concentration of a precursor dissolved in a molten salt. Thus, higher process efficiency in the UTE process is expected than that in a molten salt process. Due to higher temperature, the decomposition voltage in the UTE is also lower when compared with the molten salt process. These positive effects could probably offset the negative factor – an increased energy consumption necessary for the bath heating.

From the viewpoint of thermodynamics and process kinetics, the main factors are the equilibrium potential, the charge transfer through the interface, and the reduction overpotential, which is mainly determined by transport phenomena, that is, cation transport from electrolyte depth to electrode/electrolyte interface and vice versa – anion transport from the interface to electrolyte. When cations discharge, concentration of anions at the interface increases, and the accumulated negative charge retards the electron transfer from electrode to electrolyte.

Diffusion coefficient relates to viscosity of the electrolyte, which is determined by the electrolyte structure, for instance, polymerization degree: the higher the polymerization degree, the higher the viscosity. Addition of alkaline oxides to molten silica decreases the polymerization degree, reduces viscosity, and increases the diffusion coefficient. On the contrary, addition of acidic oxides (e.g. Al_2O_3) increases the polymerization degree and, in turn, reduces the viscosity (Toplis and Dingwell 2004). The transportation of electroactive ion is also affected by the surrounding ions in the electrolyte. A larger cluster can form, such as combinations of SiO_4^{4-} with covalent bonds, which retard the ion transport.

Ionic conductivity of molten oxide electrolyte strongly depends on temperature and composition. Early work done in 1952 by Shartsis et al. reports the data about densities and viscosities of SiO_2 in composition with Li_2O, Na_2O, and K_2O in the temperature range of 1000–1400 °C (Shartsis et al. 1952). It was determined that the density in each composition decreased with increasing alkali concentration. The thermal expansivity of the K_2O–SiO_2 system was greater than that of Na_2O–SiO_2 or Lia_2O–SiO_2 systems. The authors also concluded that viscosity was almost independent of the kind of the alkali ion introduced into the system.

Around the same time, Bockris and coworker studied viscous flow in pure silica and binary liquid silicates (Bockris and Lowe 1954; Bockris et al. 1955). The measurements have been performed using a rotated molybdenum crucible, and the studied liquid systems were composed of SiO_2 with Li_2O, Na_2O, K_2O, MgO, SrO, and BaO. The experiments have been done at temperatures up to 1800 °C. Pure silica was studied at temperatures as high as 2050 °C. The viscosity isotherms showed an exponential increase in viscosity with the silica content. The activation heat is derived by the authors from the slope of the viscosity logarithm vs. temperature ($\log \eta - 1/T$).

The authors observed that change in the silicate lattice occurs when about 10 mol% of alkali metal oxide and about 20% of alkali earth oxide is added to the silicate. They also concluded that the silicate ions obey the formula $(Si_nO_{2n+3})^{6-}$, and the main building units of the structure are the ions $Si_3O_9{}^{6-}$.

In the same year, Simnad et al. reported a study about current efficiency of silicate electrolysis, which yield oxygen as an anodic product (Simnad et al. 1954). They studied the process in FeO–SiO_2 liquid mixture depending on silica content, while varying it from 0% to 35%. The study was performed at 1400 °C using iron and platinum anodes. The current efficiency was found to be 8–12% when SiO_2 content was 0–10%. At higher SiO_2 contents, the efficiency increased linearly; it reached a saturation at about 90% when SiO_2 content was 35%. The authors concluded that the ionic current in the FeO–SiO_2 system was carried almost completely by ferrous ions. It is also important to note that platinum anodes were stable under the studied conditions: no change in electrode weight and composition has been observed. Iron anodes changed their shape due to some thermal gradient in the apparatus construction. The total electrode mass, however, did not change significantly. No change was observed when thermal gradient was removed. During the electrolysis, the iron electrode weight loss was measured to be about 0.25 g. Thus, the iron electrode showed stability in the electrolyte at 1400 °C.

Urbain et al. reported complementary results on the viscosity of liquid amorphous silica and the alumino-silicates in composition with alkaline earth metal oxides, MnO, Na_2O, K_2O, as well as silica in composition with CaO–FeO–Fe_2O_3 (Urbain et al. 1982). The viscosities have been determined experimentally at various temperatures. The authors have drawn a conclusion that viscosity measurements on liquid rock compositions require they be performed under a controlled atmosphere with knowledge of the ratio Fe^{2+}/Fe^{3+}.

Sukenaga et al. extended the viscosity studies to the system CaO–SiO_2–Al_2O_3 with the additive of oxides of Li, Na, K, Ba, and Mg (Sukenaga et al. 2006). They used the classical approach of the rotating crucible viscometer. Additionally, they studied quenched vitreous samples by the isotopes ^{27}Al and ^{29}Si magic-angle spinning (MAS)-nuclear magnetic resonance (NMR) spectroscopy. The spectra identified that the degree of polymerization of silicate anions decreased when the lithium, magnesium, or sodium oxides were added. However, the degree of polymerization of aluminosilicate anions increased with the addition of barium oxide (BaO).

Sohn and Min reviewed the relationship between the viscosity and the structure of calcium-silicate based systems (Sohn and Min 2012). The viscous behavior of the multi-component system was studied using various spectroscopic techniques, such as Fourier transform infrared (FTIR), X-ray photoelectron (XPS), and Raman. NMR technique has also been used. The review analyses also the viscous behavior and structure of binary silicate systems.

Electrolysis of molten oxides is associated with the oxygen evolution reaction $O^{2-} \rightarrow O_2$ on anode. Graphite anodes are typically used in high-temperature electrochemistry, for instance, in the industrial production of aluminum. Such electrodes are consumable; emission of CO_x greenhouse gases occurs during electrolysis, which consumes graphite. It is a challenge to develop inert anodes for UTE processes, which operate under extreme conditions: ultra-high temperature and highly destructive environment. The carbon-free electrodes should possess sufficiently high electrical conductivity and good antierosion-anticorrosion resistance. The electrode should also be stable against anodic oxidation-dissolution. The potential candidates are metals and alloys, oxides and ceramics (Caldwell et al. 2016).

Platinum as anode material has been applied in the early study of oxygen generation performed in a molten $FeO–SiO_2$ electrolyte (Simnad et al. 1954). In recent decades, noble metals, such as Pt, Rh, or Ir, have been studied as anodes operating at high temperatures (Kim et al. 2010; Wiencke et al. 2018). Wiencke et al. extracted iron at 1550 °C using Pt-10Rh anode from a molten oxide electrolyte composed of Al_2O_3, Fe_3O_4, MgO, and SiO_2. Large surface anode was used, whose surface area was much greater than the surface of the cathode. In such configuration, the anodic current density was independent of the cell potential; only the cathodic current was affected by the potential change. It is important to note that the quantity of oxygen produced on the anode was directly proportional to the Fe_3O_4 content. Sadoway's group studied performance of iridium as inert anode in acidic and alkaline oxides based on certain proportions: $SiO_2–CaO–MgO–Al_2O_3$ (acidic) and $CaO–MgO–Al_2O_3$ (alkaline) (Kim et al. 2010). The depth of corrosion attack on Ir was about 200 μm in an alkaline medium after four hours electrolysis, whereas much higher stability was observed in acidic medium. The electrode weight loss in acidic molten oxides was by ~20 times less when compared with that observed in alkaline oxides. Certainly, the cost issues limit the application of noble metals on an industrial scale.

Iron electrode for oxygen production showed rather high stability at 1400 °C in molten $FeO–SiO_2$ system (Simnad et al. 1954). A new anode material based on Cr–Fe alloys has been proposed for iron extraction from molten oxide (Allanore et al. 2013). The operating temperature of the extraction was $T = 1565\,°C$. Limited changes in dimensions of the $Cr_{90}Fe_{10}$ anode have been observed during electrolysis. Fray regarded the results as bringing closer to reality the long scientists' dream of "electrification" of iron production (Fray 2013).

At ultra-high temperatures, there is an opportunity to explore liquid metal cathodes. The metal extracted in a liquid form can further be used as a liquid cathode for subsequent extraction. The liquid product can easily be separated from the electrolyte. Another opportunity is an application of the liquid cathodes made of foreign metals, which are different from those to be extracted. The foreign metal, whose melting temperature is lower than that of the extracted metal, can be used.

Liquid metal electrodes are also used for electrolytic refining processes. Thus, Sadoway's group patented a high-temperature method for electrolytic refining of liquid metals whose melting point is above ~1000 °C (Johanson et al. 2007). The method uses a two-layer electrochemical cell, which comprises a molten electrolyte layer as an anode and another alloy layer as a cathode. The latter is prepared from an alloy, which contains the metal to be purified, and at least one metal, which is nobler than the metal to be purified. A three-layer cell was also designed. It contains, in addition, an intermediate layer – a molten electrolyte between the cathode and the anode. Molten oxides are employed as electrolytes: one electrolyte contains the oxide of the metal to be refined; another oxide electrolyte has a melting point below the operating temperature of the process.

Oishi et al. proposed three layers of Si refining concept, which is based on the application of liquid Al–Si layer as cathode, molten salt electrolyte, and liquid Cu–Si layer as an anode (Oishi et al. 2016). Low purity Si is added to liquid Cu–Si anode and Si is anodically dissolved along with the impurities less noble than Si. The Si and more noble impurities are deposited onto a liquid cathode like Al–Si. The Al–Si cathode ensures proper separation between the melt and the cathode product at relatively low temperature (1000–1300 K). After electrochemical treatment, the Si-enriched liquid is transferred to another vessel where Si is obtained by precipitation, and the remaining liquid alloy is reused as a cathode material. Si-based cells from commercial PV modules were used as feed material. The purification, however, did not meet the purity requirement for solar-grade silicon. Note that the treatment was particularly effective with respect to silver, which is used as the electrode on the front side of PV cell.

It is important that silicon purification using liquid anodes is a more energy-efficient process when compared with the conventional Siemens process, which is based on silane distillation and its decomposition at high temperatures. The energy consumption of Si purification using liquid Si–Cu anode and $CaCl_2$–SiO_2 melt was estimated to be as high as ~2 kWh kg^{-1} (Cai et al. 2012), whereas the consumption in the Siemens process is 65–160 kWh kg^{-1} (Fabri and Hesse 2012; De Wild-Scholten and Alsema 2004). Thus, there is a potential for industrial Si purification using molten electrodes. However, the purity degree is still insufficient to produce the material for PV applications.

Liquid metal batteries denote another field of application of liquid electrodes (Kim et al. 2013). Similar to the purification cells, the liquid metal battery comprises two liquid electrodes separated by a molten salt electrolyte. The difference from a purification cell is that the battery electrodes are made of liquid metals, not oxides. From a practical point of view, there is a requirement that the melting temperatures of liquid metal electrodes should be below 1000 °C. It is assumed that electrode materials with a deposition potential less than $-2.0\ V_{SHE}$ are applicable as negative electrodes and those with potential over $-1.0\ V_{SHE}$ can be used as positive electrodes. The exception is aluminum, which can serve both as a negative and a positive electrode. Liquid electrodes endow superior kinetics of charge transfer, while liquidity of the electrolyte facilitates ion transport. Typical inefficiencies of conventional batteries include factors such as the sluggish electrode kinetics and, as a result, charge transfer losses, the resistivity of electrolyte, electrodes and current collectors, the transport limitations to (from) the electrochemical reaction interface. These drawbacks can be mitigated or avoided when using liquid electrodes and molten salt electrolytes. However, elevated temperatures restrict the application of liquid metal batteries. For instance, application of these batteries in portable devices is difficult due to elevated temperatures. Another limitation is the batteries' sensitivity to motion. Note also that silicon is not considered as a potential candidate material for liquid metal batteries (Kim et al. 2013).

Molten oxide electrolysis produces oxygen gas on the anode, provided the electrode is inert. This suggests an exciting application field – oxygen production in space exploration (Sanderson 2009; Fray 2013; Schwandt et al. 2012; Wang et al. 2022). Oxygen production on-site is of great importance for space exploration due to the high cost and technical complexity of oxygen transportation from the Earth to another planet. The transportation weight could be reduced once the oxygen is produced on-site. Fortunately, Lunar or Mars soils contain oxygen (O^{2-}), which is chemically bonded with metals. The oxygen content in the soil is up to ~44 wt% (Wang et al. 2022). The oxygen-rich soil, therefore, can be used as a feedstock for oxygen production. Considerable contents of titanium oxide are found on the Moon. This is a favored situation, as deoxidation of this oxide in molten salts has comprehensively been studied (Alexander et al. 2011; Schwandt et al. 2010). Valuable titanium metal would be extracted from the soil as a byproduct of oxygen production. Other oxygen-containing minerals comprise iron, magnesium, calcium, aluminum, or silicon. Lunar rocks have been found rich in anorhite ($CaAl_2Si_2O_8$); some rocks have a high content of titanium in the form of ilmenite ($FeTiO_3$) (Crawford 2015).

Debatable is the question concerning techniques of temperature rising in order to melt the regolith. It is known that temperatures over 100 °C are reached on the Moon's surface when sunlight hits it vertically. In the Earth, temperatures up to 3000 °C can be reached when the sunlight is concentrated by mirrors from large areas into a much smaller region. Practically, temperatures of up to 1000 °C are in operation due to the limited thermal resistance of the receivers (higher temperatures can melt the receiver). Nakamura and Senior reported a solar power system for *in situ* lunar materials processing (Nakamura and Senior 2008). The authors addressed the high-temperature processes, such as hot liquid processing, welding, pyrolysis, gas–solid reactions, and gas–liquid or three-phase reactions. They pointed out that the thermal energy up to 2500 K, which is necessary for the above processes, could be achieved by utilizing solar thermal power. The proposed technique for lunar materials processing is based on collecting the solar radiation by a concentrator and transferring it to the destination area through a line made of optical fiber.

There are several distinctive features between the regolith deoxidation in molten salts (e.g. in molten $CaCl_2$ electrolyte) and electrolysis of molten regolith (molten oxides). The first approach is that the minerals are not melted; they are introduced into molten electrolyte and deoxidized directly by the electricity. In the second approach, oxygen is produced from the molten minerals. The process in molten salt requires lower temperature (500–900 °C), whereas a much higher temperature is necessary to melt the minerals. The dust regolith for molten salt processing can be dissolved in the electrolyte when selecting appropriate components, e.g. SiO_2 can be dissolved in $CaCl_2 + CaO$. The precursor can also be used as a mineral suspension in the molten salt electrolyte. Some aspects of electrolysis of the Moon regolith are also covered in Chapter 8.

In summary, one could say that molten oxide electrolysis provides an opportunity of developing sustainable technologies for metal extraction and oxygen production. Conceptually, both processes are well defined and proven. From an industrial point of view, however, these processes require deeper economic and environmental analysis. The entire value chain has to be assessed, which ranges from the feedstock to the final products and waste utilization. The energy and materials cost, as well as sustainability are particularly important. The molten oxide electrolysis is an energy intensive process due to the use of extreme temperatures. The sustainability condition requires the use of energy from renewable sources. The application of conventional inexpensive graphite electrodes is hardly acceptable due to CO_x emission; thus, the usage of such electrodes compromises the goal of carbon neutrality. The use of precious metals as electrodes is technologically justified but of little value in terms of the processing cost. Specific materials should be used also for crucibles and reactors; the molten oxide electrolytes at high temperatures are corrosive. Collection and storage of the produced oxygen require engineering solutions as well. Overall, these issues pose challenges for future researchers.

References

Alexander, D.T.L., Schwandt, C., and Fray, D.J. (2011). The electro-deoxidation of dense titanium dioxide precursor in molten calcium chloride giving a new reaction pathway. *Electrochim. Acta* 56: 3286–3295. https://doi.org/10.1016/j.electacta.2011.01.027.

Allanore, A., Yin, L., and Sadoway, D.R. (2013). A new anode material for oxygen evolution in molten oxide electrolysis. *Nature* 497: 353–356. https://doi.org/10.1038/nature12134.

Bockris, J.O.'.M. and Lowe, D.C. (1954). Viscosity and the structure of molten silicates. *Proc. R. Soc. London, Ser. A* 226: 423–435. https://doi.org/10.1098/rspa.1954.0266.

Bockris, J.O.'.M., Mackenzie, J.D., and Kitchener, J.A. (1955). Viscous flow in silica and binary liquid silicates. *Trans. Faraday Soc.* 51: 1734–1748. https://doi.org/10.1039/TF9555101734.

Cai, J., Luo, X., Haarberg, G.M., Kongstein, O.E., and Wang, S. (2012). Electrorefining of metallurgical grade silicon in molten $CaCl_2$ based salts. *J. Electrochem. Soc.* 159: D155–D158. https://doi.org/10.1149/2.041203jes

Caldwell, A.H., Lai, E., Gmitter, A.J., and Allanore, A. (2016). Influence of mass transfer and electrolyte composition on anodic oxygen evolution in molten oxides. *Electrochim. Acta* 219: 178–186. https://doi.org/10.1016/j.electacta.2016.09.132.

Crawford, I.A. (2015). Lunar resources: a review. *Prog. Phys. Geog.* 39: 137–167. https://doi.org/10.1177/0309133314567585.

De Mattei, R.C., Elwell, D., and Feigelson, R.S. (1981a). Electrodeposition of silicon at temperatures above its melting point. *J. Electrochem. Soc.* 128: 1712–1714. https://doi.org/10.1002/chin.198148039.

De Mattei, R.C., Elwell, D., and Feigelson, R.S. (1981b). US Patent, No 4, 292, 145.

De Wild-Scholten, M. and Alsema, E. (2004). Towards cleaner solar PV: environmental and health impacts of crystalline silicon photovoltaics. *Refocus* 5: 46–49. https://doi.org/10.1016/S1471-0846(04)00225-2.

Elwell, D. (1981). Electrowinning of silicon from solutions of silica in alkali metal fluoride/alkaline earth fluoride eutectics. *Sol. Energy Mater.* 5: 205–210. https://doi.org/10.1016/0165-1633(81)90032-0.

Fabri, L. and Hesse, K. (2012). Crystalline silicon feedstock preparation and analysis. *Semiconduct. Semimet.* 87: 185–261. https://doi.org/10.1016/B978-0-12-388419-0.00007-8.

Fray, D.J. (2013). Iron production electrified. *Nature* 497: 324–325. https://doi.org/10.1038/nature12102.

Johanson, K., Sadoway, D.R., Myhre, B., et al. (2007). Method for electrolytic production and refining of metals. US Patent US20070215483A1.

Kim, H.J., Paramore, J., Allanore, A., and Sadoway, D.R. (2010). Stability of iridium anode in molten oxide electrolysis for ironmaking: influence of slag basicity. *ECS Trans.* 33: 219–230. https://doi.org/10.1149/1.3484779.

Kim, H., Boysen, D.A., Newhouse, J.M. et al. (2013). Liquid metal batteries: past, present, and future. *Chem. Rev.* 113: 2075–2099. https://doi.org/10.1021/cr300205k.

Nakamura, T. and Senior, C.L. (2008). Solar thermal power for lunar materials processing. *J. Aerospace. Eng.* 21: 91–101. https://doi.org/10.1061/(ASCE)0893-1321(2008)21-2(91).

Oishi, T., Koyama, K., and Tanaka, M. (2016). Electrorefining of silicon using molten salt and liquid alloy electrodes. *J. Electrochem. Soc.* 163: E385–E389. https://doi.org/10.1149/2.0391614jes.

Sanderson, K. (2009). How to breathe on the moon. *Nature* https://doi.org/10.1038/news.2009.803.

Schwandt, C., Doughty, G.R., and Fray, D.J. (2010). The FFC-Cambridge process for titanium metal winning. *Key Eng. Mater.* 436: 13–25. https://doi.org/10.4028/www.scientific.net/KEM.436.13.

Schwandt, C., Hamilton, J.A., Fray, D.J., and Crawford, I.A. (2012). The production of oxygen and metal from lunar regolith. *Planet. Space Sci.* 74: 49–56. https://doi.org/10.1016/j.pss.2012.06.011.

Shartsis, L., Spinner, S., and Capps, W. (1952). Density, expansivity, and viscosity of molten alkali silicates. *J. Am. Ceram. Sot.* 35: 155–160. https://doi.org/10.1111/j.1151-2916.1952.tb13090.x.

Simnad, M.T., Derge, G., and George, I. (1954). Ionic nature of liquid iron-silicate slag. *JOM* 6: 1386–1390. https://doi.org/10.1007/BF03398060.

Sohn, I. and Min, D.J. (2012). Review of the relationship between viscosity and the structure of calcium-silicate-based slags in ironmaking. *Steel Res. Int.* 2012 (83): 611–630. https://doi.org/10.1002/srin.201200040.

Sukenaga, S., Saito, N., Kawakami, K., and Nakashima, K. (2006). Viscosities of $CaO–SiO_2–Al_2O_3–(R_2O$ or RO) melts. *ISIJ Int.* 46: 352–358. https://doi.org/10.2355/isijinternational.46.352.

Suzuki, Y., Inoue, Y., Yokota, M., and Goto, T. (2019). Effects of oxide ions on the electrodeposition process of silicon in molten fluorides. *J. Electrochem. Soc.* 166: D564-D568. https://doi.org/10.1149/2.0441913jes

Toplis, M.J. and Dingwell, D.B. (2004). Shear viscosities of $CaO–Al_2O_3–SiO_2$ and $Mg–Al_2O_3–SiO_2$ liquids: Implications for the structural role of aluminium and the degree of polymerisation of synthetic and natural aluminosilicate melts. *Geochim. Cosmochim. Acta* 68: 5169–5188. https://doi.org/10.1016/j.gca.2004.05.041.

Urbain, G., Bottinga, Y., and Richet, P. (1982). Viscosity of liquid silica, silicates and alumino-silicates. *Gochim. Cosmochim. Acta* 46: 1061–1072. https://doi.org/10.1016/0016-7037(82)90059-X.

Wang, M., Jiao, H., Pu, Z. et al. (2022). Ultra-high temperature molten oxide electrochemistry. *Angew. Chem. Int. Ed.* 61: https://doi.org/10.1002/anie.202206482.

Wiencke, J., Lavelaine, H., Kane, G.J. et al. (2018). Kinetics of iron electrochemical reduction into liquid metal at 1823 K in a molten oxide electrolyte. *Mater. Chem. Phys.* 212: 214–223. https://doi.org/10.1016/j.matchemphys.2018.03.020.

14

Silicon Surface Structuring

Silicon surface structuring is important in a variety of applications: solar cells (Otto et al. 2010, 2015; Liu et al. 2014) and optoelectronics (Torres-Costa and Martin-Palma 2010; Steglich et al. 2013), lithium-ion batteries (Kasavajjula et al. 2007; Ge et al. 2013; Li et al. 2015), supercapacitors (Oakes et al. 2013; Grigoras et al. 2016; Grigoras et al. 2018), photoelectrochemical hydrogen production (Oh et al. 2011 2012a,b; Ao et al. 2012), terahertz emitters (Hoyer et al. 2011), sensors (chemical (Gesemann et al. 2011), gas (Barillaro et al. 2003), and biosensors (Angelescu et al. 2003; Jane et al. 2009; Kilian et al. 2009)), biomicrofilters (Bonanno and DeLouise 2007), drug delivery *in vivo* (Anglin et al. 2008; Salonen et al. 2008; Park et al. 2009), physical bactericides (Ivanova et al. 2013), tunable electrowetting (Barberoglou et al. 2010), and platforms for laser-induced micronanopolymerization (Žukauskas et al. 2013). Most of the Si structuring methods are based on surface etching, which fall into several categories: (i) electrochemical, (ii) chemical, (iii) metal-assisted chemical (MACE), (iv) reactive ion (RIE), (v) plasma immersion and ion implantation (PIII), (vi) laser engineering, etc. These methods are referred to as "top-down" methods; they dominate the semiconductor production industry. The "bottom-up" method is also applied, such as a vapor–liquid–solid (VLS) growth method. The major features of the above structuring will be considered in this chapter. Attention will also be given to production of black silicon (b-Si), which is a nanomicrostructured silicon, whose light absorbance is close to 100% in a wide range of wavelengths.

14.1 Electrochemical Structuring, Porous Silicon

Electrochemical structuring yields the porous silicon (P-Si), which is referred to a material whose surface is structured by a network of nano- or micrometer-sized formations surrounded by a void space. Electrochemical formation of P-Si was invented by Uhlir in 1956 at Bell Laboratories when studying silicon electropolishing (Uhlir 1956). This method is based on silicon electrochemical etching (anodizing) in aqueous acidic electrolytes containing hydrofluoric acid (HF).

Lehman and Gösele studied porous silicon formation in an n-type and p-type silicon (Lehman and Gösele 1991; Lehmann 1993, 1999). They studied porous

Silicon: Electrochemistry, Production, Purification and Applications, First Edition. Eimutis Juzeliūnas.

Si layers, which were grown electrochemically on low-doped silicon in HF. The porosity features depending on factors such as doping density, electrolyte composition, and applied current density have been discussed. They also discussed the two-dimensional quantum confinement effect caused by very narrow walls between the nanopores. The quantum confinement was assumed to be the reason for the change in the Si band gap energy. The confinement is also important in order to explain the mechanism of porosity formation. A specific current density at the pore tip is essential to form the porosity. At this specific current density, the dissolution limiting stage changes from charge transfer to mass transfer limitation.

Electron acceptors are needed to perform dissolution of silicon electrodes. In electrochemical etching, electron acceptors serve as the holes (h^+), whose presence is essential to initiate silicon etching. Minority carriers are generated when illuminating Si with the sufficient photon energy $E = h\nu$, that is, electrons in p-Si and holes in n-Si. The minority carriers in n-Si, that is, h^+, make the electrode passive with respect to HF etching. One could activate the electrode when applying high anodic voltages, which exceed the breakdown potential of the passive state. Another opportunity provides electrode illumination with the photon energy sufficient to increase the electron energy from valence to conductivity band. Energy by 1.1 eV higher than that required for the recombination is needed for electron injection (Lehmann 1993). The generated holes are concentrated at the pore tips promoting in-depth dissolution rather than that of the walls. There is no need to induce h^+ generation in p-Si; electrons are minority carriers in such samples. Therefore, etching of this material could be performed in the dark and without applying high voltages. Thus, the availability of valence band holes is an essential factor for the etching; the holes are referred to as oxidizing equivalents (Anglin et al. 2008).

The Si surface properties largely depend on the applied current density. Pore dimensions, therefore, can be tailored by varying the current density. Macroporous silicon is obtained at relatively low current densities. At high current densities, a large number of holes are injected, the reaction is limited by SiO_x removal, and the surface becomes polished. In the intermediate current region, nanopillars could be formed. According to (Zhang et al. 1989; Zhang 2007), there are several distinctive regions on a current–potential curve. In the first region, where current changes exponentially, porous silicon is formed. In the second region, where constant current is observed, electropolishing of silicon occurs. The electropolishing is observed in a transition region as well. The maximum current for porous p-Si production in 1% HF solution is $j \sim 3\,\text{mA cm}^{-2}$, while electrochemical polishing occurs at $j > 5.5\,\text{mA cm}^{-2}$. Other parameters, which can be tuned to tailor surface architectures, include HF concentration, doping level, etching time, and temperature. The doping level of p-Si defines porosity because the pores randomly formed at doping sites may overlap (Otto et al. 2015). The porosity features may also be predefined by means of lithography. Photoelectrochemical Si etching has been surveyed in the following review articles (Otto et al. 2015; Liu et al. 2014).

The Si surface in the HF electrolyte is hydrogen terminated by the Si–H bonds. Anglin et al. stated that the oxidation process includes four oxidizing equivalents: two of them being produced electrochemically and two others are produced because of proton reduction by surface SiF_2 species (Anglin et al. 2008). The entire etching process includes removal of Si atoms as SiF_4 and subsequent formation occurs in the electrolyte of SiF_6^{2-}.

When performing the etching, the current flows because of silicon oxidation on the Si anode $Si \rightarrow Si^{4+} + 4e^-$ and hydrogen evolution on a cathode $2H^+ + 2e^- \rightarrow H_2$. As the presence of the valence band holes is a prerequisite, they are included in the anodic half-reaction as the oxidizing equivalents ($2h^+$) (Anglin et al. 2008):

$$Si + 6F^- + 2H^+ + 2h^+ \rightarrow SiF_6{}^{2-} + H_2 \quad (14.1)$$

The electrochemically formed porous silicon has diverse surface features. The pore morphologies and sizes vary depending on a variety of factors such as current density, HF concentration as well as type, dopant level, and crystallographic structure of the Si substrate. This provides good structural management of the product with porosity tunability opportunities. The pores from 1 nm to a few micrometers in diameter could be produced by the electrochemical method. Note that according to the IUPAC recommendations, porosity is classified with respect to the pore diameter (d) as macroporous ($d > 50$ nm), mesoporous (2 nm $< d <$ 50 nm), and microporous ($d < 2$ nm) (Rouquerol et al. 1994).

An enormous amount of studies have been done in the field of P-Si production and characterization. The anodizing method, however, remains the most common in the preparation of porous silicon. An important milestone in the field was when Canham in 1990 discovered room temperature photoluminescence of mesoporous Si layers (Canham 1990). The author used electrochemical and chemical dissolution to produce networks of wires from bulk wafers without lithography means. Mesoporous layers exhibited visible photoluminescence observable with the naked eye under green or blue laser excitation. This discovery made the P-Si a promising material for optoelectronic applications. Canham's paper received from 1990 to 2020 over 11 000 citations.

As an interesting application example of P-Si, a biosensor reported by Bonanno and DeLouse is given in Figure 14.1 (Bonanno and DeLouise 2007). The authors

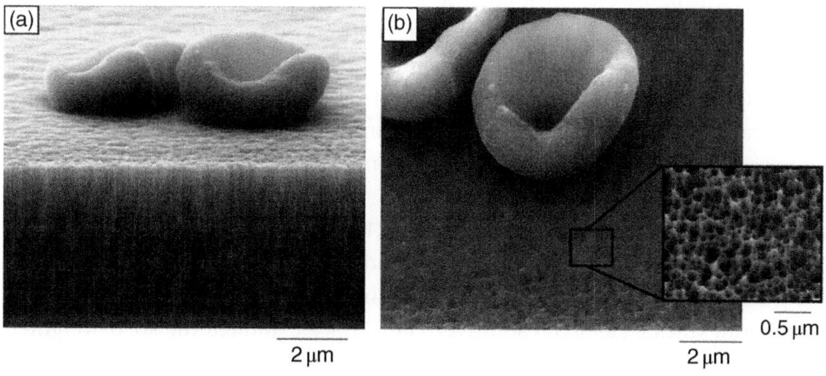

Figure 14.1 SEM images of a P-Si biosensor. (a) Cross section of the P-Si biosensor substrate with filtered out erythrocytes; (b) top view and inset, which shows a porosity of ~88 nm pores in diameter. Source: Reproduced from (Bonanno and DeLouise 2007) with permission from the publisher. © Elsevier, 2007.

developed a label-free optical biosensor for the detection of rabbit antibody immunoglobulin G (IgG) in blood with a linear detection range of 2–10 mg mL^{-1}. The antibody has been immobilized on the P-Si surface by biotin–streptavidin chemistries.

14.2 Chemical–Physical Structuring

Various silicon surface engineering methods have been developed. They fall into two major categories: "top-down" and "bottom-up." The top-down approach is based on physical or chemical etchings performed by laser treatment as well as dry reactive ion (RIE), metal-assisted chemical (MACE) or (photo)electrochemical etching. The bottom-up structuring is performed by Si deposition on patterned substrates. To this end, means of physical vapor deposition (PVD), chemical vapor deposition (CVD), or atomic layer deposition (ALD) are used. Micrometer-sized architectures are obtained by photolithographical surface pattering and subsequent VLS growth or physical–chemical etchings. For instance, vertically oriented Si microwire arrays were grown on a photolithographically pattered substrate using Au or Cu as the growth catalysts (Kayes et al. 2007; Boettcher et al. 2010, 2011). Honeycomb-structured Si was formed by means of photolithography and RIE (Baggetto et al. 2011). Pattering using porous alumina prepared by aluminum anodizing has been used to obtain Si nanostructures such as gold-decorated nanocolumns (Leinartas et al. 2011). Versatile structures are also obtained by means of electrochemistry. An emerging field is the structuring in molten salt electrolytes, which provides a variety of nanostructures, such as wires and fibers, pores, globules, etc. Furthermore, major Si structuring methods will be discussed in detail.

14.2.1 Chemical Etching

Si surface texturing by chemical etching in alkaline solutions is a state-of-the-art process applied industrially. Wide applications of this process are due to simplicity and cost-effectiveness. Random micropyramid textures are produced by chemical etching in the solution of KOH and isopropyl alcohol (IPA). The pyramidal textures show typically around 90% of light absorbance in visual and near-infrared (IR) spectral ranges (Otto et al. 2015). The enhancement of light absorbance is due to the geometric light-trapping effect.

An example of pyramid texture is shown in Figure 14.2, which was obtained in KOH-IPA solution with the potassium silicate additive (Basu et al. 2013). Monocrystalline wafers have been textured using a standard KOH-IPA industrial process with the potassium silicate (K_2SiO_3) additive. The surface exhibited the solar-weighted average reflectance from 11.4% to 12.2%, while non-textured surfaces typically reflect around 60% of the irradiation in the range of 300–1180 nm (Otto et al. 2015). The textures could also be produced by etching in tetramethylammonium hydroxide (TMAH) and isopropanol solutions (Savin et al. 2015; Khandelwal et al. 2013).

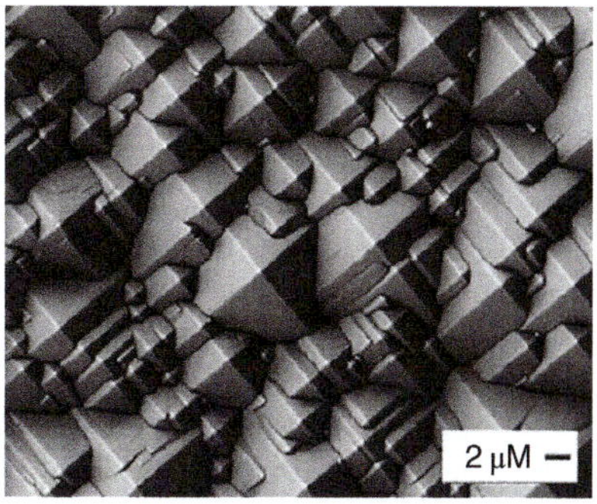

Figure 14.2 SEM image of the monocrystalline Si wafer textured in KOH and isopropyl alcohol solutions with a potassium silicate additive. Source: Reproduced from (Basu et al. 2013) with permission from the publisher. © Elsevier, 2013.

The Si pyramidal microstructures formed in an alkaline solution are effectively applied as a substrate for subsequent finer structuring. A superstructure was fabricated by Otto et al. combining KOH and shallow RIE etchings (Otto et al. 2015). The combination showed an improved absorption in the UV–vis region when compared to the individual etchings alone. The superstructure outperformed the light trapping of the surfaces structured by fs-laser treatment, Ag- and Au-assisted chemical etching (ACE) methods, or electrochemical etching. Another example of application of KOH-treated silicon substrate relates to the production of ultra-dark absorber, whose formation principle is outlined in Figure 14.3 (Agnihotri et al. 2018). Perfect light absorbers can be applied in various ways, such as energy harvesting and antireflection coatings for space and defense applications or stealth technology also known as low observable technology. Nearly zero reflection was achieved (99.4% absorption) in the range of 300–2000 nm. Carbon nanotubes

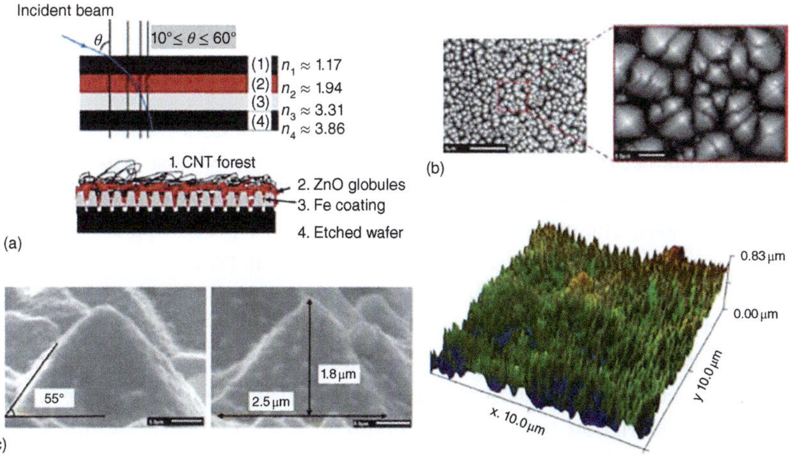

Figure 14.3 (a) Schematic representation of the assembly of an ultra-dark absorber composed of KOH-etched Si, Fe coating, and Zn globules decorated with carbon nanotubes (CNTs) on the top. Refractive indexes (*n*) are indicated in the direction of the incident beam. (b) SEM and atomic force microscopy (AFM) images show the Si surface structure after etching in KOH and (c) shows the side view of the pyramidal structure. Source: Reproduced from (Agnihotri et al. 2018), which is an open-access article. © The authors, 2018.

(CNTs), whose refractive index is $n = \sim 1.2$, were applied on the top, and subsequent, layers were applied in an increasing sequence so that zero reflection condition can be achieved.

14.2.2 Laser Engineering

Important silicon surface texturing is attained by the femtosecond (fs) or nanosecond (ns) laser engineering. During the fs-laser pulse, the thermal equilibrium in silicon is not established and its structure remains thermally cold. At the same time, some electrons are exited from the valence band to the conduction band, which disorganizes the lattice structure and leads to surface roughening or blacking. The silicon ablation rate is determined mainly by laser power.

The fs-laser treatment can be performed in various environments: vacuum, inert gas, air, corrosive gas SF_6, water, or oil. Important works were obtained from Mazur's group (Her et al. 1998, 2000). They reported an array of conical spikes produced by 500 laser pulses of 100 fs duration in SF_6 or Cl_2 gas. The spikes were up to 40 μm tall tapered to ~1 μm diameter at the tip. The authors showed that the Si surface could be structured by laser irradiation in N_2, Ne, or vacuum as well. In SF_6, not only texture with microspikes could be produced but also the surface was simultaneously doped with sulfur atoms, which increased the light absorption efficiency (Crouch et al. 2004; Lin et al. 2015). In air, mechanically robust microgrooves were produced with superior antireflection features in visual and mid-infrared wavelength regions (Vorobyev and Guo 2011). In vacuum, rather rounded spike networks – "penguin-like" structures – were fabricated, which were so named because of their similarity with colonies of penguins when they stand next to each other (Halbwax et al. 2008) (Figure 14.4). Such structures increased the photocurrents by more than 30%. Laser treatment in water using 400 nm and 100 fs pulses led to the formation of spikes 1–2 orders of magnitude smaller than those formed in gases or vacuum (Shen et al. 2004).

Silicon structural features also depend on the duration of laser pulses. The structural components became five times larger when irradiation pulse was increased from 100 fs to 30 ns, although the optical properties of both samples were similar (Crouch et al. 2004). The state of the art of ultrafast laser processing for texturing and hyperdoping of silicon was overviewed by Franta et al. (2018).

It is worth mentioning that laser texturing, when compared to chemical etchings, is a rather slow and costly process. This limits its application in mass production technologies. Also, laser process leads to material damage and requires defect removal when seeking high-quality materials. An advantage is that this physical method has no restriction to crystal orientations, while chemical etching could be sensitive to crystallographic features. Nevertheless, chemical MACE process was effectively used for multi- and microcrystalline as well as amorphous silicon (Koynov et al. 2007, 2011; Han et al. 2014). The laser treatment is in particular attractive in solar cell processing because of the possibility to perform doping and structuring in one single step, so that there is no need for pre-deposited layer of a doping source (Turner et al. 1981; Otto et al. 2015).

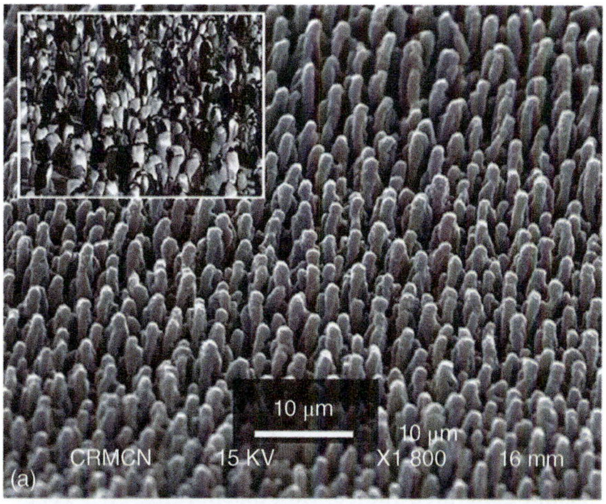

Figure 14.4 (a) SEM image of the fs laser-induced structures on silicon known as the penguin-like structure. The inset shows the real picture of penguin colony in Antarctica. Source: Reproduced from (Halbwax et al. 2008) with permission from the publisher © Elsevier, 2008. (b) King penguins early spring South Georgia photo by Darrell Gulin. The license purchased from gettyimages®.

14.2.3 Reactive Ion Etching

The reactive ion etching (RIE) is a dry technique, which typically employs SF_6 and O_2 gases. When in plasma, these gases provide F* and O* radicals. Silicon is etched by the fluorine radical forming volatile SiF_x. Fluorine also interacts with O* and forms a passivation layer of silicon oxyfluoride (SiO_xF_y), which protects the surface from further sputtering. Etching anisotropy is achieved because of different bombardment conditions by plasma ions: on horizontal plane, the passivation layer is bombarded much more intensively than that on a vertical plane. As a result, silicon needles are formed with an outstanding aspect ratio. Temperature rise favors desorption of the passive layer; therefore, etching is performed at temperatures below −100 °C and sometimes is referred to as cryogenic RIE. The RIE is more suitable for the formation of the micrometer-sized structures when high aspect ratios are not required. Note that although the RIE systems are commercially available, high cost owing to vacuum equipment limits their application for mass production. Also, the process of optimization could be difficult. Liu et al. comprehensively reviewed specifics and application of the RIE method (Liu et al. 2014).

14.2.4 Plasma Immersion Ion Implantation Etching

PIII etching is a dry method, which is slightly similar to the RIE. The approach is based on the generation of F* and O* radicals in plasma forming volatile SiF_4 gas and surface passivation by $Si_xO_yF_z$. Thereafter, high-energy ions accelerated by dc bias collide with the passive $Si_xO_yF_z$ opening to the fluorine radicals' active Si sites in order to react with them. As a result, a needle-like structure with an average reflectance of 1.79% in the range of 300–1000 nm has been attained (Xia et al. 2011). X-ray photoelectron spectroscopy (XPS) showed that the surface structures obtained by PIII were composed of silicon, carbon, oxygen, and fluorine (Xia et al. 2012). The surface changed from porous to needle-like depending on the ratio of SF_6 and O_2 gas. Si doping with nitrogen along with the nanotexturing has been proposed by the PIII method to form the p–n junction (Lim et al. 2018). Subsequent nanotexturing and surface defect passivation by hydrogen has been achieved through the plasma-generated radical flux discharging $Ar + H_2$ gas mixture. Thus, this method appears to be suitable for injecting impurity elements into Si nano–microstructures and to promote optical absorption, thereby expanding it in a wider range of wavelength.

14.2.5 Stain Etching

The stain etching method is based on the HF–HNO_3 action: HNO_3 oxidizes Si and HF removes the formed oxide. The oxidation rate and Si surface passivity may vary depending on the concentration ratio $[HF]/[HNO_3]$. Accordingly, different Si surface properties may be achieved by changing the concentration ratio: polished, textured, or porous. Silicon surfaces with light absorbance over 90% (Liu et al. 2014) as well as luminescent nanostructures were obtained by stain etching (Cullis et al. 1997). The stain etching, however, is rather slow, irreproducible, and applicable for very thin layers. An advantage is the metal-catalyzed etching. Dimova-Malinovska et al. demonstrated a rapid aluminum-assisted silicon etching in a $HF + HNO_3$ solution (Dimova-Malinovska et al. 1997). Soon thereafter, Li and Bohn showed the formation of porous silicon with luminescence properties in the $HF + H_2O_2$ mixture using ultrathin Au, Pt, or Au/Pd layers ($d < 10$ nm) as catalysts (Li and Bohn 2000).

14.2.6 Metal-Assisted Chemical Etching

Si nanostructures of high-aspect ratio can be obtained by the metal-assisted chemical etching (MACE). Highly vertical and deep structures can be formed by MACE. This wet electroless etching process is based on the Si interaction with HF in the presence of a strong oxidizer (e.g. hydrogen peroxide, H_2O_2) and a noble metal (e.g. Au and Ag) catalyst. The overall MACE reaction is

$$Si + H_2O_2 + 6HF \rightarrow 2H_2O + H_2SiF_6 + H_2 \qquad (14.2)$$

The basic principle of the MACE process is illustrated in Figure 14.5. The process is electrochemical in nature. Electrons are released from Si at the Me–Si boundary

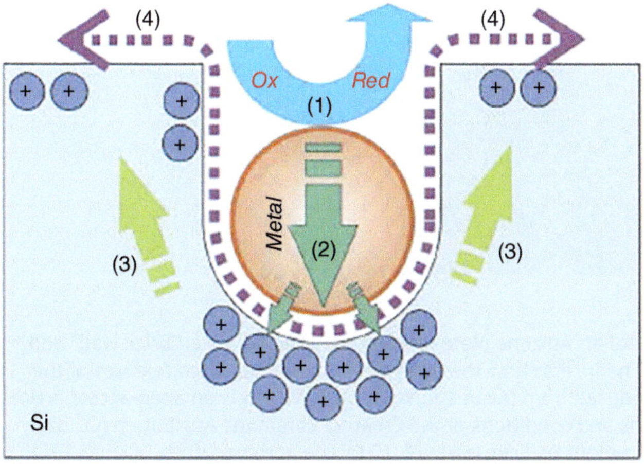

Figure 14.5 Major steps of the MACE process: the reduction of an oxidative agent on a metal particle Ox → Red (e.g. $H_2O_2 + 2H^+ \rightarrow 2H_2O$) generating the holes with the highest concentration underneath the particle (1)–(2); the migration of holes to silicon surface (3); and the removal of oxidized silicon using hydrofluoric acid (4). Source: Reproduced from (Huang et al. 2011) with permission from the publisher © WILEY-VCH, 2011.

and accepted by H_2O_2 via a metal catalyst at the Me–H_2O_2 interface. The oxidized silicon (Si^{ox}) is selectively etched by HF at the Me–Si^{ox} boundary and, as a result, the metal nanoparticle "sinks" into the silicon forming a channel – the pore. A two-layer structure was formed by Au-catalyzed etching, which consisted of a cone-like substructure and a nanoporous coverage (Algasinger et al. 2014). The pore formation mechanism depended on the Si doping level: for low-substrate resistivity (~0.1 Ω cm), a transition from the two-layer to nanoporous structure was observed. Deep analysis of the MACE process and its applications could be found in reviews written by Huang et al. (2011) and Han et al. (2014).

Metal nanoparticles on Si for the MACE process are formed by several methods. Branz and coworkers fabricated silicon nanostructures in $HAuCl_4$, HF, and H_2O_2 solution. Au nanoparticles are formed in this solution on the silicon surface, which catalyzes etching of a network of nanopores (Branz et al. 2009). The surface reflected less than 2% of the light in the wavelength range from 300 to 1000 nm. Metal nanoparticles could also be formed by depositing metal thin films using thermal evaporation or sputtering and subsequent annealing in an inert atmosphere. Noor et al. investigated the effects of annealing temperature to surface morphology and optical properties of b-Si fabricated by silver-ACE (Noor et al. 2019). The authors removed the native oxide from Si surface and deposited a 15 nm Ag film using radio frequency (RF) sputtering. The sample was then annealed at 200–260 °C in a nitrogen atmosphere and blackened by the MACE process. Annealing at 230 °C resulted in lowest reflection, which was 3% at the wavelength of 600 nm.

The noble metal catalyst can be distributed over the Si surface by means of lithography. The catalyst patterns can also be formed using the zone plate approach. The zone plate is a device used to focus the light diffractively. Such focusing optics

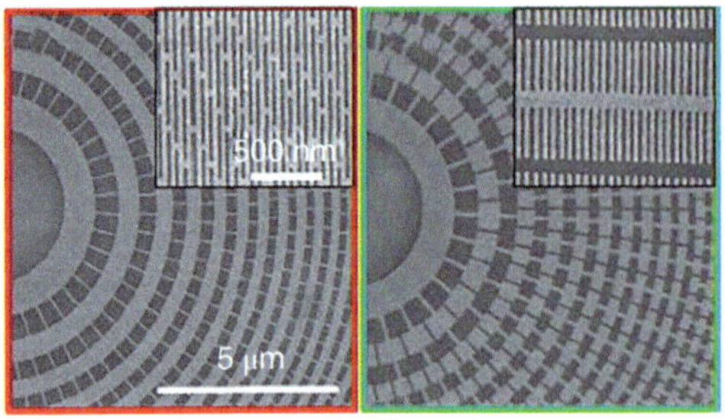

Figure 14.6 SEM images of an Au zone plate catalyst patterns on Si: (a) "brick wall" and (b) "fishbone" structures. The insets show the outermost nanometer-sized features of the zone plates. Source: Reproduced from (Akan and Vogt 2021), which is an open-access article distributed under the terms and conditions of the Creative Commons Attribution (CC BY) license (https://creativecommons.or/licenses/by/4.0/) © The authors, 2021.

is used in X-ray microscopes. Zone plates have circular grating structures, which decrease in width with the zone plate radii. The structures of the X-ray zone plates range from micron-sized features in the center to nanometer-sized features in the outer parts and, therefore, they are well suited as model patterns to MACE investigations (Akan and Vogt 2021). The scanning electron microscopy (SEM) micrographs in Figure 14.6 demonstrate the catalyst patterns of the Au zone plate.

Akan and Vogt pointed out two major challenges for the MACE nanoetching: (i) maintaining a vertical etching and (ii) keeping a good mechanical stability for deep etching with extreme aspect ratios. A variety of MACE parameters can influence the Si etching performance, such as etching temperature, Si doping, and catalyst arrangements. Figure 14.7 demonstrates the etched structures obtained at the concentration molar ratio $[H_2O_2]/[HF] = 0.68/4.7$, temperatures 8–25 °C, and when adding isopropanol as an agent, which lowers the surface tension and facilitates the transport of generated gas. The fastest etching rate (0.45 µm min^{-1}) and deepest etching (~2.4 µm) were attained at room temperature with the higher H_2O_2 concentration. However, some zones, particularly, the outermost areas of both the "brick wall" and the "fishbone" design, showed some deformations (Figure 14.7a,e). The authors came to a conclusion that the "brick wall" structures were mechanically stable up to an aspect ratio of 30 : 1; deeper etching led to a collapse of the structures.

A vast majority of metal-assisted etching has been performed applying silver or gold as catalysts, which contribute significantly to the processing cost – a key factor for industrial applications. Non-precious catalysts are attractive in terms of cost and wide availability. Thus, copper and nickel catalysts have recently attracted considerable attention (Kuo et al. 2020; Takaloo et al. 2018; Volovlikova et al. 2019; Kamarauskas et al. 2020).

Major difficulty in Cu application as a replacement for noble metals lies in Cu dissolution (corrosion) during the Si processing. Kuo et al. solved this problem by splitting the procedure into two steps: (i) electroless Cu deposition and (ii) etching in H_2O_2–HF–H_2O solution (Cu-ACE) (Kuo et al. 2020). The p-type

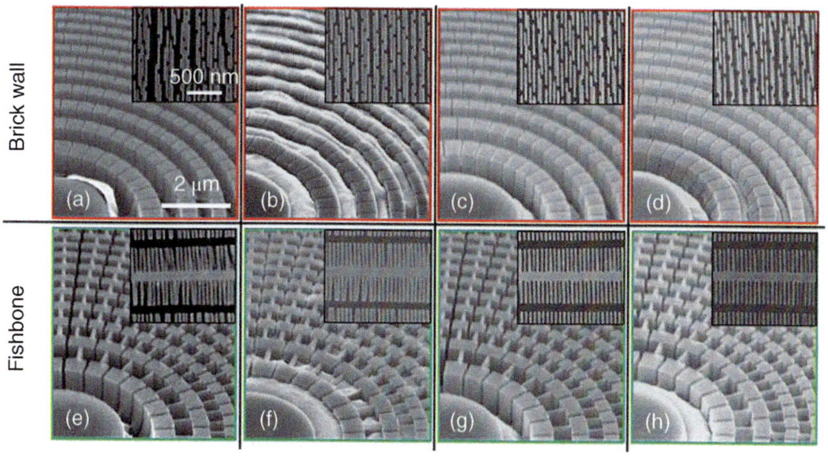

Figure 14.7 The "brick wall" and "fishbone" structures etched by MACE using the zone plate designs as shown in Figure 14.6. The etching time was six minutes. The molar concentration ratio [H_2O_2]/[HF] was 0.68/4.7 for (a), (c), (d), (e), (g), and (h). The ratio was 0.068/4.7 for (b) and (f). The solution was supplemented with 0.91 M IPA for (d) and (h). The temperature was 25 °C in most cases. The temperature was lowered to 8 °C for the samples (c) and (g). Isopropanol was added to the etching solution for the (d) and (h) samples. Source: Reproduced from (Akan and Vogt 2021), which is an open-access article distributed under the terms and conditions of the Creative Commons Attribution (CC BY) license (https://creativecommons.or/licenses/by/4.0/) © The authors, 2021.

silicon was structured by the Cu-ACE approach; the produced surfaces exhibited photoluminescence. Ultraviolet irradiation in conjunctions with Cu-ACE led to the development of a blackened surface. Thus, the authors obtained the knowledge, which is important for future development of cost-effective silicon structuring technologies.

Ni-ACE is favorable because the equilibrium potential of this metal is positive when compared to that of Si in the MACE process. Therefore, Ni acts during the etching as a cathode and Si as an anode in the Si–Ni couple (Volovlikova et al. 2019). Takaloo et al. proposed a Ni-assisted (Ni-ACE) single-step technique to produce a nanoporous antireflective Si surface (Takaloo et al. 2018). The surface structuring has been formed in conventional etching solution HF + H_2O_2 with an additive of Ni^{2+} ions ($NiSO_4$) at a slightly elevated temperature (40 °C). Both horizontal and vertical nanopores were formed, and the surfaces reflected about 2% of incident light. Two-step Ni-ACE has also been demonstrated. Gao et al. deposited the first Ni nanoparticles on the Si wafer by chemical solution deposition. Thereafter, the Ni-modified Si surface was etched in conventional etching solution HF–H_2O_2–H_2O at room temperature. An anisotropic etching has been achieved perpendicular to the surface with the light reflection as low as 2.31%. It is worth noting that there is no residual Ni within the Si pores, the metal completely corrodes in the etching solution (Volovlikova et al. 2019; Gao et al. 2018). Treideris et al. and Kamarauskas et al. applied a two-step Ni-ACE process using the Si wafer coated with a nanometer-thin Ni film as a precursor (Treideris et al. 2017; Kamarauskas et al. 2020). They demonstrated that a large area of Si wafer could be structured homogeneously attaining the reflectivity below 2% in the visible light range. They also showed that the increase in the thickness of the catalyst layer did not suppress the etching process; instead, the film thickness enables tuning of the roughness of the Si surface.

In summary, one could conclude that the MACE processes have good application prospects for mass production. In particular, one-step process using non-precious catalysts is attractive when formation of the catalyst and silicon etching are attained simultaneously. However, the MACE processes have to be further developed in order to improve the controllability of the surface morphology. In this respect, although more costly, the laser engineering seems to be a superior technique. The surface architectures and porosity are well tuned by selecting proper irradiation parameters. Note also that the Au-ACE method finds it difficult to remove inert gold residual particles from the produced pores. The Ag-ACE process faces a drawback because of silver diffusion into silicon, which changes surface doping and compromises its electric properties.

14.2.7 Vapor–Liquid–Solid Method

The chemical etchings discussed above represent the so-called "top-down" fabrication methods, which currently dominate the semiconductor production industry. An opposite "bottom-up" structuring occurs in a VLS approach. This method employs silicon growth on the silicon substrate at high temperatures from $SiCl_4$ gas. Si microwire arrays for high-efficiency photovoltaic cells were grown by VLS (Kayes et al. 2007). The entire VLS procedure is rather complex, and it includes the following main steps: (i) thermal growth of SiO_2 buffer layer on Si to separate the microislands of the metal catalyst; (ii) photolithographical patterning; (iii) removal of the oxide within the patterned resist holes by HF etching; (iv) evaporation of the catalyst, e.g. Au and Cu submicron layer; (v) lift-off of the resist; (vi) sample annealing at 900–1000 °C in H_2; and (vii) Si wire growth in $H_2 + SiCl_4$ mixture at 850–1100 °C. BCl_3 (0.25% in H_2) was added to the grown doped p-type Si wire with a resistivity of $0.05 \pm 0.01\,\Omega\,cm$, corresponding to a carrier concentration of $7 \times 10^{17}\,cm^{-3}$ (Boettcher et al. 2010). This method produced nearly defect-free arrays of Si microwires with the possibility to control the size, position, and uniformity of the arrays. Also, nanowire growth using the VLS has attracted the attention of researchers (Pinion et al. 2016). It has been shown that this method allows nanofabrication of complex sub-10 nm structures and, thus, it can potentially substitute expensive lithography.

14.2.8 Nanostructuring Based on Porous Alumina Template

Nanocolumnar Si structures could be obtained by the application of porous alumina (aluminum oxide, Al_2O_3) as a template for surface pattering. Regularly distributed pores develop at the Al_2O_3–electrolyte interface during aluminum anodizing. The surface structure consists of densely packed hexagonal columnar cells with cylindrical pores at the center, as schematically outlined in Figure 14.8.

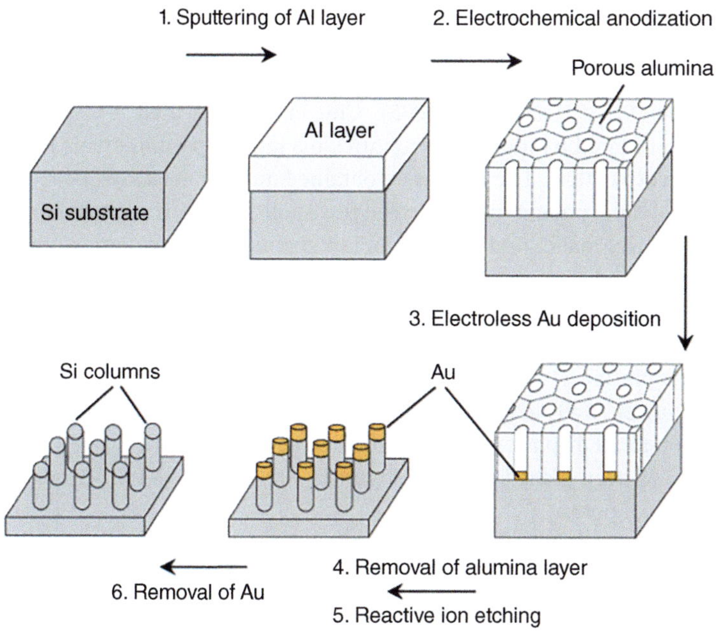

Figure 14.8 Schematic representation of the formation of nanocolumnar Si structures with and without gold decoration using Al sputtering on the Si substrate, electrochemical anodizing, electroless Au inversion within the pores, removal alumina, RIE, and removal of gold by dissolution. Porous anodic alumina consists of densely packed honeycomb cells with cylindrical pores. Reprinted from (Leinartas et al. 2011) with permission from the publisher. © Springer, 2011.

The pore size as well as the inter-pore distance could vary from a few to hundreds of nanometers depending on anodizing parameters such as applied voltage or electrolyte composition and temperature.

Studies of anodic oxide films on aluminum have a rather long history. A review about aluminum anodizing published by J. W. Diggle, T. C. Downie, and C. W. Goulding covers a period from 1930 to 1968 (Diggle et al. 1969). A great variety of electrolytes were reviewed, such as those based on boric acid, ammonium borate or tartrate, ammonium tetraborate in ethylene glycol, as well as numerous organic and inorganic acids: citric, malic, glycolic, sulfuric, phosphoric, chromic, and oxalic.

Self-assembly of the pores within anodic alumina has been widely studied. Keller, Hunter, and Robinson reported in 1953 the investigation of anodic alumina with an electron microscope (Keller et al. 1953). They observed close-packed cells hexagonal in shape with single pores in the middle. The oxides have been formed in sulfuric, oxalic, chromic, and phosphorous acid electrolytes. The pore size depends on the electrolyte used, and the wall thickness was predominantly a function of the applied voltage. Hoar and Mott discussed in 1959 the mechanism of porous anodic alumina formation, showing that diffusion of oxygen and aluminum through a compact part of the film plays an essential role (Hoar and Mott 1959). Masuda and coworkers produced ordered metal nanohole arrays using honeycomb structures of anodic alumina (Masuda and Fukuda 1995). Jessensky et al. showed that ordered hexagonal pore arrays in anodic alumina are formed by a self-organized process, which starts from randomly distributed porosity (Jessensky et al. 1998). Ono et al. produced self-ordering porous alumina films in malonic and tartaric acids applying high voltages up to 195 V (Ono et al. 2005). Chu et al. fabricated transparent alumina films with ordered nanoporous structures having the pore intervals from 130 to 980 nm (Chu et al. 2006). The films were obtained under voltages of 70–450 V in various inorganic and organic acid electrolytes, such as sulfuric, phosphoric, oxalic, glycolic, tartaric, malic, and citric. Garcia-Vergara and coworkers studied pore generation in anodic alumina using transmission and scanning electron microscopes, Rutherford backscattering spectroscopy, and nuclear reaction analysis (Garcia-Vergara et al. 2006). The authors disputed the widely accepted mechanism of field-assisted dissolution of alumina. They explained the pore formation by the high plasticity of barrier Al_2O_3 in the presence of ionic transport, which leads to the film stresses and material displacement from the barrier layer toward the cell wall region. Emphasis has been placed on the role of the stress and relief processes in the self-organization of pores.

Figure 14.9 shows an example of fabrication of gold-decorated Si nanocolumns obtained using an alumina template (Leinartas et al. 2011). Note that Si–Au nanoarchitectures are of great interest as platforms to prepare biochemical sensors. Advantages of Si nanocolumnar structures include their biocompatibility, high surface-to-bulk ratio, and oxygen- or hydrogen-terminated surfaces, which are pertinent to chemical functionalization. The structures decorated by Au particles are attractive for the detection of biomolecules because of high affinity of thiol groups that is binding them via sulfur to gold (Wanunu et al. 2005; Hostetler et al. 1999; Yang et al. 2006). For instance, the gold nanoparticle embedded in Si has been proposed for the detection of nucleic acids – DNA and RNA (Ryu et al. 2010; Martins et al. 2007).

The alumina template preparation involves two major steps: (i) deposition of an aluminum thin layer (100–500 nm) on the silicon substrate by a PVD technique, e.g. by magnetron sputtering on p-type <100> wafer, and (ii) formation of porous self-ordered alumina by anodizing the aluminum film, e.g. applying galvanostatic polarization in an oxalic acid electrolyte, which is followed by treatment in orthophosphoric acid to remove the rest of the aluminum layer at the bottom of the pores. As a result, a nanoporous alumina surface is obtained (Figure 14.9).

The formation of nanocolumns involves the following route: (i) electroless inversion of Au in the alumina pores, e.g. using the $AuCl_3 + NH_4F$ electrolyte adjusted to pH 1.5–2.0 by HF; (ii) removal of the residual alumina layer, e.g. in $K_2Cr_2O_7$ solution; and (iii) RIE with preferential sputtering of unmasked silicon, e.g. using anisotropic etching in CF_4 gas. As a result, a gold-capped nanocolumnar structure is formed (Figure 14.10).

The above discussion shows a great variety of Si structuring opportunities employing physical and chemical methods. Each of them has inherent advantages and disadvantages. In summary, it could be emphasized that most of the overviewed techniques are costly and technically sophisticated, which limits their wider application in mass production where the cost efficiency is of overriding

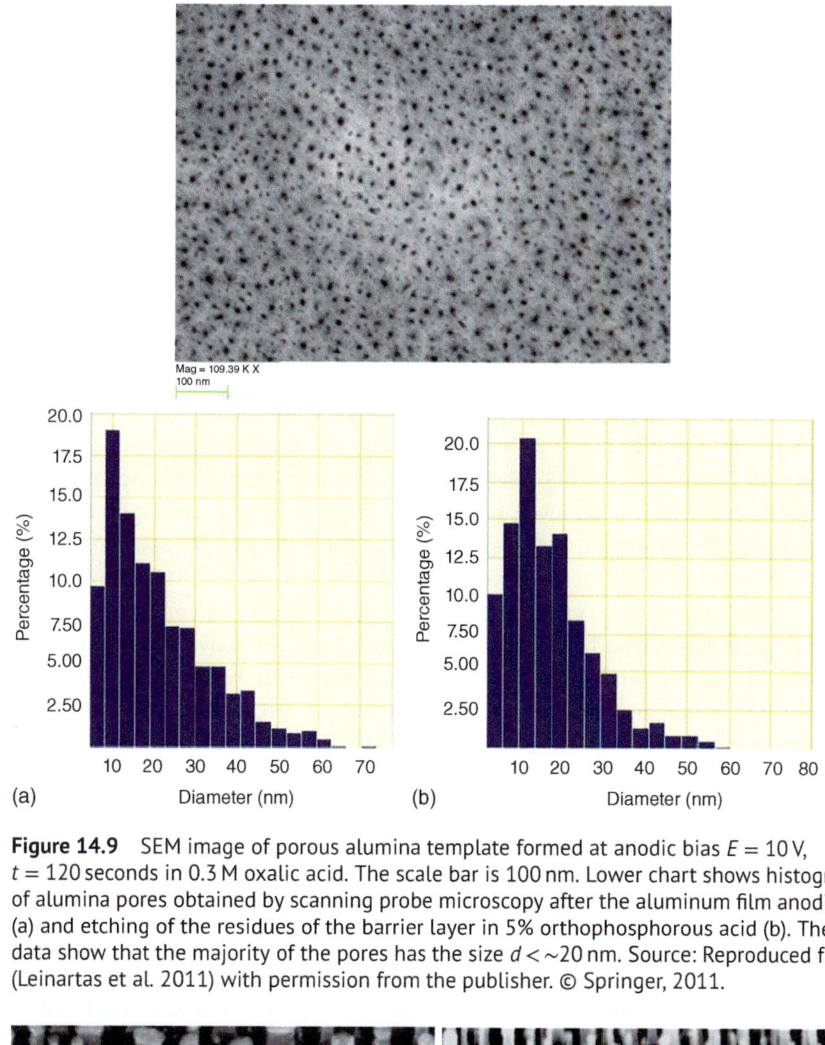

Figure 14.9 SEM image of porous alumina template formed at anodic bias $E = 10\,V$, $t = 120$ seconds in 0.3 M oxalic acid. The scale bar is 100 nm. Lower chart shows histograms of alumina pores obtained by scanning probe microscopy after the aluminum film anodizing (a) and etching of the residues of the barrier layer in 5% orthophosphorous acid (b). The data show that the majority of the pores has the size $d < \sim 20$ nm. Source: Reproduced from (Leinartas et al. 2011) with permission from the publisher. © Springer, 2011.

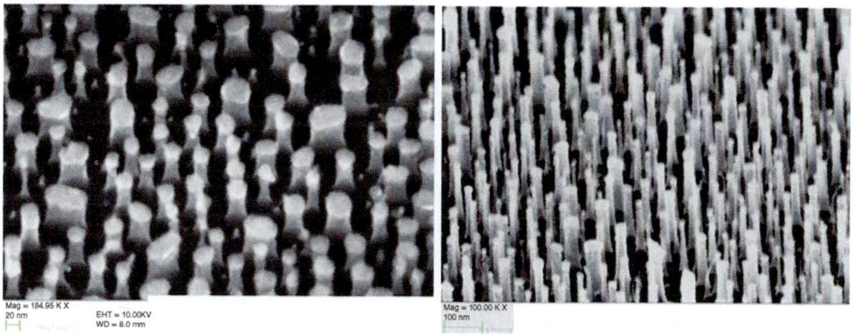

Figure 14.10 SEM images of gold-capped Si nanocolumns obtained by the synthesis route shown in Figure 14.8. The scale bar is 20 nm (a) and 100 nm (b). Source: Reproduced from (Leinartas et al. 2011) with permission from the publisher. © Springer, 2011.

importance. The industrial texturing performed by chemical etching, such as that in alkaline solutions, seems to be most attractive in terms of cost and technical simplicity. Some other etching technologies, however, involve toxic and aggressive chemicals, such as HF and in certain cases expensive catalysts, such as Au, Ag, and Pt. Economic and environmental advantages are expected from electrochemical technologies of Si surface structuring in molten salts – an emerging field, which will be discussed below.

14.3 Black Silicon

Black silicon (b-Si) is a modification of structured silicon, whose light absorbance is close to 100% in a wide range of wavelengths. The term "black" has been coined because of the visually black appearance of the material, while a conventional silicon plate is silver-gray. b-Si shows quasi-omnidirectional incidence of light. The blackening effect is achieved producing various nano–microshapes, such as wires (Garnett and Yang 2010; Kelzenberg et al. 2010; Oh et al. 2012a), cones (Zhu et al. 2008; Savin et al. 2015), pores, and globules (Oh et al. 2012b; Juzeliūnas 2012a,b). The material is of great interest in the application of silicon-based devices with superior light absorbance. In particular, b-Si is attractive to produce the photovoltaic architectures, which could operate efficiently without antireflection coatings. The other important applications include hydrogen production by photoelectrochemical water splitting (Oh et al. 2011; Oh et al. 2012a; Ao et al. 2012), physical bactericides (Ivanova et al. 2013), tunable electrowetting on dielectric (Barberoglou et al. 2010), optoelectronics (Torres-Costa and Martin-Palma 2010; Steglich et al. 2013), and platforms for laser-induced micro–nanopolymerization (Žukauskas et al. 2013). The material is also of great interest for production of sensing devices, as it has been demonstrated using the platforms of porous silicon for chemical-, gas-, and biosensors (Gesemann et al. 2011; Barillaro et al. 2003; Angelescu et al. 2003; Jane et al. 2009; Kilian et al. 2009). Black silicon application in photovoltaics was properly reviewed by Otto et al. (2015), Liu et al. (2014), and Fan et al. (2021).

In a solar cell, b-Si increases the short-circuit current and the open-circuit voltage. This is due to superior harvesting of the light and efficient absorption of photons, in particular those of the near band edge. The light trapping properties make it possible for the reduction of the cell thickness below 100 μm. Losses due to non-radiative bulk recombination are in linear proportion to the solar cell thickness; thus, reduction of solar cell thickness leads to higher efficiency of the cell. Also, reduction in cell thickness has an economic effect – less of expensive feedstock material is consumed. Other beneficial features of b-Si application include avoiding of the costs related to additional texturing and production of antireflection coatings, such as SiN_x.

An essential point in b-Si application is Si surface passivation. Fine nanostructures enhance the photocarrier recombination and reduce the efficiency of the device. Therefore, proper electronic passivation is essential to obtain efficient solar cells based on b-Si. The surface recombination velocity (S_{eff}) could be used as an indicator of the passivation quality (Savin et al. 2015). This parameter characterizes the velocity of charge carriers, at which they reach the surface to recombine (Richter et al. 2012):

$$S_{eff} = \frac{W}{2}\left(\frac{1}{\tau_{eff}} - \frac{1}{\tau_{inn}}\right) \tag{14.3}$$

where W stands for the Si wafer thickness, τ_{eff} – effective carrier lifetime, and τ_{inn} – intrinsic carrier lifetime (according to a general parameterization for intrinsic recombination developed by Richter et al. (2012)).

The increasing efficiencies of structured silicon-based solar cells have recently been reported. Fukui, Inomata, and Shirasawa reported in 1997 the cell with multicrystalline silicon textured by RIE and passivation schemes based on a silicon nitride film deposited by the plasma CVD and hydrogen annealing (Fukui et al. 1997; Inomata et al. 1997). The cell with the area as large as 225 cm^2 has been produced with a conversion efficiency of 17.1%.

High efficiency demands suppressing of surface recombination processes by proper surface passivation. SiO_2, SiN, and Al_2O_3 have been used for this purpose. In particular, passivation by a thin aluminum oxide layer formed by ALD exhibited low surface recombination. Even an ultrathin Al_2O_3 layer of 3.6 nm thickness effectively passivated crystalline silicon (Schmidt et al. 2009).

Oh et al. produced the b-Si-based cell, which has been nanostructured by the MACE process using Ag as a catalyst (Oh et al. 2012). The authors applied damage removal etching (DRE) in an alkaline solution and passivated the surface by thermal SiO_2 formation. As a result, the conversion efficiency has been achieved as high as 18.2%. It is worth mentioning that defects or damages produced during the nanotexturing play the role of recombination centers, which reduces the carriers' lifetime. The DRE procedures can help to increase the charge carrier lifetime improving the blue response of RIE- or MACE-structured surfaces (Otto et al. 2015; Oh et al. 2012). Khandelwal and coworkers improved the effective lifetime of charge carriers from 24.4 to 1092 μs when systematically removing the plasma-induced damages from the Si surface (Khandelwal et al. 2013). Repo et al. reported 18.7% efficient cells (Repo et al. 2013). They produced the cell from n-type Si with a diffused boron emitter and a RIE nanostructure. A cryogenic deep RIE was used to produce b-Si. The alumina (Al_2O_3) layer deposited by ALD has been applied to passivate the surface. The authors demonstrated that the ALD alumina layer effectively reduces the reflectivity and passivates the b-Si surface so that the b-Si-Al_2O_3 surfaces reflected less than 1% of light. Shortly thereafter, Ortega et al. reported 22% and the same group – 22.1% efficiencies (Ortega et al. 2015; Savin et al. 2015).

An important result obtained by Savin et al. is that the b-Si solar cell increased the energy production by 3% when applying better angular acceptance (Savin et al. 2015). It means that the efficiency of the cell can be maintained for a wide range of the angle of incident light. This property of b-Si surface is highly beneficial, as the angle of solar irradiation changes during the day and the year with respect to the cell surface. It is worth noting that in the case of conventional antireflection coatings, the efficiency is optimized for the perpendicular light incidence to the cell surface.

The major fabrication methods of b-Si and its optoelectronic properties have comprehensively been compared by Otto and coworkers (Otto et al. 2015). The authors studied four mostly used blackening methods: (i) inductively coupled plasma reactive ion etching (ICP-RIE) in a gas mixture of SF_6 and O_2; (ii) metal-assisted wet chemical etching (MACE) based on Au or Ag catalysts in HF and H_2O_2 solutions; (iii) macroporous Si (macP-Si) fabrication by electrochemical etching in the HF electrolyte; and (iv) femtosecond (fs) laser treatment (LSi) with or without DRE in solutions. The authors used optical absorption and minority carrier lifetime as benchmark parameters. In terms of surface passivation, the atomic layer-deposited alumina film (ALD Al_2O_3) has been selected as a benchmark. The alumina layer offered a combination of two effects: (i) a high number of negative charges that lead to field passivation and (ii) good chemical quality with low defect densities. Figure 14.11 summarizes the optical absorbance of differently prepared surface structures. The absorbance was integrated within the wide irradiation

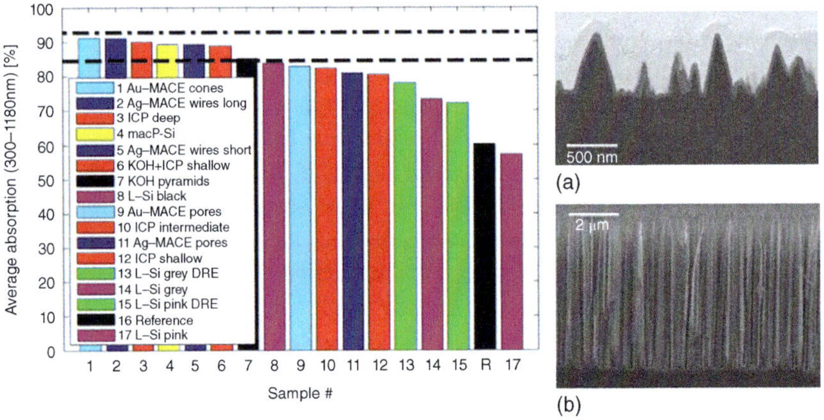

Figure 14.11 Optical absorption of b-Si prepared by different means and passivated with 35 nm Al_2O_3. The absorption is integrated within the wavelength range of 300 to 1180 nm. Sample specification: 1 – nano-cones obtained by MACE using the Au catalyst, SEM image (a) with the scale bar 500 nm; 2 – nanowire structure composed of the wires of ~6 μm length and diameters ~50–150 nm obtained by MACE and the Ag catalyst, image (b) with the scale bar 2 μm; 3 – ICP pores with the depth on the order of several micrometers; 4 – macroporous Si with micrometer-sized pores formed by electrochemical anodization in HF solution; 5 – MACE nanowire structure with peak-to-valley height 2.55 μm; 6 – KOH-treated Si with random pyramids and ICP treated as in 3 producing peak-to-valley height 0.59 μm; 7 – KOH-treated Si with random pyramids; 8 – L-Si obtained applying 500 pulses per spot, which forms up to 15 μm spikes with an aspect ratio of ~3; 9 – MACE porous surface with peak-to-valley height ~ 0.53 μm obtained using the Au catalyst; 10 – ICP intermediate structure with peak-to-valley height ~ 0.5 μm; 11 – MACE porous surface with peak-to-valley height ~ 0.55 μm obtained using the Ag catalyst; 12 – ICP shallow structure with peak-to-valley height ~ 0.59 μm; 13 – L-Si gray surface obtained applying five pulses per spot and damage removal etching (DRE); 14 – L-Si gray without DRE; 15 – L-Si pink surface obtained applying a single pulse per spot and DRE; 16 – a planar sample used as a reference one; and 17 – L-Si pink surface without DRE. Source: Reproduced from (Otto et al. 2015) with permission from the publisher © WILEY-VCH, 2014.

spectra covering the range from the middle wave UV region (300 nm) to the near-infrared (1180 nm) one.

The best absorption performance demonstrated the nano–cone structure obtained by MACE with a characteristic peak-to-valley height of ~0.65 μm. A very similar performance was demonstrated by the nanowire structure obtained by MACE using the Ag catalyst. The array was composed of nanowires ~6 μm in length and ~ 50–150 nm in diameter. Both samples showed the total integrated optical absorption over 90%. In the UV–vis range, these samples showed almost complete light absorption. For instance, sample 1 (Figure 14.11 at λ = 500 nm, that is, green light) exhibited an absorption of 99.6%. Note that absorbance of the plain unstructured surface was around 60% and that of conventionally used pyramidic structure etched in KOH was around 85%. Thus, the applied Si surface structuring suppresses substantially the light reflection.

As shown above, the light absorption capability of structured surface is dependent on surface nano–micro–architectures. The principles of wave optics are considered to characterize the antireflection and light-trapping properties of the structural elements, which are smaller than the wavelength of incident light. To this end, the correlation length (L_c) is determined, which means approximately the lateral distance between adjacent peaks and valleys (Otto et al. 2015). The light trapping performance is linked to L_c; this could be determined from a cross section (e.g. Figure 14.11) when applying statistical data analysis that is Gaussian function and standard deviation. The parameter has been determined from 30 nm for nanoporous surface (sample 11, Figure 14.11) to 360 nm for macroporous surface (sample 4, Figure 14.11). The authors concluded that surface architectures with a large L_c showed better light-trapping performance than those with low L_c. The antireflection performance was found to be best of the structure with an intermediate value L_c = 100 nm. The authors concluded that at larger values of this parameter, the structures suffer from increased backscattering. It is important that the light trapping has to be achieved without compromising the antireflection properties. The height of structural elements is of less importance for the optical properties: the structures in the range of 0.5–5 μm did not show significant influence to antireflection and scattering properties. However, the height has substantial influence on the surface recombination velocity that is to minority carrier lifetime, which affects the solar device efficiency. The height strongly depends on the degree of surface damages resultant from the surface structuring technique. Damage removal methods have to be applied to increase the efficiency of b-Si-based photodevices.

Another essential parameter of b-Si, which characterizes the maximum short circuit current, is the effective carrier lifetime (τ_{eff}). This parameter could be obtained, for instance, by contactless determination of photoconductance data as proposed by Sinton and Cuevas (1996). The method relies on contactless measurement of quasi-steady-state photoconductance (QSSPC) during a slow variation of light pulse. The method makes it possible to obtain voltammetric characteristics of silicon to be used in solar cells. The data obtained by Otto et al. (2015) showed that τ_{eff} reduces with the surface blackening. The out-of-box wafer exhibited more than 3 ms the lifetime, and the wafer textured in KOH shoved a time of around 1 ms.

The longest lifetime of dry etched samples (ICP) was in the range from 0.68 to 1.6 ms, depending on the etching characteristics. Much shorter lifetimes were determined for MACE-produced b-Si: 0.111 ms for the Au-catalyzed surface and 0.102 ms for the Ag-catalyzed one. Less than 0.01 ms exhibited the macroporous surface obtained by electrochemical anodizing. The laser-treated samples exhibited shortest lifetimes, which are around 1 μs. However, the time increased up to 5.6 μs (pink surface, sample 15, Figure 14.11) after the application of DRE. One could conclude that surface damage is a critical factor introduced by the structuring technique. Another key parameter is the surface enhancement factor, which characterizes the true surface area with respect to the projected one.

Laser processing provides an opportunity for b-Si doping along with the surface structuring. When the fs-pulses irradiate Si in a SF_6 atmosphere, a sulfur (S) doping of b-Si is attained, which leads to the surface with almost uniform absorption in a wide range from ultraviolet to near-infrared wavelengths (Crouch et al. 2004; Her et al. 2000). The S-doped surface attained by laser processing showed a white emission, which originated from intrinsic point defects in SiO_x (Fabbri et al. 2015). However, the S-doped surface is unstable to annealing, which is essential in the preparation of semiconductor devices. In this respect, phosphorous (P) doping via fs laser irradiation is attractive, as the surface retains high infrared absorption after annealing as well as the opportunity to form a semiconductor junction (Li et al. 2016). Thermostable IR absorption of gold-hyperdoped b-Si produced by fs-laser irradiation has been reported (Yu et al. 2017). The authors studied Au-supersaturated systems, which exceed by four orders of magnitude the solubility of Au in Si, so that gold silicide phase has been formed. Also, enhanced absorbance of b-Si in the short-wavelength infrared (SWIR) region was achieved using fs irradiation of crystalline Si coated by a 100 nm aluminum film (Su et al. 2018). Aluminum implantation led to up to 50% absorption enhancement in the SWIR.

An interesting modification of b-Si has been developed by Omar and coworkers, which is flexible black silicon (Omar et al. 2021). To this end, monocrystalline Si wafers were thinned down to 65 μm, dissolving them in a potassium hydroxide (KOH) solution. Then, the surfaces were modified with nanowires about 1.5 μm in length, applying the MACE process and the silver catalyst. The samples exhibited the short-circuit current densities by ~70% higher when compared to crystalline silicon. Flexible b-Si is promising to reduce the solar cell cost while maintaining superior light harvesting performance because of high light absorbance.

There are some technological limits for b-Si industrial applications. These include thermal management, mechanical strength of the nanostructures, and their structural uniformity over a large area. An important aspect is also photon loss after encapsulation of the surface structures in glass, as common in the production of photovoltaic cells. Also, most of the techniques referred to the above are costly, technically sophisticated, and involve toxic chemicals and expensive catalysts, which limits providing of competitive solutions for industrial-scale applications.

An environmentally friendly method of black silicon formation has been proposed, which is based on the Si–SiO_2 surface processing in molten calcium chloride at 900 °C (Fray and Juzeliūnas 2011; Juzeliūnas, 2012b; Liu et al. 2014).

Well-defined silicon substrates with thin SiO_2 layers thermally grown on p-type silicon were used as the precursors. Versatile surface architectures were obtained, and it is composed of micro–nanopores, spheres, and nanofiber structures. The samples exhibited specifically high surface-to-volume ratio, which showed great ability to absorb visible light. The processed surfaces reflected less than 1% of visual light. The proposed method offer attractive opportunities in silicon surface technologies and applications, for instance, in the preparation of electrodes for rechargeable lithium batteries, the antireflection coatings in solar cells or the platforms for biological and chemical sensors. Details on the electrochemical Si processing in a molten salt electrolyte will be provided and discussed in the subsequent chapter.

References

Agnihotri, P.K., Ghai, V., and Singh, H. (2018). Optically graded ultra dark absorber for visible and near-infrared wavelength range. *Sci. Rep.* 8: 12312. https://doi.org/10.1038/s41598-018-30844-5.

Akan, R. and Vogt, U. (2021). Optimization of metal-assisted chemical etching for deep silicon nanostructures. *Nanomaterials* 11: 2806. https://doi.org/10.3390/nano11112806.

Algasinger, M., Berndt, M., Koynov, S., and Stutzmann, M. (2014). Porous silicon formation during Au-catalyzed etching. *J. Appl. Phys.* 115: 164308. https://doi.org/10.1063/1.4873892.

Angelescu, A., Kleps, I., Mihaela, M. et al. (2003). Porous silicon matrix for applications in biology. *Rev. Adv. Mater. Sci.* 5 (5): 440–449.

Anglin, E.J., Cheng, L., Freeman, W.R., and Sailor, M.J. (2008). Porous silicon in drug delivery devices and materials. *Adv. Drug Delivery Rev.* 60 (11): 1266–1277. https://doi.org/10.1016/j.addr.2008.03.017.

Ao, X., Tong, X., Kim, D.S. et al. (2012). Black silicon with controllable macropore array for enhanced photoelectrochemical performance. *Appl. Phys. Lett.* 101 (11), Art. No.111901. https://doi.org/10.1063/1.4752231.

Baggetto, L., Danilov, D., and Notten, P.H.L. (2011). Honeycomb-structured silicon: remarkable morphological changes induced by electrochemical (de)lithiation. *Adv. Mater.* 23: 1563–1566. https://doi.org/10.1002/adma.201003665.

Barberoglou, M., Zorba, V., Pagozidis, A. et al. (2010). Electrowetting properties of micro/nanostructured black silicon. *Langmuir* 26 (15): 13007–13014. https://doi.org/10.1021/la101138u.

Barillaro, G., Nannini, A., and Pieri, F. (2003). APSFET: a new, porous silicon-based gas sensing device. *Sens. Actuators, B* 93 (1–3): 263–270. https://doi.org/10.1016/S0925-4005(03)00234-X.

Basu, P.K., Sarangi, D., Shetty, K.D., and Boreland, M.B. (2013). Liquid silicate additive for alkaline texturing of mono-Si wafers to improve process bath lifetime and reduce IPA consumption. *Sol. Energy Mater. Sol. Cells* 113: 37–43. https://doi.org/10.1016/j.solmat.2013.01.037.

Boettcher, S.W., Spurgeon, J.M., Putnam, M.C. et al. (2010). Energy-conversion properties of vapor-liquid-solid-growth silicon wire-array photocathodes. *Science* 327: 185–187. https://doi.org/10.1126/science.1180783.

Boettcher, S.W., Warren, E.L., and Putnam, M.C. (2011). Photoelectrochemical hydrogen evolution using Si microwire arrays. *J. Am. Chem. Soc.* 133: 1216–1219. https://doi.org/10.1021/ja108801m.

Bonanno, L.M. and DeLouise, L.A. (2007). Whole blood optical biosensor. *Biosens. Bioelectron.* 23: 444–448. https://doi.org/10.1016/j.bios.2007.05.008.

Branz, H.M., Yost, V.E., Ward, S. et al. (2009). Nanostructured black silicon and the optical reflectance of graded-density surfaces. *Appl. Phys. Lett.* 94: 231121. https://doi.org/10.1063/1.3152244.

Canham, L.T. (1990). Silicon quantum wire array fabrication by electrochemical and chemical dissolution of wafers. *Appl. Phys. Lett.* 57: 1046. https://doi.org/10.1063/1.103561.

Chu, S.Z., Wada, K., Inoue, S. et al. (2006). Large-scale fabrication of ordered nanoporous alumina films with arbitrary pore intervals by critical-potential anodization. *J. Electrochem. Soc.* 153: B384–B391. https://doi.org/10.1149/1.2218822.

Crouch, C.H., Carey, J.E., Warrender, J.M. et al. (2004). Comparison of structure and properties of femtosecond and nanosecond laser-structured silicon. *Appl. Phys. Lett.* 84 (11): 1850–1852. https://doi.org/10.1063/1.1667004.

Cullis, A.G., Canham, L.T., and Calcott, P.D.J. (1997). The structural and luminescence properties of porous silicon. *J. Appl. Phys.* 82 (3): 909–965. https://doi.org/10.1063/1.366536.

Diggle, J.W., Downie, T.C., and Goulding, C.W. (1969). Anodic oxide films on aluminum. *Chem. Rev.* 69: 365–405. https://doi.org/10.1021/cr60259a005.

Dimova-Malinovska, D., Sendova-Vassileva, M., Tzenov, N., and Kamenova, M. (1997). Preparation of thin porous silicon layers by stain etching. *Thin Solid Films* 297 (1,2): 9–12. https://doi.org/10.1016/S0040-6090(96)09434-5.

Fabbri, F., Lin, Y.-T., Bartoni, G. et al. (2015). Origin of the visible emission of black silicon microstructures. *Appl. Phys. Lett.* 107: 021907. https://doi.org/10.1063/1.4926912.

Fan, Z., Cui, D., Zhang, Z. et al. (2021). Recent progress of black silicon: from fabrications to applications. *Nanomaterials* 11: 41. https://doi.org/10.3390/nano11010041.

Franta, B., Mazur, E., and Sundaram, S.K. (2018). Ultrafast laser processing of silicon for photovoltaics. *Int. Mater. Rev.* 63 (4): 227–240. https://doi.org/10.1080/09506608.2017.1389547.

Fray, D.J. and Juzeliūnas, E. (2011). Method for texturing silicon surface to create black silicon for photovoltaic applications. Int. Patent WO 2011/161479 A1; International publication date: 29 December 2011. Appl. No: PCT/GB2011/051213, UK. Priority date 26 June 2010.

Fukui, K., Inomata, Y., and Shirasawa, K. (1997). Surface texturing using reactive ion etching for multicrystalline silicon solar cells. In: *Conference Record of the Twenty Sixth IEEE Photovoltaic Specialists Conference*, 47–50. Anaheim, CA: https://doi.org/10.1109/PVSC.1997.653921.

Gao, K., Shen, H., Liu, Y. et al. (2018). Fabrication of black silicon by Ni assisted chemical etching. *Mater. Res. Express* 5: 015020. https://doi.org/10.1088/2053-1591/aaa1fb.

Garcia-Vergara, S.J., Iglesias-Rubianes, L., Bianco-Pinzon, C.E. et al. (2006). Mechanical instability and pore generation in anodic alumina. *Proc. R. Soc. A.* 462: 2345–2358. https://doi.org/10.1098/rspa.2006.1686.

Garnett, E.C. and Yang, P. (2010). Light trapping in silicon nanowire solar cells. *Nano Lett.* 10: 1082–1087. https://doi.org/10.1021/nl100161z.

Ge, M., Fang, X., Rong, J., and Zhou, C. (2013). Review of porous silicon preparation and its application for lithium-ion battery anodes. 24 (42): 422001. https://doi.org/10.1088/0957-4484/24/42/422001.

Gesemann, B., Wehrspohn, R., Hackner, A., and Müller, G. (2011). Large-scale fabrication of ordered silicon nanotip arrays used for gas ionization in ion mobility spectrometers. *IEEE Trans. Nanotechnol.* 10 (1): 50–52, Art. No. 5484434. https://doi.org/10.1109/TNANO.2010.2053046.

Grigoras, K., Keskinen, J., Grönberg, L. et al. (2016). Porous silicon for energy storage at microscale: supercapacitors. *ECS Trans.* 75 (1): 97–104. https://doi.org/10.1149/07501.0097ecst.

Grigoras, K., Ahopelto, J., and Prunnila, M. (2018). Porous silicon supercapacitors. In: *Handbook of Porous Silicon: Second Edition*, vol. 2–2, 1517–1529. https://doi.org/10.1007/978-3-319-71381-6_116.

Halbwax, M., Samet, T., Delaporte, P. et al. (2008). Micro and nano-structuration of silicon by femtosecond laser: application to silicon photovoltaic cells fabrication. *Thin Solid Films* 516: 6791–6795. https://doi.org/10.1016/j.tsf.2007.12.117.

Han, H., Huang, Z., and Lee, W. (2014). Metal-assisted chemical etching of silicon and nanotechnology applications. *Nano Today* 9: 271–304. https://doi.org/10.1016/j.nantod.2014.04.013.

Her, T.H., Finlay, R.J., Wu, C. et al. (1998). Microstructuring of silicon with femtosecond laser pulses. *Appl. Phys. Lett.* 73: 1673. https://doi.org/10.1063/1.122241.

Her, T.H., Finlay, R.J., Wu, C., and Mazur, E. (2000). Femtosecond laser-induced formation of spikes on silicon. *Appl. Phys. A* 70: 383–385. https://doi.org/10.1007/s003390051052.

Hoar, T.P. and Mott, N.F. (1959). A mechanism for the formation of porous anodic oxide films on aluminium. *J. Phys. Chem. Solids* 9: 97–99. https://doi.org/10.1016/0022-3697(59)90199-4.

Hostetler, M.J., Templeton, A.C., and Murray, R.W. (1999). Dynamics of place-exchange reactions on monolayer-protected gold cluster molecules. *Langmuir* 15: 3782–3789. https://doi.org/10.1021/la981598f.

Hoyer, P., Matthäus, G., Blumröder, U. et al. (2011). Induced terahertz emission as a probe for semiconductor devices. *Appl. Phys. Lett.* 22, Art. No. 221112. https://doi.org/10.1063/1.3664769.

Huang, Z., Geyer, N., Werner, P. et al. (2011). Metal-assisted chemical etching of silicon. A review. *Adv. Mater.* 23: 285–308. https://doi.org/10.1002/adma.201001784.

Inomata, Y., Fukui, K., and Shirasawa, K. (1997). Surface texturing of large area multicrystalline silicon solar cells using reactive ion etching method. *Sol. Energy Mater. Sol. Cells* 48: 237–242. https://doi.org/10.1016/S0927-0248(97)00106-2.

Ivanova, E., Hasan, J., Webb, H.K. et al. (2013). Bactericidal activity of black silicon. *Nat. Commun.* 4, Art. No. 2838. https://doi.org/10.1038/ncomms3838.

Jane, A., Dronov, R., Hodges, A., and Voelcker, N.H. (2009). Porous silicon biosensors on the advance. *Trends Biotechnol.* 27 (4): 230–239. https://doi.org/10.1016/j.tibtech.2008.12.004.

Jessensky, O., Müller, F., and Gösele, U. (1998). Self-organized formation of hexagonal pore structures in anodic alumina. *J. Electrochem. Soc.* 145: 3735–3740. https://doi.org/10.1149/1.1838867.

Juzeliūnas, E., Cox, A., Abdelkader, A., and Fray, D.J. (2012a). Electrochemical formation of porous silicon layers in molten salt. In: *Fray International Symposium - On Metals and Materials Processing in Clean Environment*, Materials Recycling, Silicon Photovoltaic Cells, Boron & Borates, vol. 4 (ed. F. Kongoli), 491–496. Flogen Star OUTREACH. ISBN: 978-0-9879974-1-8.

Juzeliūnas, E., Cox, A., and Fray, D.J. (2012b). Electro-deoxidation of thin silica layer in molten salt – Globular structures with effective light absorbance. *Electrochim. Acta* 68: 123–127. https://doi.org/10.1016/j.electacta.2012.02.054.

Kamarauskas, M., Treideris, M., Agafonov, V. et al. (2020). Black silicon quality control by conditions of nickel-assisted etching of crystalline silicon surfaces in photovoltaic devices. *Lith. J. Phys.* 60: 57–66. https://doi.org/10.3952/physics.v60i1.4164.

Kasavajjula, U., Wang, C., and Appleby, A.J. (2007). Nano- and bulk-silicon-based insertion anodes for lithium-ion secondary cells. *J. Power Sources* 163: 1003–1039. https://doi.org/10.1016/j.jpowsour.2006.09.084.

Kayes, B.M., Filler, M.A., Putnam, M.C. et al. (2007). Growth of vertically aligned Si wire arrays over large areas (>1 cm^2) with Au and Cu catalysts. *Appl. Phys. Lett.* 91: 103110. https://doi.org/10.1063/1.2779236.

Keller, F., Hunter, M.S., and Robinson, D.L. (1953). Structural features of oxide coatings on aluminum. *J. Electrochem. Soc.* 100: 411–419. https://doi.org/10.1149/1.2781142.

Kelzenberg, M.D., Boettcher, S.W., Petykiewicz, J.A. et al. (2010). Enhanced absorption and carrier collection in Si wire arrays for photovoltaic applications. *Nat. Matter.* 9: 239–244. https://doi.org/10.1038/nmat2727.

Khandelwal, R., Plachetka, U., Min, B. et al. (2013). A comparative study based on optical and electrical performance of micro- and nano-textured surfaces for silicon solar cells. *Microelectron. Eng.* 111: 220–223. https://doi.org/10.1016/j.mee.2013.03.162.

Kilian, K.A., Böcking, T., and Gooding, J.J. (2009). The importance of surface chemistry in mesoporous materials: lessons from porous silicon biosensors. *Chem. Commun.* 6: 630–640. https://doi.org/10.1039/b815449j.

Koynov, S., Brandt, M.S., and Stutzmann, M. (2007). Black multi-crystalline silicon solar cells. *Phys. Status Solidi RRL* 1 (2): R53–R55. https://doi.org/10.1002/pssr.200600064.

Koynov, S., Brandt, M.S., and Stutzmann, M. (2011). Black thin film silicon. *J. Appl. Phys.* 110: 043537. https://doi.org/10.1063/1.3626900.

Kuo, K.-H., Ku, W.-H., and Lee, B.T.-H. (2020). Photoluminescent or blackened silicon surfaces synthesized with copper-assisted chemical etching: for energy applications. *ECS J. Solid State Sci. Technol.* 9: 024006. https://doi.org/10.1149/2162-8777/ab682f.

Lehmann, V. (1993). The physics of macropore formation in low doped n-type silicon. *J. Electrochem. Soc.* 140: 2836–2843. https://doi.org/10.1149/1.2220919.

Lehmann, V. (1999). The physics of macropore formation in low-doped p-type silicon. *J. Electrochem. Soc.* 146: 2968–2975. https://doi.org/10.1149/1.1392037.

Lehmann, V. and Gösele, U. (1991). Porous silicon formation: a quantum wire effect. *Appl. Phys. Lett.* 58: 856–858. https://doi.org/10.1063/1.104512.

Leinartas, K., Miečinskas, P., Selskis, A. et al. (2011). Formation of gold-capped silicon nanocolumns on silicon substrate. *J. Solid State Electrochem.* 15: 2419–2425. https://doi.org/10.1007/s10008-011-1442-4.

Li, X. and Bohn, P.W. (2000). Metal-assisted chemical etching in HF/H_2O_2 produces porous silicon. *Appl. Phys. Lett.* 77 (16): 2572–2574. https://doi.org/10.1063/1.1319191.

Li, X., Yan, C., Wang, J. et al. (2015). Stable silicon anodes for lithium-ion batteries using mesoporous metallurgical silicon. *Adv. Energy Mater.* 5 (4): 401556. https://doi.org/10.1002/aenm.201401556.

Li, C.H., Zhao, J.H., Yu, X.Y. et al. (2016). Fabrication of black silicon with thermostable infrared absorption by femtosecond laser. *IEEE Photon. J.* 8: 6805809. https://doi.org/10.1109/JPHOT.2016.2617403.

Lim, J.W.M., Huang, S., Xu, L. et al. (2018). Ultra-low reflective black silicon photovoltaics by high density inductively coupled plasmas. *Sol. Energy* 171: 841–850. https://doi.org/10.1016/j.solener.2018.07.032.

Lin, Y.-T., Mangan, N., Marbach, S. et al. (2015). Creating femtosecond-laser-hyperdoped silicon with a homogeneous doping profile. *Appl. Phys. Lett.* 106: 062105. https://doi.org/10.1063/1.4907988.

Liu, X., Coxon, P.R., Peters, M. et al. (2014). Black silicon: fabrication methods, properties and solar energy applications. *Energy Environ. Sci.* 7: 3223–3263. https://doi.org/10.1039/C4EE01152.

Martins, R., Baptista, P., Raniero, L. et al. (2007). Amorphous/nanocrystalline silicon biosensor for the specific identification of unamplified nucleic acid sequences using gold nanoparticle probes. *Appl. Phys. Lett.* 90: 023903. https://doi.org/10.1063/1.2431449.

Masuda, H. and Fukuda, K. (1995). Ordered metal nanohole arrays made by a two-step replication of honeycomb structures of anodic alumina. *Science* 268: 1466–1468. https://doi.org/10.1126/science.268.5216.1466.

Noor, N.A., Mohamad, S.K., Hamil, S.S. et al. (2019). Effects of annealing temperature towards surface morphological and optical properties of black silicon fabricated by silver-assisted chemical etching. *Mater. Sci. Semicond. Process.* 91: 167–173. https://doi.org/10.1016/j.mssp.2018.11.006.

Oakes, L., Westover, A., Mares, J.W. et al. (2013). Surface engineered porous silicon for stable, high performance electrochemical supercapacitors. *Sci. Rep.* 3, Art. No. 3020. https://doi.org/10.1038/srep03020.

Oh, J., Deutsch, T.G., Yuan, H.-C., and Branz, H.M. (2011). Nanoporous black silicon photocathode for H_2 production by photoelectrochemical water splitting. *Energy Environ. Sci.* 4 (5): 1690–1694. https://doi.org/10.1039/C1EE01124C.

Oh, I., Kye, J., and Hwang, S. (2012a). Enhanced photoelectrochemical hydrogen production from silicon nanowire array photocathode. *Nano Lett.* 12 (1): 298–302. https://doi.org/10.1021/nl203564s.

Oh, J., Yuan, H.-C., and Branz, H.M. (2012b). An 18.2%-efficient black-silicon solar cell achieved through control of carrier recombination in nanostructures. *Nat. Nanotechnol.* 7: 743–748. https://doi.org/10.1038/nnano.2012.166.

Omar, H.D., Hashim, M.S., and Pakhuruddin, M.Z. (2021). Surface morphological and optical properties of flexible black silicon fabricated by metal-assisted chemical etching. *Opt. Laser Technol.* 136: 106765. https://doi.org/10.1016/j.optlastec.2020.106765.

Ono, S., Saito, M., and Asoh, H. (2005). Self-ordering of anodic porous alumina formed in organic acid electrolytes. *Electrochim. Acta* 51: 827–833. https://doi.org/10.1016/j.electacta.2005.05.058.

Ortega, P., Calle, E., Gastrow, G. et al. (2015). High-efficiency black silicon interdigitated back contacted solar cells on p-type and n-type c-Si substrates. *Prog. Photovoltaics Res. Appl.* 23: 1448–1457. https://doi.org/10.1002/pip.2664.

Otto, M., Kroll, M., Kasebier, T. et al. (2010). Conformal transparent conducting oxides on black silicon. *Adv. Mater.* 22: 5035–5038. https://doi.org/10.1002/adma.201002515.

Otto, M., Algasinger, M., Branz, H. et al. (2015). Black silicon photovoltaics. *Adv. Optical Mater.* 3 (2): 147–164. https://doi.org/10.1002/adom.201400395.

Park, J.-H., Gu, L., Maltzahn, G. et al. (2009). Biodegradable luminescent porous silicon nanoparticles for *in vivo* applications. *Nat. Mater.* 8: 331–336. https://doi.org/10.1038/nmat2398.

Pinion, C.W., Christensen, J.D., and Cahoon, J.F. (2016). Understanding the vapor-liquid-solid mechanism of Si nanowire growth and doping to synthetically encode precise nanoscale morphology. *J. Mater. Chem. C* 4: 3890–3897. https://doi.org/10.1039/C5TC03898G.

Repo, P., Haarahiltunen, A., Sainiemi, L. et al. (2013). Effective passivation of black silicon surfaces by atomic layer deposition. *IEEE J. Photovoltaics* 3: 90. https://doi.org/10.1109/JPHOTOV.2012.2210031.

Richter, A., Glunz, S.W., Werner, F. et al. (2012). Improved quantitative description of Auger recombination in crystalline silicon. *Phys. Rev. B* 86: 165202. https://doi.org/10.1103/PhysRevB.86.165202.

Rouquerol, J., Everett, D.H., Haynes, J.M. et al. (1994). IUPAC, Recommendations for the characterization of porous solids. *Pure Appl. Chem.* 55: 1739–1758. https://doi.org/10.1351/pac199466081739.

Ryu, S.W., Kim, C.H., Han, J.W. et al. (2010). Gold nanoparticle embedded silicon nanowire biosensor for application of label-free DNA detection. *Biosens. Bioelectron.* 25: 2182–2185. https://doi.org/10.1016/j.bios.2010.02.010.

Salonen, J., Kaukonen, A.M., Hirvonen, J., and Lehto, V.-P. (2008). Mesoporous silicon in drug delivery applications. *J. Pharm. Sci.* 97 (2): 632–653. https://doi.org/10.1002/jps.20999.

Savin, H., Repo, P., Gastrow, G. et al. (2015). Black silicon solar cells with interdigitated back-contacts achieve 22.1% efficiency. *Nat. Nanoechnol.* 10: 624–629. https://doi.org/10.1038/nnano.2015.89.

Schmidt, J., Veith, B., and Brendel, R. (2009). Effective surface passivation of crystalline silicon using ultrathin Al_2O_3 films and $Al_2O_3/SiN_{\underline{x}}$ stacks. *Phys. Stat. Sol.-Rap. Res. Lett.* 3: 287. https://doi.org/10.1002/pssr.200903272.

Shen, M.Y., Crouch, C.H., Carey, J.E., and Mazur, E. (2004). Femtosecond laser-induced formation of submicrometer spikes on silicon in water. *Appl. Phys. Lett.* 23: 5694–5696. https://doi.org/10.1063/1.1828575.

Sinton, R.A. and Cuevas, A. (1996). Contactless determination of current–voltage characteristics and minority-carrier lifetimes in semiconductors from quasi-steady-state photoconductance data. *Appl. Phys. Lett.* 69: 2510. https://doi.org/10.1063/1.117723.

Steglich, M., Zilk, M., Schrempel, F. et al. (2013). Improvement of Ge-on-Si photodiodes by black silicon light trapping. *Appl. Phys. Lett.* 102 (11), Art. No. 111110. https://doi.org/10.1063/1.4798322.

Su, Y., Zhan, X., Zang, H. et al. (2018). Direct and stand-off fabrication of black silicon with enhanced absorbance in the short-wavelength infrared region using femtosecond laser treatment. *Appl. Phys. B* 124, No 223. https://doi.org/10.1007/s00340-018-7092-9.

Takaloo, A.V., Kolahdouz, M., Poursafar, J. et al. (2018). Fabrication of low reflective nanopore-type black Si layer using one-step Ni-assisted chemical etching for Si solar cell application. *Mater. Res. Express* 5: 035905. https://doi.org/10.1088/2053-1591/aab2ee.

Torres-Costa, V. and Martin-Palma, R.J. (2010). Application of nanostructured porous silicon in the field of optics. A review. *J. Mater. Sci.* 45 (11): 2823–2838. https://doi.org/10.1007/s10853-010-4251-8.

Treideris, M., Reza, A., Kamarauskas, M. et al. (2017). Fashioning "black" silicon by nickel-film assisted chemical etching. In: *Proceedings of the 33rd European Photovoltaic Solar Energy Conference and Exhibition*, 687–689. https://doi.org/10.4229/EUPVSEC20172017-2AV.2.36.

Turner, G.B., Tarrant, D., Pollock, C. et al. (1981). Solar cells made by laser-induced diffusion directly from phosphine gas. *Appl. Phys. Lett.* 39: 967. https://doi.org/10.1063/1.92628.

Uhlir, A. (1956). Electrolytic shaping of germanium and silicon. *Bell System Tech. J.* 35: 333–347. https://doi.org/10.1002/j.1538-7305.1956.tb02385.x.

Volovlikova, O.V., Gavrikov, S.A., Lazarenko, P.I. et al. (2019). Influence of etching regimes on the reflectance of black silicon films formed by Ni-assisted chemical etching. *Key Eng. Mater.* 806: 24–29. https://doi.org/10.4028/www.scientific.net/KEM.806.24.

Vorobyev, A.Y. and Guo, C. (2011). Direct creation of black silicon using femtosecond laser pulses. *Appl. Surf. Sci.* 257: 7291–7294. https://doi.org/10.1016/j.apsusc.2011.03.106.

Wanunu, M., Popovitz-Biro, R., Cohen, H. et al. (2005). Coordination-based gold nanoparticle layers. *J. Am. Chem. Soc.* 127: 9207–9215. https://doi.org/10.1021/ja050016v.

Xia, Y., Liu, B., Liu, J. et al. (2011). A novel method to produce black silicon for solar cells. *Sol. Energy* 85: 1574–1578. https://doi.org/10.1016/j.solener.2011.03.012.

Xia, Y., Liu, B., Zhong, S., and Li, C. (2012). X-ray photoelectron spectroscopic studies of black silicon for solar cell. *J. Electron Spectros. Relat. Phenomena* 184: 589–592. https://doi.org/10.1016/j.elspec.2011.10.004.

Yang, K., Wang, H., Zou, K., and Zhang, X. (2006). Gold nanoparticle modified silicon nanowires as biosensors. *Nanotechnol.* 17: S276–S279. https://doi.org/10.1088/0957-4484/17/11/S08.

Yu, X.Y., Zhao, J.H., Li, C.H. et al. (2017). Gold-hyperdoped black silicon with high IR absorption by femtosecond laser irradiation. *IEEE Trans. Nanotechnol.* 16: 502–506. https://doi.org/10.1109/TNANO.2017.2693691.

Zhang, X. (2007). Porous silicon: morphology and formation mechanism. *J. Electrochem. Soc.* 151: 65–133. https://doi.org/10.1007/978-0-387-31701-4_2.

Zhang, X.G., Collins, S.D., and Smith, R.L. (1989). Porous silicon formation and electrochemical polishing of silicon by anodic polarization in HF solutions. *J. Electrochem. Soc.* 136: 1561. https://doi.org/10.1149/1.2096961.

Zhu, J., Yu, Z., Burkhard, G.F. et al. (2008). Optical absorption enhancement in amorphous silicon nanowire and nanocone arrays. *Nano Lett.* 9: 279–282. https://doi.org/10.1021/nl802886y.

Žukauskas, A., Malinauskas, M., Kadys, A. et al. (2013). Black silicon: substrate for laser 3D micro/nano-polymerization. *Opt. Express* 21 (6): 6901–6909. https://doi.org/10.1364/OE.21.006901.

15

Electrochemical Si Surface Structuring and Formation of Black Silicon in High-Temperature Molten Salts

Electrochemical Si structuring has been known for decades since the invention of electrochemical formation of porous silicon at Bell laboratories in 1956 (Uhlir 1956). This method is based on silicon electrochemical etching in aqueous acidic electrolytes containing hydrofluoric acid (HF). The foremost information on chemical etching of Si in $HF + HNO_3$ solutions was published by Robbins and Schwartz (1959), Archer (1960), and Memming and Schwandt (1966). The principles of laser engineering Birnbaum has proposed in 1965 (Birnbaum 1965), and the principles of the reactive ion etching (RIE) was published in 1979 by Schwartz and Schaible (1979). Compared to these early works, the electrochemical structuring of Si in molten salts is at its early stage of development. Nohira et al. (2003) studied pinpoint and bulk electrochemical reduction of insulating silicon dioxide to silicon, and their first paper on electrochemically formed Si structures was published in 2003. The study on Si surface electrochemical texturing in molten $CaCl_2$ appeared in 2010 (Juzeliūnas et al. 2010). Below, we will discuss the electrochemical processing in high-temperature molten salts, which produces versatile surface nano/micro-architectures on silicon surface.

15.1 Anodic and Cathodic Processing in Molten $CaCl_2$

Distinctive macroporosity can be obtained when applying anodic polarization of crystalline Si in molten $CaCl_2$. The process requires the temperatures around 850 °C. Figure 15.1 shows the scanning electron microscopy (SEM) images of typical macroporous surface structures on a micrometer scale obtained when p-Si was electrolyzed at 1.0 V vs. graphite in molten $CaCl_2$. The specific feature of the applied approach was that only silicon contacted the electrolyte; that is, no contribution to the electrochemical process was imposed from the contacting electrode (Mo). The latter played the role of the current collector. Most of the experiments reported in the literature on silicon electrochemistry in molten salts usually apply the heterogeneous electrodes, e.g. Mo, Ni, or W in contact with the $Si–SiO_2$ interface (Juzeliūnas and Fray 2020).

Silicon: Electrochemistry, Production, Purification and Applications, First Edition. Eimutis Juzeliūnas.

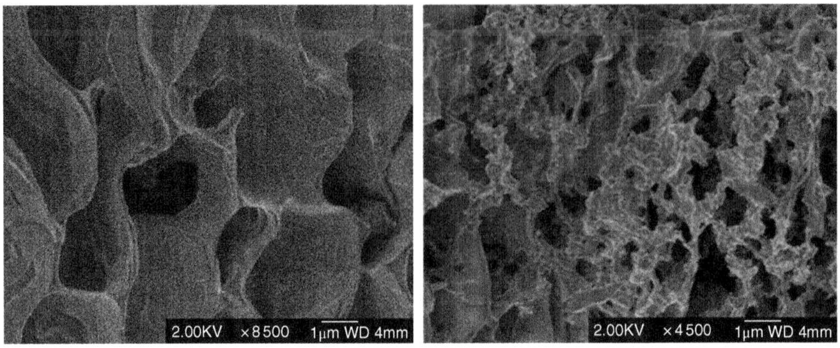

(a) (b)

Figure 15.1 SEM images of the silicon electrolyzed at 1.0 V vs. graphite for five minutes in molten $CaCl_2$ at 850 °C. Scale bars are 1 μm. Crystalline p-Si <100> orientation was used. The diameter of the wafer was 2″, geometric area – 22.8 cm^2, thickness – 275 ± 25 μm, and resistivity 1–30 Ω cm^{-1}. The wafers were thermally coated with a 10 nm oxide film after removal of the native oxide. The specimens were cut from the wafer and attached to a molybdenum (Mo) wire (0.5 mm) frame. The precursor was deoxygenated in the electrolyte at −1.0 V vs. graphite and then polarized anodically. Before anodic polarization, the sample was partially withdrawn from the electrolyte so that Si solely contacted the melt with no contribution from the Mo current collector. Source: E. Juzeliūnas.

Macroporous structures are also attained in molten $CaCl_2$ using cathodic polarization. The structuring takes place because of surface etching by reduced calcium. The melting point of calcium is $T_m = 839$ °C. The operating temperature in molten $CaCl_2$ electrolyte is typically about $T = 850$ °C. Therefore, in such electrolyte, calcium metal could be reduced in a liquid phase forming liquid silicides. Numerous silicides are known, such as $CaSi_2$, $Ca_{14}Si_{19}$, Ca_3Si_4, CaSi, Ca_5Si_3, and Ca_2Si (Ropp 2013). The melting points of these compounds are in the range of 700–935 °C. Figure 15.2a demonstrates the porous structure, which developed on the silicon substrate in the cathodic region of calcium deposition. The sample has been polarized down to $E = -0.9$ V vs. graphite for three hours. A rather thick macroporous layer (~10 μm) is obtained, which indicates in-depth silicon etching. The pore size is on the order of micrometers; thus, the pores can be ascribed to macropores. According to the International Union of Pure and Applied Chemistry (IUPAC) classification, the micropores are those below 2 nm in diameter, the mesopores are in the range of 2–50 nm, and the macropores are greater than 50 nm. Initial formation of porosity is shown in Figure 15.2b.

The applied precursor in Figure 15.2b was p-Si with thermally produced 2 μm SiO_2 layer. The oxide layer is electrochemically converted to silicon; as a result, a nanofiber structure forms on top of the Si sample. The cross section identifies the development of the pits, whose size is on the order of hundreds of nanometers. The void spaces with a frozen liquid at the bottom signify an etching process. X-ray diffraction (XRD) study identified Si, $CaSi_2$, and Ca_2Si within the surface structure. The obtained surfaces absorbed up to 95% of the incident light in a wide range of wavelengths from 450 to 1000 nm. The samples also exhibited the photoresponsiveness, which was distinctively different from that typical of p-Si wafer (Juzeliūnas et al. 2016).

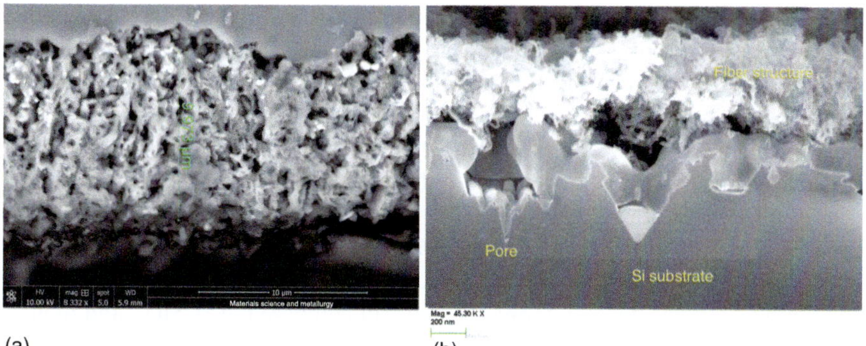

(a) (b)

Figure 15.2 (a) SEM image of the cross section of the macroporous layer formed on the silicon substrate (upper part of the image) using cathodic polarization in molten $CaCl_2$ at $E = -0.9$ V vs. graphite for three hours. Precursor – p-Si wafer with 100 nm thermal oxide and (b) cross section of the surface obtained at $E = -1.0$ V (one minute) of the precursor with a 2 μm silica layer. A nanofiber structure on the top covers the pores with frozen liquid inside, which identifies the etching process. Source: Reproduced with permission from Juzeliūnas et al. (2016), © ELSEVIER.

15.2 Microcolumnar and Amorphous Structures

A microcolumnar Si structure was the first observable result of the electrochemical silica deoxidation in molten $CaCl_2$. Nohira et al. (2003) used a quartz plate as a precursor and performed electro-deoxidation in molten $CaCl_2$. They concluded by direct observation that the silica reduction occurs at polarization of 1.1 V or below with respect to equilibrium potential of calcium electrode $E^0_{Ca/Ca^{2+}}$. The depth of the reduced zone on SiO_2 substrate was about 100–200 μm. It consisted of the columns of diameter 5–10 μm, and these were composed of 90 at.% of Si and 8 at.% of oxygen. The authors took a closer look at the structure in their subsequent works (Yasuda et al. 2005a,b,c). The authors reduced an amorphous quartz precursor to silicon at 850 °C and confirmed its crystallinity by XRD. A conclusion followed from these data that Si atoms rearrange during silica reduction. The reduced silicon appeared in a microcolumnar structure with a lot of vacant spaces between the columns. The vacant spaces have been formed because of great volume decrease when converting SiO_2 to Si: the molar volume of SiO_2 is 22.6 $cm^3\,mol^{-1}$ and that of Si is 12.1 $cm^3\,mol^{-1}$. The Si columns showed a hexagonal prismatic structure.

The mechanism of $SiO_2 \rightarrow Si$ conversion is rather complex. It includes such multiplex steps as charge transfer at a three-phase interface (3PI), oxygen diffusion within the surface structure (e.g. pores), Si crystallization, and 3D spreading over the Si–SiO_2 interface. An electric current conductor is necessary to attach to the SiO_2 surface in order to supply electrons and induce $SiO_2 \rightarrow Si$ conversion. To this end, a molybdenum wire is typically used, which also plays the role of current collector (Juzeliūnas and Fray 2020). Attractiveness of molybdenum as a contacting electrode is that this metal does hardly react with calcium, which can be electrochemically produced in the $CaCl_2$ electrolyte. Molten $CaCl_2$ has been widely used as an electrolyte, whose economic importance includes lower cost and wide availability. Also, oxygen ions (O^{2-}) and calcium metal are well soluble in this molten salt.

Si reduction is initiated at the site where three phases meet each other: the contacting electrode (Mo), the Si precursor (SiO_2), and the electrolyte (molten salt). Such sites are referred to as the three phase interlines (3PI). Once sufficiently negative voltage is exerted, silicon is reduced ($Si^{4+} + 4e^- \rightarrow Si$) and oxygen ions ($O^{2-}$) are released at 3PI. The reduced Si is well conductive at high temperature. Specific conductance of Si at 850 °C was evaluated as high as $\sim 50\,\Omega^{-1}\,cm^{-1}$ (Burton and Madjid 1969). Because of conductivity, the reduced Si plays the role of the contacting electrode, thereby forming a new 3PI composed of Si, SiO_2, and molten $CaCl_2$. As a result, silicon layer spreads over the SiO_2 surface. At the same time, in-depth penetration of the reaction into SiO_2 phase occurs.

An important step of the $SiO_2 \rightarrow Si$ conversion mechanism is Si crystallization. Development of a crystalline Si has been observed from an amorphous silica precursor in molten $CaCl_2$ (Yasuda et al. 2005c; Juzeliūnas et al. 2010). Breaking of the O—Si—O bonds and formation of Si dangling bonds occur when Si^{4+} is reduced to Si^0. At the first instances, Si atoms are arranged randomly, that is in an amorphous structure. At higher temperatures, a transition from amorphous Si to thermodynamically stable crystalline phase occurs. Crystal grain nucleation in amorphous silicon has comprehensively been studied by Spinella and Lombardo (1998). They applied Si annealing at high temperatures and studied morphological evolution from amorphous to crystalline phase. They demonstrated the importance of structural and thermodynamic properties on the starting phase of nucleation. It is assumed that amorphous Si is transformed into a crystalline structure at temperatures over 470 °C (Spinella and Lombardo 1998; Yasuda et al. 2005c; Olson and Roth 1988). Olson and Roth discussed that the solid-phase transitions in amorphous Si can occur at temperatures below the melting point (Olson and Roth 1988). They also showed that the crystallization rates, among other factors, could strongly be enhanced by certain impurities. It seems convincing that Si in molten $CaCl_2$ at 850 °C is reduced in an amorphous structure, which later on transforms into a crystalline structure. It is important that the reaction $Si^{4+} + 4e^- \rightarrow Si$ and Si crystallization are considered to be much faster than diffusion of the produced O^{2-}. The oxygen diffusion is retarded because of the effects of porosity or capillary. The diffusion is considered to be a rate-determining step (Yasuda et al. 2005c).

Temperature lowering below that of molten $CaCl_2$ led to less crystallinity, which is consistent with the above considerations. In LiCl–KCl–$CaCl_2$ eutectic at 500 °C, amorphous and a sponge-like nanocrystalline structure with a particle size less than 50 nm was formed (Yasuda et al. 2005c). The authors explained the structural features by dependence of crystallization rate on temperature. Amorphous Si is transformed into stable crystalline Si at $T \sim 500\,°C$ as that in LiCl–KCl–$CaCl_2$ eutectic. The silicon crystallization rate at such temperature is as low as $10^{-12}\,m\,s^{-1}$; therefore, only minor structural changes were observed. Most of the silicon reduced from amorphous SiO_2 remained amorphous, and some of the crystalline fraction has been produced with the crystal size less than 10 nm. In molten $CaCl_2$, the temperature is much higher ($T \sim 850\,°C$) and the Si crystallization rate is by 6 orders of magnitude also higher ($10^{-6}\,m\,s^{-1}$). The reduced silicon, therefore, forms a crystalline structure.

15.3 Electrodeoxidation of Thin SiO_2 Layers

Versatile Si surface architectures can be formed using the electrochemical processing of silicon using a thin silica layer as a precursor (Juzeliūnas et al. 2010, 2011, 2016, 2012a,b; Juzeliūnas and Fray 2020; Liu et al. 2014; Coxon et al. 2015).

Thermal oxidation provides an opportunity to prepare a well-defined thin layer of SiO_2 on Si for electrochemical processing in molten salt. Basics of the Si thermal oxidation were considered in detail in Chapter 11. This method is based on forced oxygen diffusion into Si wafer at high temperature and its reaction with silicon forming the oxide. Generally, the Deal–Grove model describes the relationship between the layer thickness and its growing time (11.3) and (11.4). The model has been proven for both oxygen and water oxidation over the temperature range of 700–1300 °C and the oxide thickness of 30–2000 nm. The growth of very thin oxides ($<\sim$25 nm) deviates from the model: more rapid growth occurs in silicon nanostructures, such as nanowires (NWs) (Liu et al. 2016).

Apart from thermal oxidation, there are also other methods of SiO_2 film formation, such as various sputtering techniques, chemical vapor deposition (CVD), or atomic layer deposition using silane and organosilicon precursors and atmospheric pressure plasma oxidation of silicon (Kakiuchi et al. 2007). Mostly, a homogeneous Si–SiO_2 interface is formed by thermal oxidation – the oxygen penetrates directly the Si crystal lattice so that there are no other structural or elemental components at the interface. By contrast, when the SiO_2 layer is deposited, it forms on the Si surface, which always has some inhomogeneity because of the presence of surface states, dangling bonds, hydrogen termination, adsorbed oxygen, etc. Thus, the Si–SiO_2 interface in that case is less homogeneous.

Thermal oxidation has been applied to form Si–SiO_2 precursors using crystalline Si wafers as the substrates (Juzeliūnas et al. 2010, 2011, 2016, 2012; Juzeliūnas and Fray 2020, Coxon et al. 2015). The oxide with the thickness ranging from 10 to 2000 nm has been studied. When sufficiently negative potentials are applied, several processes occur within the precursor: (i) deoxidation of silica layer $SiO_2 \rightarrow Si$; (ii) silica dissolution in the form of silicates with an opportunity of electrochemical Si redeposition; and (iii) calcium metal reduction and silicide formation. As a result of the cathodic processing, a variety of surface architectures forms at the top, e.g. pits and pores, globular formations, and nano–microfiber structure (Figures 15.2 and 15.3).

Deoxidation of a thin silica layer leads to a fine fiber structure, composed of intertwined Si NWs (Figure 15.3). Formation of such structure should be consistent with the regularities considered above for a bulk quartz precursor, of which deoxidation led to a microcolumnar structure (Yasuda et al. 2005c). The experiments in

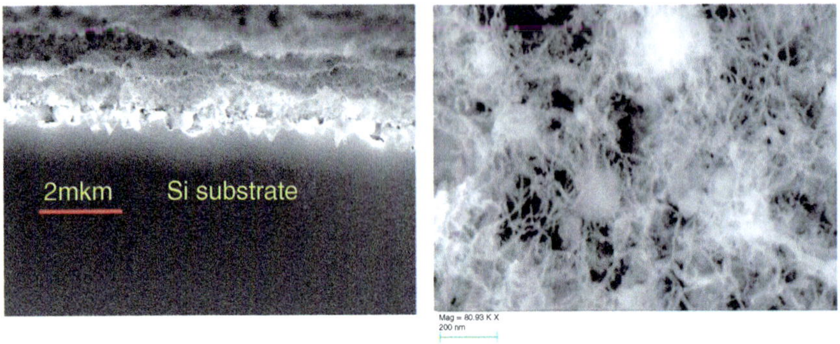

(a) (b)

Figure 15.3 SEM image of the cross section on a micrometer scale (a) and top view on a nanometer scale (b) of the surface obtained when deoxidizing amorphous a 2 μm SiO_2 layer on p-Si <100> in molten $CaCl_2$ at 850 °C. The electrode polarization was −0.9 V vs. graphite pseudoreference electrode. The scale bars are 2 μm (a) and 200 nm (b). Source: (b) Reproduced with permission from (Juzeliūnas et al. 2012a,b), © ELSEVIER.

Figures 15.2 and 15.3 were performed in molten $CaCl_2$ at 850 °C. Thin silica film reduction $SiO_2 + 4e^- \rightarrow Si$ is initiated at the three-phase interface Mo–SiO_2–$CaCl_2$, and the reduced Si at high temperature takes further the function of conductive electron provider to the SiO_2 layer. As a result, the reaction zone spreads over the surface. In-depth penetration of the reaction also occurs until the silica layer is completely reduced. The oxygen ions (O^2) leaves the reaction zone and diffuse into electrolyte bulk, whereas the reduced Si crystallizes in the form of filaments, which are as thick as tens of nanometers. The thickness of the formations is limited by a minor amount of the precursor material. Thus, the formations originating from thin film are by several orders of magnitude thinner when compared to those originating from a bulk precursor.

15.4 Globular Structures

Calcium oxide is inevitably present in a molten $CaCl_2$ electrolyte because of the hygroscopic nature of the salt. The oxide is present even when the salt is dehydrated under vacuum at elevated temperatures. Calcium oxide can interact with silica producing silicates ($SiO_3{}^{2-}$, $SiO_4{}^{4-}$, and $Si_2O_7{}^{6-}$), which dissolve in the electrolyte. Because of this, thin SiO_2 layer produced on Si dissolves in the electrolyte. There are also indications that Si is not stable in molten $CaCl_2$: it can dissolve in the electrolyte. Owing to these phenomena, the electrolyte is supplemented by the electroactive species of silicon, providing an opportunity for Si redeposition on the Si substrate, and this creates the conditions for Si surface structuring.

Figure 15.4 shows distinctive globular formations, which were formed from Si–SiO_2 precursor. The samples with a 2 μm oxide layer has been pre-exposed for 20 minutes (a–c) in the electrolyte and then polarized cathodically. Submicron globules are observed, which tend to agglomerate into distinctive globular textures.

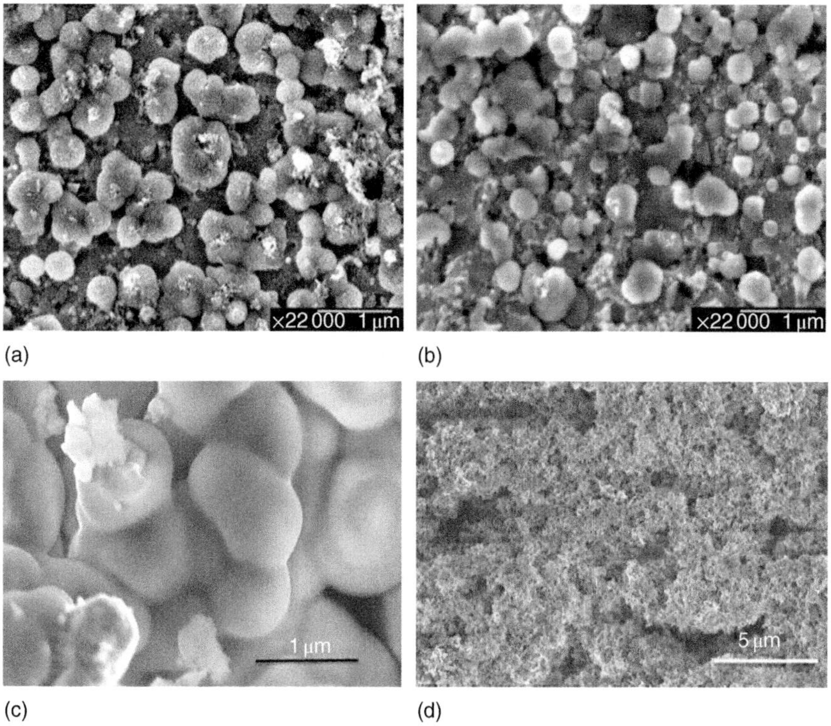

Figure 15.4 SEM micrographs of Si surfaces produced from p-Si thermally coated with a 2 μm SiO_2 layer, pre-exposed in molten $CaCl_2$ for 20 minutes at 850 °C, and polarized to $E = -0.07$ V (a–b) and 0.23 V (c) vs. Ca/Ca^{2+}. Image (d) shows the same precursor pre-exposed in the electrolyte for one minute and polarized at $E = -0.07$ V vs. Ca/Ca^{2+}. The samples after the processing in molten salt were kept for 12–17 hours in 0.5 M HCl. The scale bars are 1 μm (a–c) and 5 μm (d). Source: E. Juzeliūnas and P. Coxon.

The XRD study identifies an amorphous structure of the thermally formed oxide, which is evident from a broad diffraction peak at low angles $2\Theta = 20.08°$ (Figure 15.5). The peak is not present when the sample is processed in molten salt. Instead, several new sharp peaks appear typical of crystalline silicon. These data indicate removal of the amorphous silica layer and electrochemical deposition of crystalline silicon.

It is important to note that Peng et al. observed globular surface structures when depositing a tin-doped silicon on graphite from KCl–KF electrolyte at 650 °C with 1 mol% K_2SiF_6 additive as a silicon source (Peng et al. 2017). The major stages of the morphology development are shown in Figure 15.6. The growth process includes the nucleation and the islands formation, which is followed by island aggregation to a film with globular structure.

Globular shapes were also obtained when pre-exposing Si with a native oxide film in the $CaCl_2$ electrolyte with subsequent electrolyzing at $E = -0.07$ V vs. $E_{Ca/Ca^{2+}}$. Si with a very thin oxide layer supplies the electrolyte with a minor amount of Si^{4+} species during the pre-exposure period, which are then electrochemically

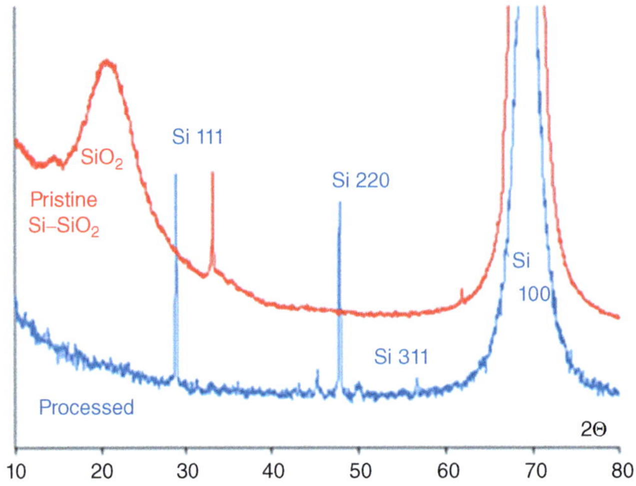

Figure 15.5 XRD data for p-Si with a 2 μm SiO_2 layer (red) and that processed in molten $CaCl_2$ at 850 °C (blue line) at −0.9 V vs. graphite pseudoreference electrode. Source: Reproduced from Juzeliūnas et al. (2012) with permission from the publisher. © Elsevier, 2012.

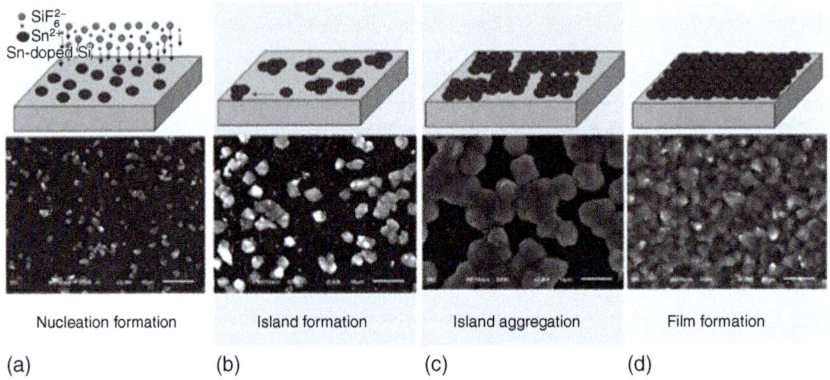

Figure 15.6 Schematic diagrams along with SEM micrographs of the surfaces obtained when depositing tin-doped silicon on graphite from the KCl-KF electrolyte with 1 mol% of K_2SiF_6 and 0.02 wt% of tin metal at a current density of 5 mA cm^{-2} and different times: (a) 5 minutes; (b) 15 minutes; (c) 30 minutes; and (d) more than 90 minutes. Scale bars 10 μm. Source: Reproduced with permission from Peng et al. (2017), John Wiley & Sons.

reduced. The figure identifies very initial stages of the reduction process (Figure 15.7).

It is important that the formation of globules was detected not only on precursor surfaces but also in the bulk of a large-scale powder precursor. Micron and submicron globules were formed when deoxidizing silica powders in a container and electrolyzed in molten LiCl–Li_2O electrolyte at 923 K (Lee et al. 2008).

Formation of globules is also typical for electrochemical synthesis of NWs. An essential point for such process is the addition of some metal additive into

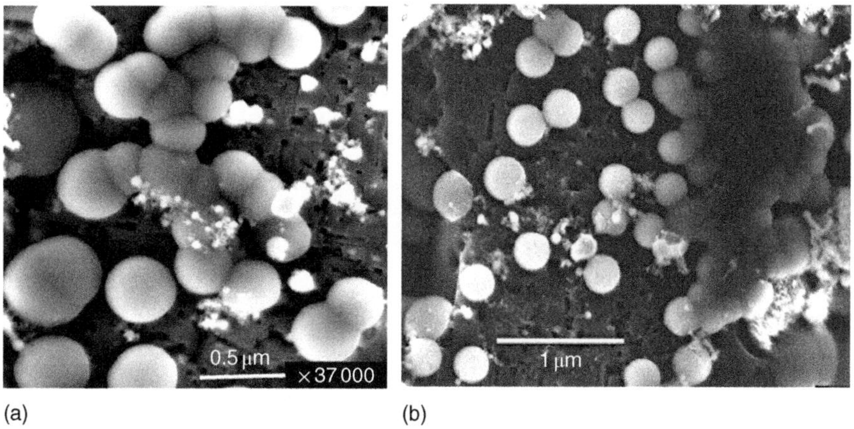

Figure 15.7 SEM images show the nucleation and aggregation of nanospheres on p-Si when using the silicon precursor with a native oxide layer. The precursor was shortly (one minute) pre-exposed in molten $CaCl_2$ at 850 °C and polarized down to $E = -0.9$ V vs. graphite pseudoreference electrode. The sample after the processing was immediately transferred to ambient temperature, cooled down, and kept in deionized water to remove residual $CaCl_2$, which is a well soluble salt. The scale bars are 0.5 μm (a) and 1 μm (b). Source: E. Juzeliūnas.

silicon, which reduces the melting point of the Si–Me system when compared to pure silicon. As a result, liquid droplets form, which promote the one-dimensional growth of silicon. Nishimura et al. discussed the formation of such droplets when reducing Si from the Au and SiO_2 nanoparticle mixture (Nishimura et al. 2011).

The experiments discussed above imply that formation of globular structures is rather typical for Si reduction in molten salts. Formation of globules is promoted by the Si–Me eutectics formed with metal additives in the electrolyte. It is important that the metals could be present in the molten salt as uncontrollable impurities. For instance, if stainless steel reactor is used, iron and nickel can contaminate the molten salt electrolyte at high temperatures.

Formation of globular structures is of practical relevance. Such structures enhance the light harvesting performance of the Si surface by significant reduction of light reflection to a certain percent level. The antireflective performance of spherically structured layer on Si has been modeled and experimentally studied by Lin et al. (2014). The authors demonstrated efficient light absorbance of the surfaces modified by a layer composed of polystyrene nanospheres, which is 450 nm in diameter. Note that such structural dimensions are similar to those obtained in molten salt as shown in Figures 16.5 and 16.7. The monolayer has been modeled and experimentally produced on p-Si (100) as Figure 15.8a, b demonstrate. The spherical structures greatly enhanced light-harvesting performance over the broadband of crystalline solar cells by suppressing the surface reflection and enhancing forward scattering. The nanostructured surface improved external quantum efficiencies by 13% and light conversion efficiency by 21.6% at $\lambda = 800$ nm irradiation. It follows that the nanospherical textures are promising light harvesters being applied as the antireflection coatings for optoelectronic devices.

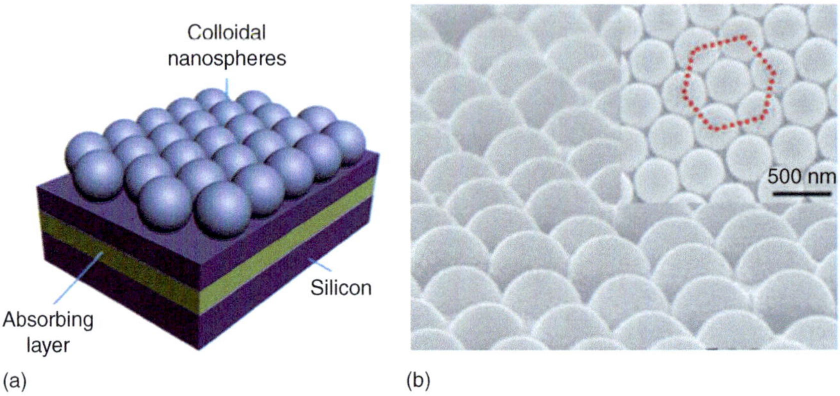

Figure 15.8 (a) Schematic representation of solar cell with colloidal nanospheres, which effectively absorb the light. (b) Polystyrene nanospheres produced on p-Si (100) from colloidal solution. Source: (b) Reproduced with permission from Lin et al. (Lin et al. 2014), ELSEVIER.

15.5 Black Silicon from Molten Salts

Black silicon (b-Si) is a nanostructured material, which effectively absorbs the light in a wide range of wavelengths. The material is of great importance in numerous fields, in particular, in sustainable energy applications: photovoltaics, photoelectrochemical hydrogen generation from water, or energy storage devices. Techniques of black silicon production include femtosecond laser treatment, dry reactive (cryogenic) ion etching, metal-assisted chemical etching, photoelectrochemical etching, vapor–liquid–solid growth, etc. These techniques are costly and technically rather sophisticated. Also, toxic chemicals and expensive catalysts are frequently applied, which limits the provision of competitive solutions for industrial-scale applications. Furthermore, formation of b-Si by means of electrochemistry in molten salts will be addressed. The electrochemical approach is attractive because of technical simplicity and environmental friendliness. The invention of electrochemical b-Si formation in molten salt has been patented worldwide (Fray and Juzeliūnas 2011).

Figure 15.9 shows an example of electrochemically produced *b*-Si in molten $CaCl_2$ at 900 °C temperature. A p-type Si wafer was thermally coated with a micron-sized SiO_2 layer, pre-exposed in molten $CaCl_2$, and polarized negatively. After the processing, the sample has been washed for several hours in 0.5 M HCl to remove possible metallic contaminants. As commonly known, Si is stable in most mineral acids; an exception makes the hydrofluoric acid (HF). A visually green–black surface is obtained with a reflectivity less than 10%.

Figure 15.10 demonstrates a rather dense globular structure obtained when processing the Si–SiO_2 precursor electrochemically. The produced globules tend to agglomerate, confining void spaces and forming a macroporous texture.

The globular textures effectively absorb the light (Figure 15.11). Their reflectivity was measured at about 8–11% in visible and near-infrared wavelength regions.

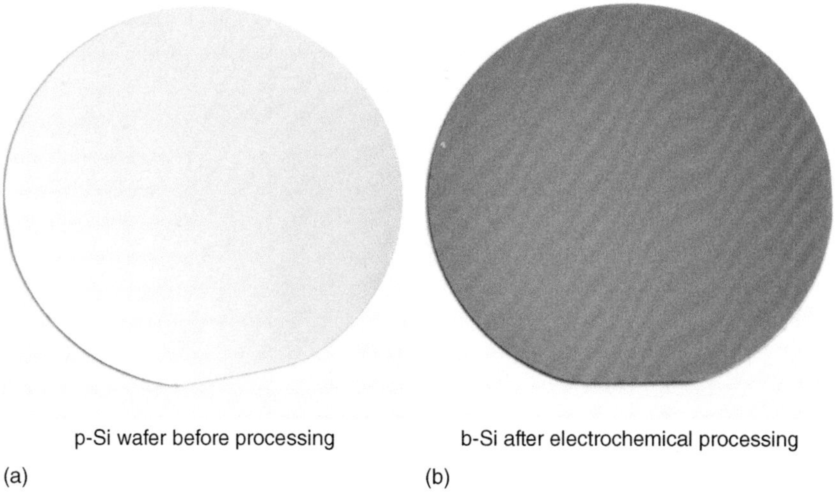

Figure 15.9 (a) p-Si wafer thermally coated with a 2 μm SiO_2 layer, which has been used as a precursor to produce b-Si; (b) b-Si with 93% light absorbance (λ = 600–800 nm) produced from the precursor when polarizing it at −0.9 V vs. graphite pseudoreference electrode in the molten $CaCl_2$ electrolyte at T = 900 °C. Boron-doped wafers were used, whose diameter was 50.8 ± 0.381 mm (2″ ± 0.015″) and thickness was 256–306 μm. Source: E. Juzeliūnas.

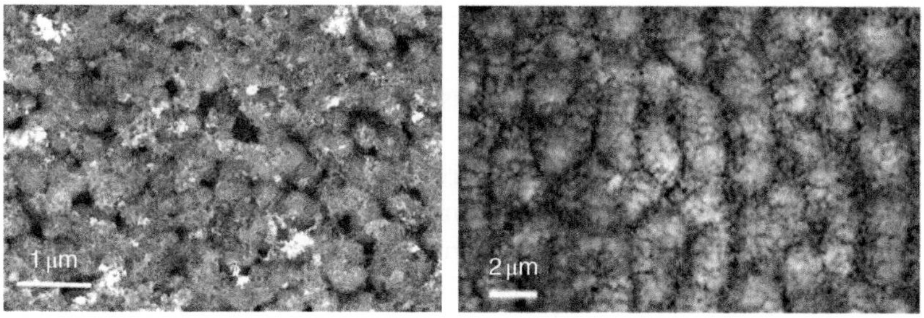

Figure 15.10 (a) SEM micrograph of a porous globular texture obtained from a 2 μm SiO_2 precursor on silicon polarized in molten $CaCl_2$ (850 °C) at E = −1.0 V vs. graphite. Image (b) shows the micrograph of the texture obtained by fs-laser treatment of the Si surface. Source: Reproduced with permission from Vorobyev and Guo (2011), ELSEVIER.

High absorption efficiency is also observed in the ultraviolet (UV) spectral region, which covers far-UV (122 nm ≤ λ < 200 nm), medium-UV (200 nm ≤ λ < 300 nm), and near-UV (300 nm ≤ λ < 400 nm) regions of the electromagnetic spectrum.

Interestingly, the morphology of the electrochemically processed surface is similar to that obtained by the laser treatment (Figure 15.10b) (Vorobyev and Guo 2011). The authors structured the Si surface by femtosecond (fs) laser and produced an array of microgrooves. The produced surface reduced reflectance less than 5% in the visible region. The antireflection effect also extended to a mid-infrared range with a reduction potential at terahertz and even millimeter wavelength range. Thus, both structuring methods, electrochemical and fs-laser engineering, provided

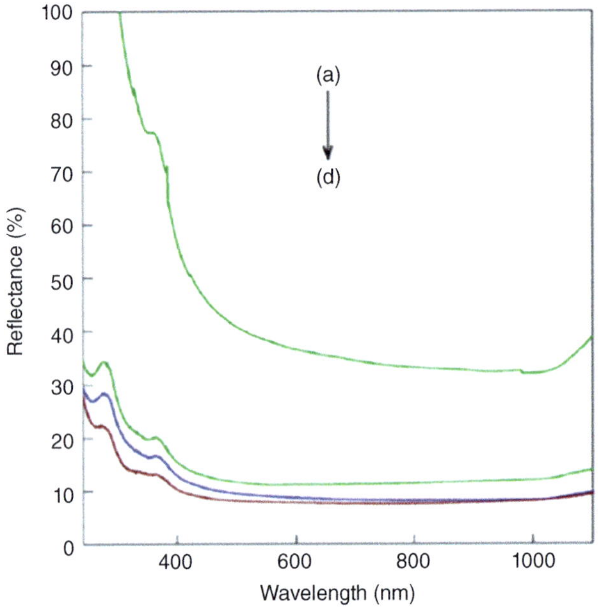

Figure 15.11 Surface reflectance spectra for bare Si wafer and that treated electrochemically under the conditions given in Figure 15.10 for different times: (a) bare Si wafer, (b) 1 hour, (c) 17 minute, and (d) 3 minute. Experimental contribution of G. Niaura is greatly acknowledged. Source: Reproduced from Juzeliūnas et al. (2012) with permission from the publisher. © Elsevier, 2012.

similar surface textures with similar light absorbance. The globular Si structuring is promising as antireflection coatings.

Furthermore, the dependence of surface structure formation on applied potential is discussed. The arrays of aligned rod-like formations were obtained at 0.63 $V_{Ca^{2+}/Ca}$ (Figure 15.12a). At 0.23 $V_{Ca^{2+}/Ca}$, a NW structure forms, which is composed of curved and intertwined filaments, whose thickness varies from several to tens of nanometers (Figure 15.12b). According to the measured value of the equilibrium potential ($E^0_{Ca/Ca^{2+}}$), calcium metal cannot be reduced at the applied polarizations; the reduction can be expected at $E < E^0_{Ca^{2+}/Ca}$, assuming $E < E^0_{Ca^{2+}/Ca} = 0.0$ V. However, studies showed that Ca can be reduced on the Si substrate even at the potentials $E > E^0_{Ca^{2+}/Ca}$ due to the formation of Ca–Si alloy (Xiao and Wang 2014; Xiao and Wang 2014; Nohira et al. 2003; Gao 2005; Yasuda et al. 2005a,b,c). The process is referred to as underpotential deposition (UPD). Thus, some Ca deposition during the electrochemical processing shown in Figure 15.12 cannot be disclaimed. To remove possible contaminations, sample leaching in 0.5 M HCl has been undertaken. The EDS analysis showed that the calcium content on sample surface after the leaching was less than 1 at.%.

It is remarkable that the Si NWs (Figure 15.12b) were converted to silica because of water washing or reaction with ambient air. The EDS analysis showed oxygen content on the surface to be ~60 at.% and that of silicon ~32 at.%. This concentration ratio corresponds to silica atomic composition.

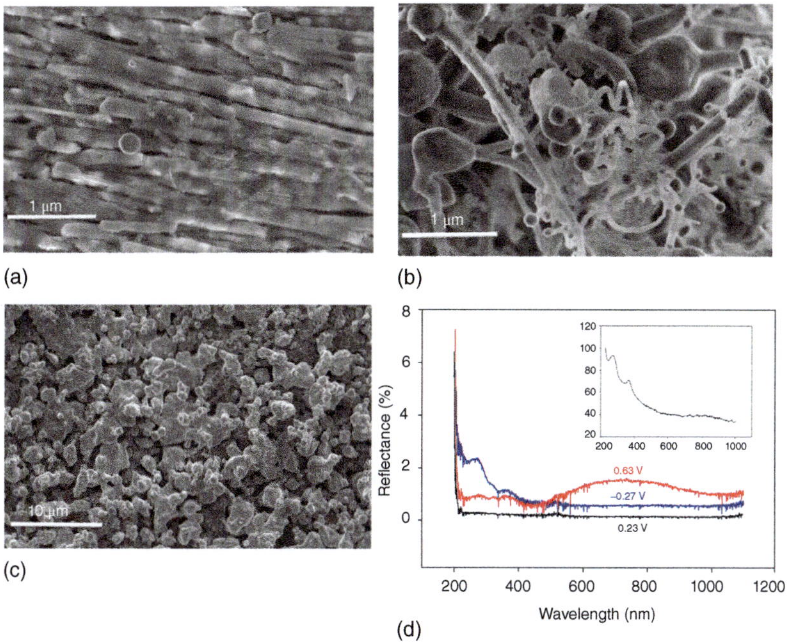

Figure 15.12 SEM views of the surfaces obtained in molten $CaCl_2$ at 850 °C from Si to 0.3 μm SiO_2 precursor polarized for 20 minutes (vs. Ca^{2+}/Ca): (a) 0.63 V, (b) 0.23 V, and (c) −0.27 V. Reflectance (%) graph (d) was obtained in a wide range of UV-Viz-IR wavelength. The spectra for bare silicon (inset) reflecting more than 30% of the light, and those for the surfaces obtained at different polarizations (a–c) reflect less than 2% of the light. Source: Authors: P. Coxon, E. Juzeliūnas, and P. Kalinauskas.

Figure 15.12c shows the surface structure obtained at polarization −0.27 V, at which calcium reduction $Ca^{2+} + 2e \rightarrow Ca$ occurs along with the silica reduction reaction. The latter occurs as $Si^{4+} + 4e \rightarrow Si$ from SiO_2 and/or from the dissolved silicates (SiO_3^{2-}, SiO_4^{4-}, and $Si_2O_7^{6-}$). The obtained surface exhibits a network structure composed of micrometer-sized agglomerations, which confine void spaces – pores. Prolonged polarization at this region led to surface etching and formation of a thick porous structure. A macroporous layer of ~10 μm thickness was formed during three hours electrolysis. The porosity effect is explained by silicon etching because of the formation of various liquid calcium silicide, whose melting points are in the range of 700–935 °C (Ropp 2013).

The processed surfaces greatly suppressed the light reflectance in a wide range of UV–viz–infrared (IR) wavelengths (Figure 15.12d). The surfaces were visually black. The NW sample processed at 0.23 V (Figure 15.12b) showed nearly zero reflectance (0.2%) within the entire studied spectrum $220\,nm < \lambda < 1150\,nm$. The macroporous sample processed at −0.27 V reflected less than 1% of the visible and IR irradiation ($400\,nm < \lambda < 1000\,nm$), whereas UV reflection of this sample was about 2% ($250\,nm < \lambda < 400\,nm$). The rod-like structure formed at 0.63 V also effectively suppressed the reflection below 2% within the entire range of wavelength.

We have discussed in this chapter the principal opportunities of b-Si formation by means of molten salt electrochemistry. The field is at a very early stage of investigations. Very little is known about electrochemical reactions and mechanisms of such processing. Numerous experimental factors, which influence the structural properties, are to be elucidated. This would open opportunities for better structural management and tunable morphology. In particular, formation of homogeneous structural arrays and their continuity over larger surface area are necessary to further investigate.

15.6 Electrochemical Synthesis of Nanowires: Implications for Li-Ion Batteries

Si NWs are of importance in electronics, solar energy conversion, batteries, and sensor applications. Of particular interest is the application of Si-NWs in Li-ion batteries (LIBs).

LIBs have been commercialized by Sony in 1991. These types of batteries are widely used in everyday life, for instance, in mobile phones, computers, or electric vehicles. LIBs could also be used for stationary applications to store energy from solar or wind power. Thus, LIBs are promising because of their substantial contribution to low-carbon economy and fossil fuel-free society. The Nobel Prize in Chemistry 2019 was awarded to J. B. Goodenough, M. S. Whittingham, and A. Yoshito "for the development of lithium-ion batteries."

In summary, LIBs are composed of a positive electrode (cathode) made of layered $LiCoO_2$ and a negative electrode (anode) made of graphite. The electrodes are separated by a nonaqueous electrolyte supplemented with a polymer separator, which prevents the flow of electrons inside the battery. When charging, the Li ions leave the $LiCoO_2$ structure and travel through the electrolyte and the separator diaphragm. At the graphite anode, the ions combine with electrons and intercalate into the graphite structure. A vice versa process occurs during battery discharging. Thus, lithium ions are reversibly intercalated–deintercalated from the electrodes, ideally with a theoretical 100% Coulombic efficiency. The term "rocking chair" has been coined to describe the shuttling principle of Li^+ ions. The typical $LiCoO_2$–graphite electrodes provide an output of 3.7 V and an energy density of 150–200 Wh kg^{-1} (Tarascon and Armand 2001).

Although Li-based technologies proved their outstanding position, manufacturers are still in active search for high energy density and high-capacity electrode materials when compared with the conventionally used carbon and $LiCoO_2$. Silicon has the highest specific capacitance to accommodate Li atoms: each Si atom could accommodate 4.4 Li atoms producing $Li_{22}Si_5$ alloy. The specific insertion capacity of silicon is 4200 mA h g^{-1}; this is by 1 order of magnitude higher when compared to the conventional graphite electrode and also higher than various nitride and oxide materials. At the same time, the lithiation potential of silicon is similar to that of graphite that is in the range of 0 to 0.4 V vs. Li/Li^+ (Zhou et al. 2019). However, unfavorable disintegration of the electrode occurs because of the nearly 400% volume change during lithium insertion and extraction. The expansion–contraction phenomenon leads to fast capacity fading, which precludes practical applications of Si electrodes. Thus, apart from the new advantageous compositions, the restoration of electrical properties after damage (self-healing) is of paramount importance in the development of Si-based energy storage devices.

The importance of self-healing technologies is emphasized in the European roadmap. The European BATTERY 2030+ Roadmap, a large-scale long-term research initiative, emphasizes that batteries are the key technologies in abatement of carbon dioxide emission from the transport, power, and industry sectors (Edström 2020). An ultra-high performance is expected from R&D initiatives, which would lead to energy and power characteristics close to theoretical limits, as well as an outstanding lifetime, safety, and environmental sustainability. Self-healing of electrodes is projected as a key smart functionality, which aims to prevent, retard, or reverse the battery degradation, thereby enhancing its reliability, lifetime, and safety.

Nanostructuring of the electrode material is one of the ways to release mechanical strains and satisfy the above requirements. Various nanostructures have been investigated with an aim to apply for battery electrodes: nanotubes (Wu et al. 2012; Park et al. 2009), hollow nanospheres (Yao et al. 2011), yolk–shell structures (Liu et al. 2012), and core–shell NWs with crystalline Si cores for mechanical support and amorphous shells for lithium storage (Cui et al. 2009).

The effective way of increasing the electrode's stability is by providing the Si-NW structuring (Chan et al. 2008; Zamfir et al. 2013; Dasgupta et al. 2014; Arbiol and Xiong 2015). Si-NWs do not only damp the electrode's mechanical deformations but it also increases the process rate, thereby reducing the Li^+ diffusion pathway to the host structure. The major advantages of the Si-NWs include the following features (Dasgupta et al. 2014): (i) nanoscale formations make it possible for better accommodation of the volume changes because of lithiation–delithiation with better stability against structural fracture when compared to bulk or micron-sized architectures; (ii) NWs ensure better utilization of active material because of the large surface area of the electrode–electrolyte interface; and (iii) NWs shorten the charge (both electrons and ions) transportation distances.

Si-NWs are mainly prepared by the vapor–liquid–solid (VLS) microfabrication, which belongs to a family of CVD methods (Chan et al. 2008; Laïk et al. 2008; Boettcher et al. 2010; Dasgupta et al. 2014). Si structures are grown from a vapor phase composed of the silicon tetrachloride (tetrachlorosilane) and the hydrogen: $SiCl_4 + 2H_2 \rightarrow Si + 4HCl$. In order to induce one-dimensional Si growth, a catalytic phase of liquid alloy in the form of nucleation seeds is produced on the silicon substrate. To this end, the Au–Si binary system is widely used. Si wafer is coated by a Au nanofilm (<10 nm) using sputtering or thermal evaporation and annealed at temperatures over the Au–Si eutectic point. The melting point of pure silicon is ~1414 °C, and it reduces when Au is introduced. A temperature minimum of 363 ± 2 °C is observed when the Au:Si ratio is 4 : 1 (19.0 at.% of Si). During annealing, Au–Si droplets are produced, whose size depends on the Au film thickness. The nano–micrometer-sized particles tend to minimize their surface energy by forming spheres or half spheres. The activation energy to deposit Si on these droplets is much lower when compared to the energy needed to deposit on the solid surface. Thus, the temperature $T > 800\,°C$ is necessary for the CVD deposition on the solid Si surface, whereas the liquid Au–Si alloy can adsorb the gas precursor at much lower temperatures, $T > 363\,°C$. Adsorption of the vapor and Si production is catalyzed on the droplet. The melting point of the Au–Si alloy droplet increases with the Si content, which leads to Si crystallization at the Si–droplet interface. As a result, the droplet rises and 1D nano-Si grows. The physical properties of NWs could be controlled by tuning the physical properties of the liquid-phase formations. Note also that lithography techniques could be used to pattern the substrate surface and to form droplets of desirable diameter and position.

It is important to note that nano-sized droplets could have even lower melting point as predicted by the bulk phase diagrams because of size-dependent depression of the melting point. Electron diffraction studies on nanocrystals showed a large depression in the melting temperature with the size reduction, which was attributable to a larger fraction of number of atoms on the surface, in other words, to increase the surface-to-volume ratio (Goldstein et al. 1992). The melting point depression of 0.1–10 nm-thick indium films on silicon was determined by nanocalorimetrical measurements (Zhang et al. 2000). A 110 °C depression was detected for the particles with a radius of 2 nm. Liang et al. modeled size-dependent solution phase diagrams for binary systems and showed that the temperatures of solidus and liquidus curves reduce with the size of the components (Liang et al. 2003). According to these calculations, the melting point of pure Si dropped down from ~1414 °C for bulk sample to ~900 °C for 5 nm sized particles.

While the CVD methods for production of Si-NWs are widely used in research practice, they have limited practical application. These methods produce low yield and have limited large-scale availability for manufacturing of NWs for mass production. Reduction of processing cost is another issue to be addressed.

Electrochemical formation of Si-NWs includes such advantages as facile and template-free fabrication as well as the usage of inexpensive precursors, for instance, silica or silicates. There are also possibilities of electrosynthesis of Si-NW compositions with other elements such as carbon or various metals.

Silicon nano–microwires are rather typical products of electrochemical silica–silicon conversion in molten salts. Microcolumnar Si structures were observed in the early studies of electrochemical silica deoxidation. Nohira et al. first demonstrated electro-deoxidation of amorphous quartz in molten $CaCl_2$ at 850 °C and observed the microcolumns 5–10 μm in diameter, which were composed of 90 at.% of Si and 8 at.% of oxygen (Nohira et al. 2003; Yasuda et al. 2005a). The depth of the reduced zone on the quartz substrate was about 100–200 μm. The silicon produced from the amorphous substrate showed crystallinity, as confirmed by XRD. Deoxidation of a thin layer of amorphous silica produced a crystalline nanofiber structure (Juzeliūnas et al. 2012). The diameter of the produced filaments was on the order of tenth of nanometers (Figure 15.3).

Yang et al. produced Si-NWs from silica nanoparticles pressed into a porous pellet and sintered. The conversion process was achieved by electrolysis in molten $CaCl_2$

at 900 °C (Yang et al. 2009). The intertwined NWs were produced, which featured irregularly curved shapes. While the feedstock had an amorphous structure, the product was crystalline with a diamond cubic structure, as XRD data identified. An important feature of the NW structure is photoluminescence with a peak at about 750 nm. The peak was at a slightly lower wavelength when compared to the NWs on bulk Si because of the differences of the interfaces between the Si core and the SiO_2 sheath. This optical property is of practical value, for instance, for *in vivo* sensing or therapeutics (Anglin et al. 2008). Si nanostructures are attractive host matrixes for drug delivery *in vivo* under controlled amount, identity, and release rate of drug payloads. Anglin et al. reviewed advanced drug delivery devices and materials and pointed that the optical properties of Si photonic structures possess a valuable feature – self-reporting for *in vivo* monitoring (Anglin et al. 2008).

Xiao et al. studied the morphology of the silicon produced by reduction of porous silica and quartz pellets in molten $CaCl_2$ (Xiao and Wang 2014). The morphology of the reduced silicon depended on the precursor and the applied potential. Curved and entangled Si-NWs with diameter from 50 to 200 nm were produced from porous silica pellet at $-0.65\,V_{Ag/AgCl}$. A mixture of interconnected NWs and nanoparticles ($d < 50$ nm) was obtained within the polarization range −0.65 V to −0.75 V. Formation of nanoparticles became dominant at lower polarizations. Electrolysis of quartz pellet yielded interconnected nanoparticles irrespective of the imposed potential.

The studies referred above were performed in a molten $CaCl_2$ electrolyte at temperatures over 800 °C. The nanowire structures could also be synthesized at lower temperatures employing molten salt eutectics. Randomly oriented Si-NWs were produced on graphite in a molten KCl–KF electrolyte at 650 °C (Peng et al. 2017). The electrolyte was supplemented with 1 mol% of K_2SiF_6 as a silicon source. Dong et al. reduced silicon from calcium silicate in a ternary eutectic melt $CaCl_2$–$MgCl_2$–NaCl at temperature 650 °C (Dong et al. 2017). Note that calcium silicate is an abundant and inexpensive material. It is soluble in the melt producing the anions (e.g. SiO_3^{2-}, SiO_4^{4-}, and $Si_xO_yCl_z^{n-}$), which can be reduced to Si. The authors synthesized Si-NWs 80–300 nm in diameter and up to hundred micrometers in length. An important role is played by CaO introduced in the melt. The oxide facilitates diffusion of the oxygen ions O^{2-} released from SiO_2 during Si conversion. The oxygen ions diffuse to the anode where they are discharged. The CaO also facilitates dissolution of SiO_2 opening the route for Si production from quartz, sand, or common glass, which is composed of SiO_2, CaO, and Na_2O. Coal ash could also be used as a feedstock. The study performed by Dong and coworkers demonstrated the effectiveness of Si-NWs in application as anodes in Li-ion battery (Dong et al. 2017).

The electrochemical method makes it possible to combine silicon in NWs with other elements such as carbon, copper, or nickel. Thus, Si–C NWs have been produced by facile electrosynthesis in molten $CaCl_2$ (Zou et al. 2017). Homogeneous NWs were produced from the SiO_2–C precursor without any template and catalyst. The precursor was prepared in a form of pellet from nanoscale (50–100 nm) and microscale (~2 μm) silica and carbon powder. The pellet prepared from the SiO_2–C mixture at a molar ratio of 1 : 1 was sandwiched between porous Ni foils and electrolyzed in molten $CaCl_2$ at a constant potential (~3.1 V) using a graphite anode. The conversion basically included solid-to-solid silica reduction $SiO_2 \rightarrow Si$, which is rather a slow process. It took about 15 hours to convert 2.0 g of the precursor into NWs. The conversion (the current) efficiency was also rather low, just about ~30%. The obtained Si product contained low amount of residual carbon. The authors applied another approach – Si electrodeposition. The process included such steps as dispersing SiO_2 and C particles in molten $CaCl_2$, the electrochemical Si codeposition with C, and the carbonization ($Si + C \rightarrow SiC$). A mixture of Si and SiC NWs was obtained because of insufficient carbon particles surrounding the Si nuclei. Tailoring of NW diameter could be achieved by tuning experimental parameters. The EDS spectrum identified carbon and silicon energy peaks of the NW. However, the question arises as to whether this composition relates to a mixture of individual elements or to carbide in which elements are covalently bonded.

The electrochemical synthesis of silicon–metal composite NWs has also been proposed. The principle is conceptually similar to that of the "bottom-up" Si structuring by VLS, which was discussed above. Below we discuss that such metal-assisted concept also applies to molten salts (Munisami and Bard 2010; Zhou et al. 2018; Fang et al. 2015; Nishimura et al. 2011).

Longitudinal NW growth has been promoted by preparing the NW precursor from Au and SiO_2 nanoparticles (Nishimura et al. 2011). Si-NPs (nanoparticles) and Au-NPs were mixed, pressed into pellets, and sintered under vacuum at 1273 K. The experiments have been performed in molten $CaCl_2$ electrolyte at 1123 K (850 °C). The electrochemically reduced Si from SiO_2 forms an Au–Si alloy at the tip of the NW in the form of a liquid droplet. It is known that the melting point of Si drops down when adding gold to the system. The temperature minimum is at 363 °C when Au content is 19 at.%. As in the case of the VLS process, the temperature depression makes it possible for the formation of liquid Au–Si droplets on the tip of reduced silicon. With the reduction progress, the Au–Si droplet is enriched by Si and the liquidus temperature increases toward an operational one. At the same time, Si is deposited at the tip because of the oversaturation of the liquid alloy. The rate of Au–Si formation in a liquid form is balanced with the rate of Si precipitation from the oversaturated alloy. The process is preferential at the tip of the NW because the surface energy is lowest at these locations.

Elongated NWs were produced according to the above metal-assisted mechanism when compared to the wires obtained from pure SiO_2. The NWs appeared promising for the production of anodes for LIBs. The particles of the Au catalyst are inert and can increase the conductivity of the Si-NW electrode.

Zhou et al. synthesized Si and Cu–Si (Cu_9Si/Si) NWs in equimolar $CaCl_2$ and NaCl mixture melted at 700 °C (Tarascon and Armand 2001). They added NH_4HCO_3 as a pore-forming agent to $SiO_2 + Cu$ powder blend, pressed into pellets, and sintered at 500 °C. The Si-NWs obtained from a Cu-free precursor were of irregularly curved shape, several hundreds of nanometers thick, and several micrometers long. On the contrary, the Si–Cu NWs exhibited a regular straight shape; they were by ~10 μm longer and by 60–100 nm narrower than the Si counterparts. The mechanism of formation of Si–Cu NWs bore a resemblance with that of the VLS approach or the Au–Si process discussed above. When Si is reduced at the Cu–SiO_2–$CaCl_2$ interface, Si dissolves in the Cu matrix, thereby forming the Cu–Si alloy. Si precipitation is induced when Si concentration exceeds the solubility limit. This causes an oriented growth of NW, actually composed of Cu_9Si/Si. As the anode, the Cu–Si composite exhibited improved capacitance and rate performance because of the enhanced conductivity and straight structure of the NWs.

Si–$NiSi_2$ NWs were also synthesized using the NiO/SiO_2 nanopowder precursor (Fang et al. 2015). The synthesis was performed in molten $CaCl_2$ at 900 °C at the concentration ratio $[SiO_2]/[NiO] = 10\,M$. NiO acted as seeds of Si-NW growth. The NWs were straight with the diameter being from 60 to 330 nm and the length up to 30 μm. However, the produced NWs were rather the composite materials than the products of pure silicon. The nano-sized nickel silicide particles were embedded in Si-NWs. Also, transmission electron spectroscopy (TEM) energy dispersive spectroscopy (EDX) identified substantial contents of carbon and copper within the NWs. The authors believed that incorporation of Ni in the Si-NWs is beneficial for Li-ion battery anodes as inactive $NiSi_2$ particles could buffer volumetric changes during the lithiation/delithiation process.

References

Anglin, E.J., Cheng, L., Freeman, W.R., and Sailor, M.J. (2008). Porous silicon in drug delivery devices and materials. *Adv. Drug Delivery Rev.* 60: 1266–1277. https://doi.org/10.1016/j.addr.2008.03.017.

Arbiol, J. and Xiong, Q. (ed.) (2015). *Semiconductor Nanowires: Materials, Synthesis, Characterization and Applications*. Elsevier Ltd.

Archer, R.J. (1960). Stain films on silicon. *J. Phys. Chem. Solids* 14: 104–110. https://doi.org/10.1016/0022-3697(60)90215-8.

Birnbaum, M. (1965). Semiconductor surface damage produced by ruby lasers. *J. Appl. Phys.* 36: 3688. https://doi.org/10.1063/1.1703071.

Boettcher, S., Spurgeon, J.M., Putnam, M.C. et al. (2010). Energy-conversion properties of vapor–liquid–solid-grown silicon wire-array photocathodes. *Science* 327: 185–187. https://doi.org/10.1126/science.1180783.

Burton, L.C. and Madjid, A.H. (1969). Coulomb screening in intrinsic medium-gap semiconductors and the electrical conductivity of silicon at elevated temperatures. *Physiol. Rev.* 185: 1127–1132. https://doi.org/10.1103/PhysRev.185.1127.

Chan, C.K., Peng, H., Liu, G. et al. (2008). High-performance lithium battery anodes using silicon nanowires. *Nat. Nanotechnol.* 3: 31–35. https://doi.org/10.1038/nnano.2007.411.

Coxon, P.R., Coto, M., Juzeliūnas, E., and Fray, D.J. (2015). The use of electro-deoxidation in molten salts to reduce the energy consumption of solar grade silicon and increase the output of PV solar cells. *Prog. Nat. Sci. Mater.* 25: 583–590. https://doi.org/10.1016/j.pnsc.2015.11.006.

Cui, L.-F., Ruffo, R., Chan, C.K. et al. (2009). Crystalline-amorphous core-shell silicon nanowires for high capacity and high current battery electrodes. *Nano Lett.* 9: 491–495. https://doi.org/10.1021/nl8036323.

Dasgupta, N.P., Sun, J., Liu, C. et al. (2014). Semiconductor nanowires – synthesis, characterization, and applications. *Adv. Mater.* 26: 2137–2184. https://doi.org/10.1002/adma.201305929.

Dong, Y., Slade, T., Stolt, M.J. et al. (2017). Low-temperature molten-salt production of silicon nanowires by the electrochemical reduction of $CaSiO_3$. *Angew. Chem. Int. Ed.* 56: 14453–14457. https://doi.org/10.1002/anie.201707064.

Edström, K., (ed.) (2020). Battery 2030+ Roadmap. Inventing the sustainable batteries of the future. Domniko, R., Fichtner, M., Otuszewski, T., et al. https://battery2030.eu/research/roadmap

Fang, S., Wang, H., Yang, J. et al. (2015). Electrochemical preparation of silicon nanowires from porous NiO/SiO_2 blocks in molten $CaCl_2$. *Mater. Lett.* 160: 1–4. https://doi.org/10.1016/j.matlet.2015.07.082.

Fray, D.J. and Juzeliūnas, E. (2011) Method for texturing silicon surface to create black silicon for photovoltaic applications. Patent WO 2011/161479 A1, 29 December 2011; China patent ZL201180030064.0, 22 April 2015; Japan patent 5, 689,530, 6 February 2015. UK patent GB2494566, 28 May 2014; US patent 8, 790,953 B2, 29 July 2014. Brazil patent Br 11 2012 032955.3; European patent 2586064 (Germany, France, UK, Norway).

Goldstein, A.N., Echer, C.M., and Alivisatos, A.P. (1992). Melting in semiconductor nanocrystals. *Science* 256: 1425–1427. https://doi.org/10.1126/science.2565062.1425.

Juzeliūnas, E. and Fray, D.J. (2020). Silicon electrochemistry in molten salts. *Chem. Rev.* 120: 1690–1709. https://doi.org/10.1021/acs.chemrev.9b00428.

Juzeliūnas, E., Cox, A., and Fray, D.J. (2010). Silicon surface texturing by electro-deoxidation of a thin silica layer in molten salt. *Electrochem. Commun.* 12: 1270–1274. https://doi.org/10.1016/j.elecom.2010.06.034.

Juzeliūnas, E., Cox, A., and Fray, D.J. (2011). Silicon surface texturing by electro-deoxidation of a thin silica layer. In: *The Minerals, Metals and Materials Society (TMS)*, Proceedings: Materials Processing and Energy Materials, vol. 1, 709–716. Wiley https://doi.org/10.1002/9781118062111.ch82.

Juzeliūnas, E., Cox, A., and Fray, D.J. (2012a). Electro-deoxidation of thin layer in molten salt – Globular structures with effective light absorbance. *Electrochim. Acta* 68: 123–127. https://doi.org/10.1016/j.electacta.2012.02.054.

Juzeliūnas, E., Cox, A., Abdelkader, A., and Fray, D.J. (2012b). Electrochemical formation of porous silicon layers in molten salt. In: *Materials Recycling, Silicon Photovoltaic Cells, Boron & Borates*, Fray International Symposium - On Metals and Materials Processing in Clean Environment, vol. 4 (ed. F. Kongoli), 491–496. Flogen Star OUTREACH. ISBN: 978-0-9879974-1-8.

Juzeliūnas, E., Coxon, P.R., Fray, D.J. et al. (2016). Photoelectrochemical activity of macroporous silicon formed in molten calcium chloride. *Chemija* 27: 11–16. http://mokslozurnalai.lmaleidykla.lt/chemija/2016/1/7224.

Kakiuchi, H., Ohmi, H., Harada, M. et al. (2007). Formation of silicon dioxide layers at low temperatures (150–400 °C) by atmospheric pressure plasma oxidation of silicon. *Sci. Technol. Adv. Mater.* 8: 137–141. https://doi.org/10.1016/j.stam.2006.12.006.

Laïk, B., Eude, L., Pereira-Ramos, J.-P. et al. (2008). Silicon nanowires as negative electrode for lithium-ion microbatteries. *Electrochim. Acta* 53: 5528–5532. https://doi.org/10.1016/j.electacta.2008.02.114.

Lee, S.-C., Hur, J.-M., and Seo, C.-S. (2008). Silicon powder production by electrochemical reduction of SiO_2 in molten LiCl–Li_2O. *J. Ind. Eng. Chem.* 14: 651–654. https://doi.org/10.1016/j.jiec.2008.04.010.

Liang, L.H., Liu, D., and Jiang, Q. (2003). Size-dependent continuous binary solution phase diagram. *Nanotechnol.* 14: 438–442. https://doi.org/10.1088/0957-4484/14/4/306.

Lin, G.-J., Wang, H.-P., Lien, D.-H. et al. (2014). A broadband and omnidirectional light-harvesting scheme employing nanospheres on Si solar cells. *Nano Energy* 6: 36–43. https://doi.org/10.1016/j.nanoen.2014.03.004.

Liu, N., Wu, H., McDowell, M.T. et al. (2012). A yolk-shell design for stabilized and scalable Li-ion battery alloy anodes. *Nano Lett.* 12: 3315–3321. https://doi.org/10.1021/nl3014814.

Liu, X., Coxon, P.R., Peters, M. et al. (2014). Black silicon: fabrication methods, properties and solar energy applications. *Energy Environ. Sci.* 7: 3223–3263. https://doi.org/10.1039/c4ee01152j.

Liu, M., Jin, P., Xu, Z. et al. (2016). Two-dimensional modeling of the self-limiting oxidation in silicon and tungsten nanowires. *Theor. App. Mech. Lett.* 6: 195–199. https://doi.org/10.1016/j.taml.2016.08.002.

Memming, R. and Schwandt, G. (1966). Anodic dissolution of silicon in hydrofluoric acid solution. *Surf. Sci.* 4: 109–124. https://doi.org/10.1016/0039-6028(66)90071-9.

Munisami, T. and Bard, A.J. (2010). Electrodeposition of Si from organic solvents and studies related to initial stages of Si growth. *Electrochim. Acta* 55: 3797–3803. https://doi.org/10.1016/j.electacta.2010.01.097.

Nishimura, Y., Nohira, T., Kobayashi, K., and Hagiwara, R. (2011). Formation of Si nanowires by direct electrolytic reduction of porous SiO_2 pellets in molten $CaCl_2$. *J. Electrochem. Soc.* 158: E55–E59. https://doi.org/10.1149/1.3575635.

Nohira, T., Yasuda, K., and Ito, Y. (2003). Pinpoint and bulk electrochemical reduction of insulating silicon dioxide to silicon. *Nat. Mater.* 2: 397–401. https://doi.org/10.1038/nmat900.

Olson, G.L. and Roth, J.A. (1988). Kinetics of solid phase crystallization in amorphous silicon. *Mat. Sci. Rep.* 3: 1–77. https://doi.org/10.1016/S0920-2307(88)80005-7.

Park, M.-H., Kim, M.G., Joo, J. et al. (2009). Silicon nanotube battery anodes. *Nano Lett.* 9: 3844–3847. https://doi.org/10.1021/nl902058c.

Peng, J., Yin, H., Zhao, J. et al. (2017). Liquid-tin-assisted molten salt electrodeposition of photoresponsive n-type silicon films. *Adv. Funct. Mater.* 1703551. https://doi.org/10.1002/adfm.201703551.

Robbins, H. and Schwartz, B. (1959). Chemical etching of silicon. *J. Electrochem. Soc.* 106: 509–511. https://doi.org/10.1017/S1551929500060739.

Ropp, R. (2013). *Encyclopedia of the Alkaline Earth Compounds*, 375. Elsevier. ISBN: 978-0-444-59550-8.

Schwartz, G.C. and Schaible, P.M. (1979). Reactive etching of silicon. *J. Vac. Sci. Technol.* 16: 410–413. https://doi.org/10.1116/1.569962.

Spinella, C. and Lombardo, S. (1998). Crystal grain nucleation in amorphous silicon. *J. Appl. Phys.* 84: 5383. https://doi.org/10.1063/1.368873.

Tarascon, J.M. and Armand, M. (2001). Issues and challenges facing rechargeable lithium batteries. *Nature* 414: 359–367. https://doi.org/10.1142/9789814317665_0024.

Uhlir, A. (1956). Electrolytic shaping of germanium and silicon. *Bell System Tech. J.* 35: 333–347. https://doi.org/10.1002/j.1538-7305.1956.tb02385.x.

Vorobyev, A.Y. and Guo, C. (2011). Direct creation of black silicon using femtosecond laser pulses. *Appl. Surf. Sci.* 257: 7291–7294. https://doi.org/10.1016/j.apsusc.2011.03.106.

Wu, H., Chan, G., Choi, J.W. et al. (2012). Stable cycling of double-walled silicon nanotube battery anodes through solid–electrolyte interphase control. *Nat. Nanotechnol.* 7: 310–315. https://doi.org/10.1038/nnano.2012.35.

Xiao, W. and Wang, D. (2014). The electrochemical reduction process of solid compounds in high temperature molten salts. *Chem. Soc. Rev.* 43: 3215–3228. https://doi.org/10.1039/c3cs60327j.

Yang, J., Lu, S., Kan, S. et al. (2009). Electrochemical preparation of silicon nanowires from nanometer silica in molten calcium chloride. *Chem. Commun.* 3273–3275. https://doi.org/10.1039/b902029b.

Yao, Y., McDowell, M.T., Ryu, I. et al. (2011). Interconnected silicon hollow nanospheres for lithium-ion battery anodes with long cycle life. *Nano Lett.* 11: 2949–2954. https://doi.org/10.1021/nl201470j.

Yasuda, K., Nohira, T., and Ito, Y. (2005a). Effect of electrolysis potential on reduction of solid silicon dioxide in molten $CaCl_2$. *J. Phys. Chem. Solids* 66: 443–447. https://doi.org/10.1016/j.jpcs.2004.06.037.

Yasuda, K., Nohira, T., Ogata, Y.H., and Ito, Y. (2005b). Direct electrolytic reduction of solid silicon dioxide in molten LiCl–KCl–$CaCl_2$ at 773 K. *J. Electrochem. Soc.* 152: D208–D212. https://doi.org/10.1149/1.2042910.

Yasuda, K., Nohira, T., Amezawa, K. et al. (2005c). Mechanism of direct electrolytic reduction of solid SiO_2 to Si in molten $CaCl_2$. *J. Electrochem. Soc.* 152: D69–D74. https://doi.org/10.1149/1.1864453.

Zamfir, M.R., Nguyen, H.T., Moyen, E. et al. (2013). Silicon nanowires for Li-based battery anodes: a review. *J. Mater. Chem. A.* 1: 9566–9586. https://doi.org/10.1039/C3TA11714F.

Zhang, M., Efremov, M. Yu., Schiettekatte, F. et al. (2000). Size-dependent melting point depression of nanostructures: nanocalorimetric measurements. *Phys. Rev. B* 62: 10548. https://doi.org/10.1103/PhysRevB.62.10548.

Zhou, Z., Zhang, Y., Hua, Y. et al. (2018). Molten salt electrolytic synthesis of silicon-copper composite nanowires with enhanced performances as lithium ion battery anode. *J. Alloys Compd.* 751: 307–315. https://doi.org/10.1016/j.jallcom.2018.04.128.

Zhou, G., Xu, L., Hu, G. et al. (2019). Nanowires for electrochemical energy storage. *Chem. Rev.* 119: 11042–11109. https://doi.org/10.1021/acs.chemrev.9b00326.

Zou, X., Ji, L., Lu, X., and Zhou, Z. (2017). Facile electrosynthesis of silicon carbide nanowires from silica/carbon precursors in molten salt. *Sci. Rep.* 7: 9978. https://doi.org/10.1038/s41598-017-10587-5.

16

Silicon Compositions – Perspectives for Semiconductor Production

16.1 Silicon Carbide

Silicon carbide (SiC), also known as carborundum, is characterized by its high degree of chemical, thermal, and mechanical stability. These properties make the material outstanding in numerous fields of applications such as abrasive and cutting tools, automobile parts, power and hostile environment electronics, etc. SiC is a semiconductor; n-type is doped by nitrogen or phosphorous, while p-type is doped by boron, beryllium, aluminum, or gallium. In particular, porous nanostructured SiC, because of its wide band gap and large surface area, is an outstanding material in numerous fields: catalytic supports, light detecting devices, photocatalytic hydrogen production, gas and chemical sensors, and in a variety of biomedical applications (Wu et al. 2015; Shcherban 2017; Ponraj et al. 2016; Prakash et al. 2015). Carbon-carbide compositions (C–SiC) applied in silicon solar cells showed a number of advantageous properties such as an increased refractive index, hardness, and improved performance of solar cells up to 1.5 times (Klyui et al. 2002).

Electrochemical synthesis of Si–C systems is at an early stage of development. Attempts have been undertaken to produce Si–C composites by facile electrosynthesis in molten salts using silica and carbon powders as the precursors (Zou et al. 2017). The proposed synthesis route is depicted in Figure 16.1. By this route, SiC was obtained in the form of nanowires. The precursor was prepared from SiO_2 and C powder mixture pressed into a pellet. The SiO_2 in the pellet is deoxidized electrochemically by applying a corresponding voltage. Silicon carbonization follows the electrochemical process. Advantages of the method include one-step process and operation without a template and any catalyst. Another concept relies on the electrodeposition from the molten salt electrolyte, which contains SiO_2 and C as electroactive species (Figure 16.1b,c). Appearance of the nanowires (e.g. diameter) could be controlled by tuning the experimental parameters.

Both opportunities, that is, direct SiO_2 deoxidation and Si electrodeposition, have been studied (Zou et al. 2017). The authors succeeded in producing homogeneous SiC nanowires. A pellet was prepared from SiO_2–C powders at a molar ratio (1 : 1), which was sandwiched between porous Ni foils and electrolyzed in molten $CaCl_2$ at a constant potential (~3.1 V) using a graphite anode. The electrochemical processes

Silicon: Electrochemistry, Production, Purification and Applications, First Edition. Eimutis Juzeliūnas.

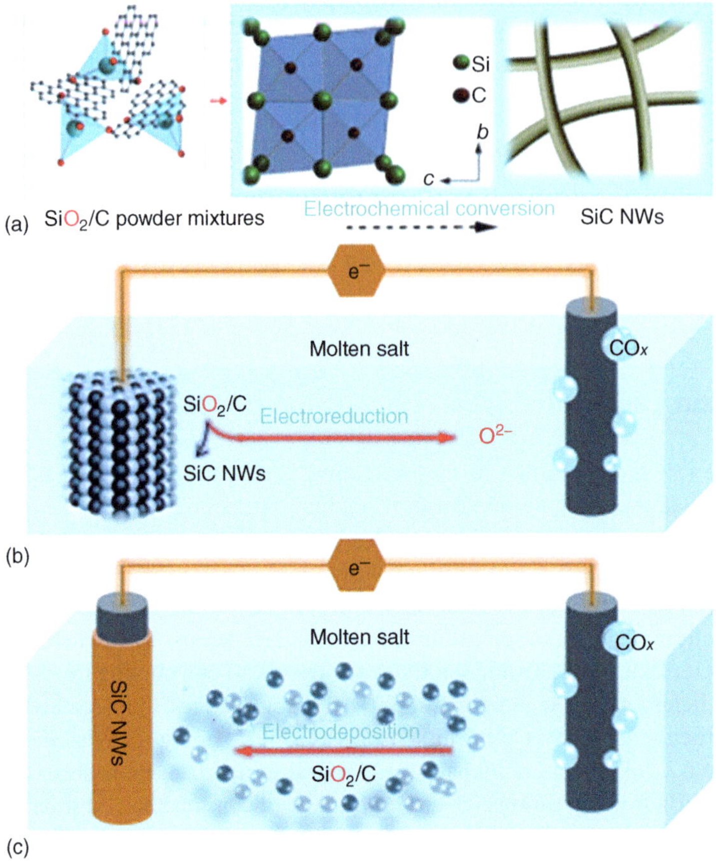

Figure 16.1 Illustration of the synthesis route of SiC nanowires in a molten salt electrolyte: (a) – schematics of the direct electrochemical conversion from SiO_2/C mixture to SiC; (b) – electrochemical reduction of SiO_2/C precursor along with electroreduction from the molten salt electrolyte; oxygen ions O^{2-} are produced, which are converted to CO_x on the graphite anode; (c) – electrodeposition process from the electrolyte, which contains SiO_2 and C particles as electroactive species. Source: From Zou et al. (2017), ELSEVIER. Licensed under CC BY 4.0.

included solid-to-solid conversion $SiO_2 \rightarrow Si$. The process was rather slow: it took about 15 hours to convert 2.0 g of SiO_2/C mixture into SiC NWs (nanowires). The current efficiency was quite low as well, being only about 30%. The Si product contained a small amount of residual carbon. Another approach, that is, Si electrodeposition from the electrolyte, was also applied. To this end, SiO_2 and C particles were dispersed in the molten $CaCl_2$ electrolyte. The SiO_2 particles dissolve in the molten $CaCl_2$ electrolyte to produce silicates. The process comprises silicon reduction from silicates ($Ca_ySi_xO_{2x+z}$), which is followed by codeposition with carbon particles and carbonization ($Si + C \rightarrow SiC$). The authors obtained a mixture of Si and SiC nanowires because of the deficiency of carbon particles in the vicinity of Si nuclei.

Silicon electrodeposition from the SiO_2 particles comprises the steps of silica dissolution, conversion to silicate, electrochemical reduction to silicon, and releasing of the oxygen ions (Zou et al. 2017):

$$x\mathrm{SiO_2} + y\mathrm{Ca^{2+}} + \mathrm{O^{2-}} \rightarrow \mathrm{Ca}_y\mathrm{Si}_x\mathrm{O}_{(2x+z)}\ (x = 1, 2; y = 1, 2, 3; z = 1, 2, 3) \quad (16.1)$$

$$\mathrm{Ca}_y\mathrm{Si}_x\mathrm{O}_{(2x+z)} \rightarrow y\mathrm{Ca^{2+}} + \mathrm{Si}_x\mathrm{O}_{(2x+z)}{}^{2y-} \quad (16.2)$$

$$\mathrm{Si}_x\mathrm{O}_{(2x+z)}{}^{2y-} + 4xe^- \rightarrow x\mathrm{Si} + (2x + z)\mathrm{O^{2-}} \quad (16.3)$$

Carbon-carbide composite (C–SiC) was electrochemically synthesized on a silicon substrate in molten $CaCl_2$ (Juzeliūnas et al. 2018). The precursor was prepared as p-Si with a 300 nm SiO_2 layer, and graphite was used as an anode. Development of the SiC–C composite has been observed at sufficiently negative potentials, at which calcium metal is reduced. A highly developed surface structure has been produced (Figure 16.2), which is promising in the applications where surface-to-volume ratio is of importance. Energy-dispersive X-ray spectroscopy (EDX) identified that the structure is composed of carbon, silicon, and oxygen. It should be noted that the configuration of the contacting electrode (CE) is important for the structural uniformity over the surface. The configuration of the CE is important in terms of uniformity of the potential distribution over silicon surface. The CE geometry may result in some variation in the produced surface properties, as the potential in the vicinity of the Mo–Si contacting points could differ from the potential at

Figure 16.2 Cross section of the Si sample, which shows the surface formed on silicon (dark area) in molten $CaCl_2$ at −1.25 V vs. graphite pseudoreference electrode. A p-Si wafer with a 300 nm SiO_2 layer was used as a precursor. The scale bar is 4 μm. A molybdenum wire contacting electrode and a graphite anode have been applied. Source: Reproduced with permission from Juzeliūnas et al. (2018), ELSEVIER.

distant locations. EDX identified some surface variations in the composition of carbon, silicon, and oxygen. The concentrations of these elements varied from place to place over the surface (in at.%): C – 22.3 to 28.5; Si – 49.2 to 53.4; and O – 23.2 to 30.5.

X-ray diffraction (XRD) showed that the obtained Si–C–SiC composition had a crystalline structure. The composition contained some calcium carbonate. It is important to note that carbonate could be removed from the structure by acid leaching, whereas other components are stable in most acidic solutions.

Information on the oxidation states of the elements could be obtained by X-ray photoelectron spectroscopy (XPS) (Figure 16.3). The method is based on the measurement of the kinetic energy of photoelectrons.

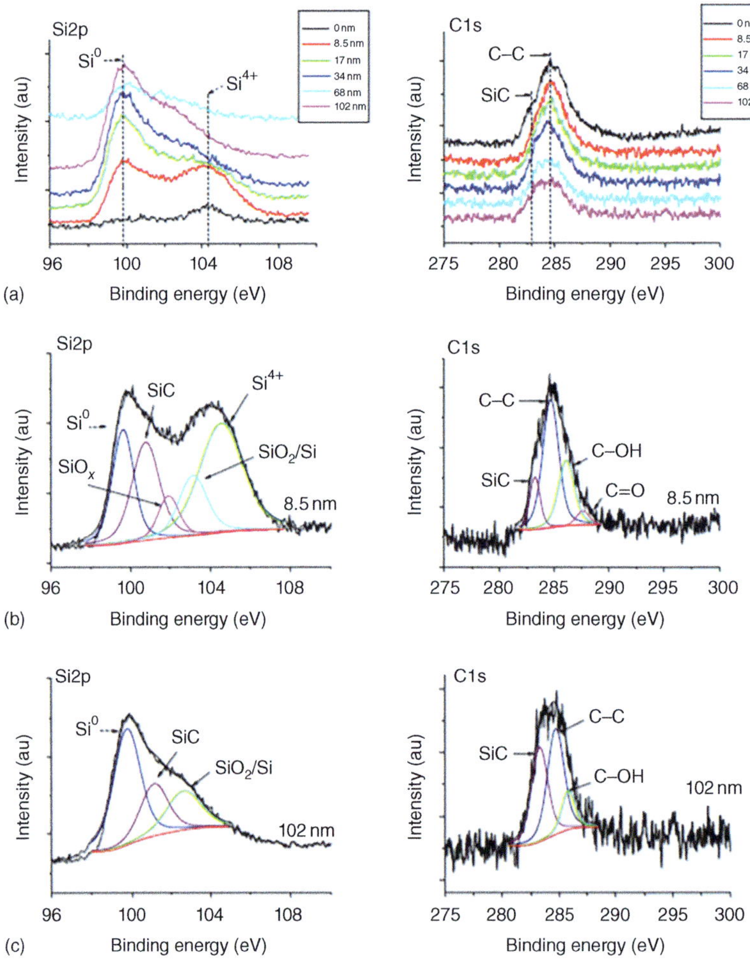

Figure 16.3 Silicon and carbon XPS spectra: (a) at a depth of up to 102 nm obtained when etching the surface by ionized argon, (b) the fitting data showing SiC at 8.5 nm, and (c) 102 nm depth. Source: Reproduced from Juzeliūnas et al. (2018) with permission from the publisher. ©Elsevier, 2018.

The XPS energy peaks identify the oxidation degree of the elements or those in a non-oxidized state. Figure 16.3b,c shows the C1s spectra and fitting curves with the energy peaks at 283.2 and 284.6 eV, which identify Si—C and C—C bonds, respectively (Murota et al. 2012).

The XPS measurements could be performed in conjunction with the surface etching by sample bombardment using ionized argon (Ar^+). In this case, the in-depth concentration profiles are obtained. The produced Si–C–SiC structures exhibited an in-depth variation of the elemental content. Thus, the concentration ratio [SiC]/[C] = 0.2 was found at a depth of 8.5 nm; the ratio increased more than three times in deeper levels, e.g. [SiC]/[C] = 0.7 was detected at a depth of 100 nm. Such an increase is understandable: deeper levels correspond to a longer interaction between the carbon deposits and the Si substrate. Analogous to the XRD data, the XPS curves identified minor amounts of carbonate.

The Si2p energy spectra in Figure 16.3a identify oxidized silicon on the pristine surface. In deeper levels, both oxidized (Si^{4+}) and non-oxidized silicon (Si) are present. No oxidized silicon was found at a depth $d \sim 100$ nm; this means the thickness of the silica film is $d < 100$ nm. The investigations performed by XRD and in-depth XPS confirmed the electrochemical synthesis of the Si–C–SiC structure.

The mechanism of the SiC–C structure formation includes several steps. When the Si–SiO_2 precursor is immersed in molten $CaCl_2$, the silica layer dissolves in whole or in part, forming silicate $SiO_2 \rightarrow SiO_3^{2-}$ as discussed above (16.1 and 16.2). When the voltage is exerted, silicon is reduced either directly from silica by a solid-to-solid mechanism $SiO_2 \rightarrow Si$ or through the electrodeposition from the silicate $SiO_3^{2-} \rightarrow Si$. Both reactions produce oxygen ions O^{2-}. In parallel, calcium metal is produced $Ca^{2+} \rightarrow Ca$. The released O^{2-} diffuses to the graphite anode, interacts with them, and produces the carbon oxide:

$$C + xO^{2-} \rightarrow CO_x \tag{16.4}$$

Some part of the produced gas is emitted outside the cell; however, some part of it dissolves in the electrolyte and diffuses back to the cathode. Finally, CO_x is converted to carbon either by Ca metal or by the electrochemical reduction.

The Si–SiC–C structure exhibited distinctive features of photoresponsiveness. Figure 16.4 compares the photoresponse of the p-Si substrate and the response of the carbon-carbide-modified surface. The plots show photopotential variation at an open circuit when irradiation is switched on and off at certain time intervals. The p-Si sample was with a native oxide film, whose thickness was estimated to be around 1 μm.

The initial potential value of the p-Si electrode is less when compared to the potential of the carbon-modified sample. This implies higher passivity of the carbon-modified sample. A sharp positive peak appears when the sample is irradiated, after which the potential returns to a value in the dark. A vice versa condition is observed when the illumination is terminated: the potential drops down and then returns to the initial value. As commonly known, under illumination of p-Si, minority carriers – electrons – are generated, whereby producing electron–hole (e^-/h^+) pears. The photo-excited Si surface becomes positively charged. A buried

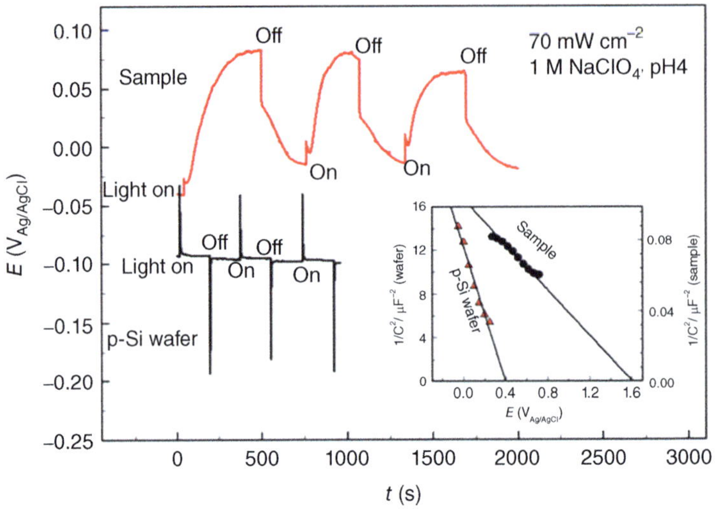

Figure 16.4 Photopotential variation at an open circuit in 1 M $NaClO_4$ (pH 4) when chopping the white light illumination ($N = 70\,mW\,cm^{-2}$) for the carbon-modified sample (upper curve) and p-Si wafer with a native oxide layer ($d \sim 1\,nm$). The inset shows the Mott–Schottky graph obtained for the p-Si wafer and the carbon-modified sample. Source: Reproduced from Juzeliūnas et al. (2018) with permission from the publisher. ©Elsevier, 2018.

solid-state junction Si–SiO_2 exists because of the presence of the native silica film. The band bending in such a case is determined by the contact between Si and the oxide. The film separates Si from the solution or, more precisely, from the Helmholtz layer situated at the oxide/electrolyte interface. Thus, the photo-generated charge carriers are captured within the silicon. Some of the carriers can be injected into the oxide. With the illumination time, the photoelectrons tend to counterbalance the interfacial positive charge returning the potential to an initial value in the dark. When illumination is terminated, the holes at the interface recombine, the positive charge no longer exists and, at the first instance, the sample remains negatively charged by the electrons previously injected into the oxide film. As a result, the potential reduces significantly. With time, the electrons spread within the sample, recombines with the defects, etc., and the electrode potential returns to the initial value. Another concept relies on the non-specific adsorption–desorption of the electrolyte anions, which counterbalance the photo-induced positive Si surface charge.

A distinctive photo-behavior exhibits the carbon-modified sample (Figure 16.4, the curve in red). When illuminated, a positive photopotential shift of ~100 mV is observed, which indicates a p-type semiconductor. When irradiation is terminated, the potential drops down to the initial value in the dark. The surface structure of the modified electrode contains elemental carbon, which does not isolate the electrode; carbon is a good conducting material. In such case, the Helmholtz layer becomes relevant, which forms at the sample–electrolyte boundary. The photoresponse, therefore, is typical of the p-type semiconductor without a passive layer. For instance, a similar shape shows hydrogen-terminated p-Si when the native oxide is removed in hydrochloric acid.

The Mott–Schottky characteristics (Figure 16.4, inset) confirm a *p*-type semiconductivity of the Si–C–SiC structure. This is evident from an increase in electrode capacitance with the electrode potential. The modified electrode showed higher capacitances measured at an open-circuit potential ($C = 3.49 \times 10^{-6}$ F for modified electrode vs. $C = 2.65 \times 10^{-7}$ F for Si). The charge carrier density $N = 1.6 \times 10^{14}$ cm^{-3} was calculated from the plot slope for p-Si using the dielectric constant $\varepsilon_{Si} = 12$. The analogous figure for the carbon-modified sample was much higher: $N = 7.5 \times 10^{16}$ cm^{-3}. This value was calculated, assuming that $\varepsilon_{Si+C+SiC} \approx \varepsilon_{Si}$ ($\varepsilon_{Si} = 11.9$–12.1 (Madelung 1996), $\varepsilon_{SiC} = 9.72$–10.03 (Patrick and Choyke 1970), and $\varepsilon_{graphite} = 12$–$15$ (Speight 2003)). The obtained charge carrier density is rather high; it is over 1 order of magnitude higher than that reported for the carbon nanospheres prepared by chemical vapor deposition ($N = 3.46 \times 10^{15}$ cm^{-3} (Boudjemaa et al. 2015)).

An important feature of the synthesized carbon-modified surface is its high capability to absorb the light. The surfaces absorbed over 90% of the light in a wide range of wavelength from 400 to 1100 nm.

Thus, molten salt electrochemistry shows great potential in producing carbon-modified silicon structures with important functional properties. Till date, however, the method is at the very early stage of development.

16.2 Silicides

Metal silicides are compound semiconductors, which gained attention as materials with excellent optical properties, high hydrogen storage capacity, and superconductivity at certain temperatures and pressures. The intermetallic silicides are also potential candidates for replacing graphite anodes in Li-ion batteries. The band gap of metal silicides could be extended to the visible region by tailoring the elemental composition and the phases (Sakanaka et al. 2015). Metal silicides are typically formed by means of reactive deposition epitaxy (RDE) at high temperatures. Another possibility provides sputter deposition of a metal layer on the Si substrate with subsequent annealing or melting. For example, one could mention the preparation of $BaSi_2$ by arc melting in an Ar atmosphere (Nakamura et al. 2002), the formation of magnesium silicide from magnetron radio frequency-sputtered Mg films and high-temperature treatment (Kogut and Record 2013), and deposition of iron and cobalt silicides by the RDE (Koga et al. 2000; Lim et al. 2005). In these approaches, metal is deposited from a gas phase at high temperatures. To form a calcium silicide ($CaSi_x$) layer, deposition of Ca from a gas phase is also feasible; however, it is difficult to prepare a uniform film because of high vapor pressure of Ca, which can evaporate from the silicon substrate. Electrochemical methods provide an opportunity.

The electrochemical formation of calcium silicides in molten $CaCl_2$ has been discussed by many authors (Xiao and Wang 2014; Nohira et al. 2003; Gao et al. 2005;

Yasuda et al. 2005a,b,c; Juzeliūnas et al. 2012, 2016; Sakanaka et al. 2015). This process plays a significant role in silicon electrode kinetics, silica deoxidation, and formation of surface nano-microarchitectures (Juzeliūnas et al. 2012, 2016, 2018). Sakanaka et al. produced electrochemically Ca–Si films at different cathodic and anodic polarizations vs. the Ca/Ca^{2+} electrode (Sakanaka et al. 2015). The authors obtained multiphase Ca–Si structures on n-Si in a $CaCl_2$–KCl eutectic melt at 923 K (~650 °C) (Figure 16.5). The samples were preserved by rinsing with ethylene glycol. It is important that calcium was deposited not only at negative polarizations with respect to the Ca/Ca^{2+} electrode $\Delta E < E^0_{Ca/Ca^{2+}}$ (e.g. −0.1 V in Figure 16.5a) but also at potentials above the equilibrium potential $\Delta E > E^0_{Ca/Ca^{2+}}$, that is, under underpotential deposition (UPD) conditions (e.g. 0.1 and 0.5 V in the Figure 16.5).

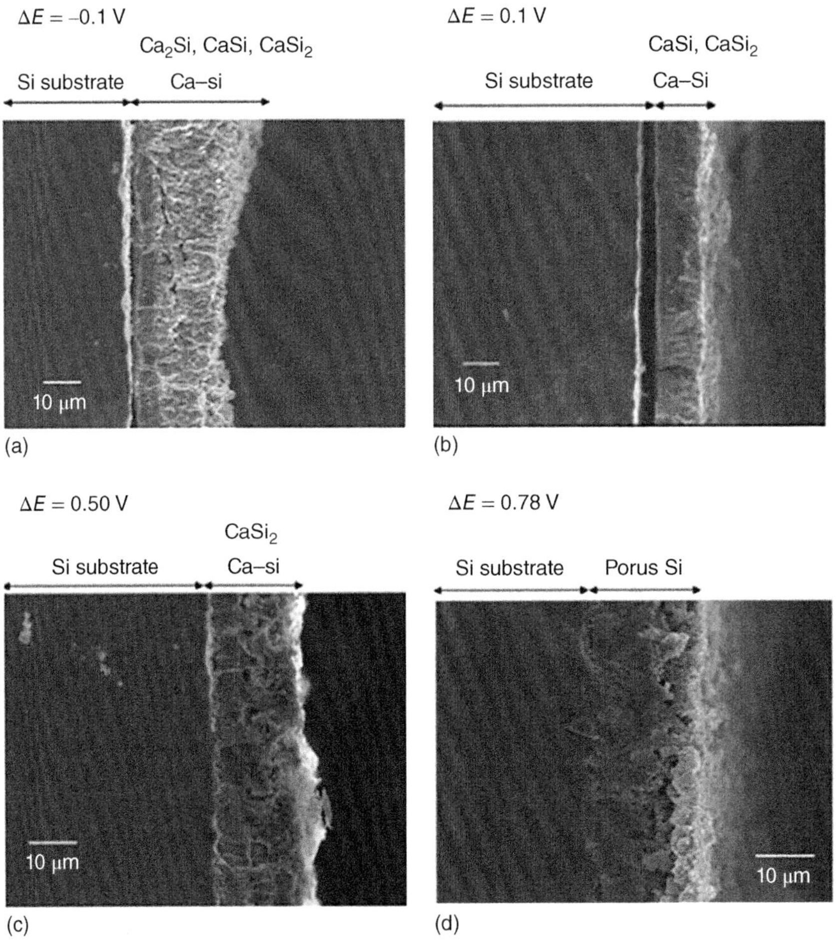

Figure 16.5 Scanning electron microscopy (SEM) image of the cross sections of the Si substrate and Ca–Si structures formed in the $CaCl_2$–KCl molten electrolyte at 923 K and polarizations $\Delta E < E^0_{Ca/Ca^{2+}}$ (a) and $\Delta E > E^0_{Ca/Ca^{2+}}$ (b–d) vs. the Ca/Ca^{2+} reference electrode. Source: Reproduced from Sakanaka et al. (2015), ELSEVIER.

No calcium deposition was observed at $\Delta E = 0.78$ V. It is remarkable that a porous Si structure of about 20 μm thickness has been formed at this polarization. A similar macroporous structure around 10 μm thickness was obtained by the authors who performed electrolysis at −0.9 V vs. graphite in molten $CaCl_2$ (Juzeliūnas et al. 2016). Thus, the electrolysis can be tuned to a regime either for silicide formation or for macroporous surface structuring.

A multiphase Ca–Si alloy deposited at $-0.1\ V_{Ca/Ca^{2+}}$ was converted into CaSi, $CaSi_2$, and Si by tuning the potential of anodic electrolysis. A direct band gap of 3.1 eV was confirmed for the $CaSi_2$ film by reflectance measurements (Sakanaka et al. 2015).

Ca–Si alloys were also prepared in a molten $CaCl_2$–LiCl electrolyte (Sun et al. 2015). The cathodic voltage of about 2.6 V was applied, which yielded calcium and silicon at 850 °C. Six nines purity of silicon (6N-Si) was achieved by applying subsequent zone melting.

As discussed above, a metal silicide layer is typically formed by depositing metals on the Si substrate and subsequent thermal treatment. Electrochemical methods make a vice versa process possible, that is, Si deposition on a metal substrate. Thus, silicon was implanted in iron and nickel electrodes in molten LiCl–KCl–LiF–K_2SiF_6 electrolyte at 450 °C and polarization $E < 1.6$ V vs. Li/Li^+ (Ueda et al. 1995). A Si–Fe layer of about 10–30 μm thickness was obtained by Si plating on Fe in the SiO_2–Li_2O–MgO electrolyte at ~1000 °C (Sasaki and Maeda 2015). Also, the $MoSi_2$ layer was formed by Si deposition on the Mo substrate. Another example provides the siliconizing of steel by pulse electrodeposition from KCl–NaCl–NaF–SiO_2 molten salts, followed by diffusion annealing. A steel layer containing 6.5 mass% of Si has been obtained (Yang et al. 2013).

References

Boudjemaa, A., Mokrani, T., Bachari, K., and Coville, N.J. (2015). Electrochemical and photo-electrochemical properties of carbon spheres prepared via chemical vapor deposition. *Mater. Sci. Semicond. Process.* 30: 456–461. https://doi.org/10.1016/j.mssp.2014.10.050.

Gao, P., Jin, X.B., Wang, D.H. et al. (2005). A quartz sealed Ag/AgCl reference electrode for $CaCl_2$ based molten salts. *J. Electroanal. Chem.* 579: 321–328. https://doi.org/10.1016/j.elechem.2005.03.004.

Juzeliūnas, E., Cox, A., and Fray, D.J. (2012). Electro-deoxidation of thin layer in molten salt – Globular structures with effective light absorbance. *Electrochim. Acta* 68: 123–127. https://doi.org/10.1016/j.electacta.2012.02.054.

Juzeliūnas, E., Coxon, P.R., Fray, D.J. et al. (2016). Photoelectrochemical activity of macroporous silicon formed in molten calcium chloride. *Chemija* 27: 11–16. http://mokslozurnalai.lmaleidykla.lt/chemija/2016/1/7224.

Juzeliūnas, E., Fray, D.J., Kalinauskas, P. et al. (2018). Electrochemical synthesis of photoactive carbon-carbide structure on silicon in molten salt. *Electrochem. Commun.* 90: 6–10. https://doi.org/10.1016/j.elecom.2018.03.003.

Klyui, N. I.; Litovchenko, V. G.; Rozhin, A. G.; Dikusha, V. N.; Kittler, M.; Seifert, W. (2002) Silicon solar cells with antireflection diamond-like carbon and silicon carbide films. *Sol. Energy Mater. Sol. Cells*, 72, 597–603. https://doi.org/10.1016/S0927-0248(01)00209-4

Koga, T., Bright, A., Suzuki, T. et al. (2000). Growth of β-$FeSi_2$ and FeSi layers by reactive deposition using Sb-related intermetallic compounds. *Thin Solid Films* 369: 248–252. https://doi.org/10.1010/S0040-6090(00)00817-8.

Kogut, I. and Record, M.-C. (2013). Growth of magnesium silicide thin films on Si(100), Si(111) and SOI substrates during rapid thermal processing. *Intermetallics* 32: 184–193. https://doi.org/10.1016/j.intermet.2012.08.029.

Lim, C.W., Shin, C.-S., Gall, D. et al. (2005). Growth of $CoSi_2$ on Si(001) by reactive deposition epitaxy. *J. Appl. Phys.* 97: 044909. https://doi.org/10.1063/1.1774263.

Madelung, O. (ed.) (1996). *Data, Semiconductors – Basic*, 2e, 20. Berlin: Springer-Verlag.

Murota, J., Kikuchi, T., Hasegawa, J., and Sakuraba, M. (2012). Strain control of Si and $Si_{1-x-y}Ge_xC_y$ layers in Si/$Si_{1-x-y}Ge_xC_y$/Si heterostructures by low-pressure chemical vapor deposition. *ECS Trans.* 50: 245–254. https://doi.org/10.1149/05009.0245ecst.

Nakamura, T., Suemasu, T., Takakura, K., and Hasegawa, F. (2002). Investigation of the energy band structure of orthorhombic $BaSi_2$ optical and electrical measurements and theoretical calculations. *Appl. Phys. Lett.* 81: 1032. https://doi.org/10.1063/1.1498865.

Nohira, T., Yasuda, K., and Ito, Y. (2003). Pinpoint and bulk electrochemical reduction of insulating silicon dioxide to silicon. *Nat. Mater.* 2: 397–401. https://doi.org/10.1038/nmat900.

Patrick, L. and Choyke, W.J. (1970). Static dielectric constant of SiC. *Phys. Rev. B: Condens. Matter* 2 (6): 2255–2256. https://doi.org/10.1103/PhysRevB.2.2255.

Ponraj, J.S., Dhanabalan, S.C., Attolini, G., and Salviati, G. (2016). SiC nanostructures toward biomedical applications and its future challenges. *Crit. Rev. Solid State Mater. Sci.* 41: 430–446. https://doi.org/10.1080/10408436.2016.1150806.

Prakash, J., Venugopalan, R., Tripathi, B.M. et al. (2015). Chemistry of one dimensional silicon carbide materials: principle, production, application and future prospects. *Prog. Solid State Chem.* 43: 98–122. https://doi.org/10.1016/j.progsolidstchem.2015.06.001.

Sakanaka, Y., Goto, T., and Hachiya, K. (2015). Electrochemical formation of Ca-Si in molten $CaCl_2$-KCl. *J. Electrochem. Soc.* 162: D186–D191. https://doi.org/10.1149/2.1021504jes.

Sasaki, H. and Maeda, M. (2015). Siliconizing of iron and molybdenum by electrochemical reduction of silicon in molten SiO_2–Li_2O–MgO. *J. Alloys Compd.* 641: 64–68. https://doi.org/10.1016/j.jallcom.2015.04.037.

Shcherban, N.D. (2017). Review of synthesis, structure, physical and chemical properties and functional characteristics of porous silicon carbide. *J. Ind. Eng. Chem.* 50: 15–28. https://doi.org/10.1016/j.jiec.2017.02.002.

Speight, J.G. (2003). *Perry's Standard Tables and Formulas for Chemical Engineers*, 67. New York: The McGraw-Hill Companies Inc.

Sun, D.-Y., He, L.-W., Lian, J.-Q. et al. (2015). Process for solar grade silicon production by molten salt electrolysis and zone melting. *J. Mater. Eng.* 43: 73–78. https://doi.org/10.3969/j.issn.1001-4381.2013.06.001.

Ueda, T., Goto, T., and Ito, Y. (1995). The implantation of Si and formation of silicide by molten salt electrochemical process. *J. Surf. Finish. Soc. Jpn.* 46: 1173–1179. https://doi.org/10.4139/sfj.46.1173.

Wu, R., Zhou, K., Yue, C.Y. et al. (2015). Recent progress in synthesis, properties and potential applications of SiC nanomaterials. *Prog. Mater Sci.* 72: 1–60. https://doi.org/10.1016/j.pmatsci.2015.01.003.

Xiao, W. and Wang, D. (2014). The electrochemical reduction process of solid compounds in high temperature molten salts. *Chem. Soc. Rev.* 43: 3215–3228. https://doi.org/10.1039/c3cs60327j.

Yang, H.-L., Shang, L., Tang, G.-Z. et al. (2013). Molten salt pulse electrodeposition of silicon on low silicon steel. *Mater. Trans.* 54: 1006–1011. https://doi.org/10.2320/matertrans.M2012257.

Yasuda, K., Nohira, T., and Ito, Y. (2005a). Effect of electrolysis potential on reduction of solid silicon dioxide in molten $CaCl_2$. *J. Phys. Chem. Solids* 66: 443–447. https://doi.org/10.1016/j.jpcs.2004.06.037.

Yasuda, K., Nohira, T., Ogata, Y.H., and Ito, Y. (2005b). Direct electrolytic reduction of solid silicon dioxide in molten LiCl–KCl–$CaCl_2$ at 773 K. *J. Electrochem. Soc.* 152: D208–D212. https://doi.org/10.1149/1.2042910.

Yasuda, K., Nohira, T., Amezawa, K. et al. (2005c). Mechanism of direct electrolytic reduction of solid SiO_2 to Si in molten $CaCl_2$. *J. Electrochem. Soc.* 152: D69–D74. https://doi.org/10.1149/1.1864453.

Zou, X., Ji, L., Lu, X., and Zhou, Z. (2017). Facile electrosynthesis of silicon carbide nanowires from silica/carbon precursors in molten salt. *Sci. Rep.* 7: 9978. https://doi.org/10.1038/s41598-017-10587-5.

17

Silicon Photo-Electrodes for Water Splitting and Their Protection

17.1 Relevance, Basic Principles, and Semiconductor Materials for Photo-Electrodes

After the intensive exploitation of nonrenewable sources over the past century, the current civilization faces threats from climate change and fossil resource depletion. The use of alternative energy is an essential condition for better quality of life. The future energy should be renewable, with minimum or zero greenhouse emission, and economically sustainable.

Collection of the solar energy in chemical bonds offers a way for sustainable energy economy. Such solar-to-chemical conversion is similar to that achieved by photosynthesis in plants. Conceptually, the production of hydrogen in combination with solar energy occurs with zero greenhouse gas emissions. When captured in chemical form (e.g. hydrogen, H_2), the energy can be converted to power in fuel cells with specifically high efficiency (up to 85%) or in combustion engines. It should also be possible to combine hydrogen generation with a fuel cell to generate electricity while replacing the water that has been consumed.

Photoelectrochemical conversion of CO_2 to fuels is also of great interest. Solar-energy-based CO_2 recycling into transportable hydrocarbon fuel would reduce the CO_2 levels and, at the same time, provide the hydrocarbon as a feedstock for chemical synthesis or as a fuel. Photocatalytic conversion of CO_2 to hydrocarbons was achieved on photosensitive semiconductor powders suspended in water under ultraviolet (UV) illumination (Inoue et al. 1979). The following conversion products were obtained: formic acid (CH_2O_2), formaldehyde (CH_2O), methyl alcohol (CH_3OH), and methane (CH_4)

$$CO_2 + H_2O \rightarrow CH_4, CH_3OH, CH_2O, CH_2O_2 + O_2 \tag{17.1}$$

Varghese et al. reported high-rate solar photocatalytic CO_2 conversion to hydrocarbon fuels in the presence of water vapor (Varghese et al. 2009; Roy et al. 2010). The carbon dioxide and water vapor were converted to methane using nitrogen-doped TiO_2 nanotube arrays. Platinum and copper have been used as co-catalysts. The experiments were performed outside under the sunlight. The photocatalytic process also yielded hydrogen and carbon monoxide as intermediate products. A hydrocarbon production rate of 160 $\mu g\, g^{-1}\, h^{-1}$ was achieved using Cu

Silicon: Electrochemistry, Production, Purification and Applications, First Edition. Eimutis Juzeliūnas.

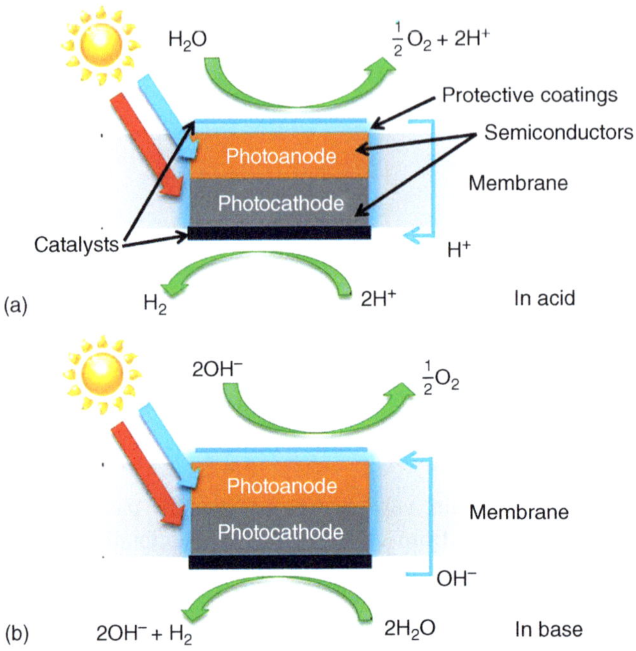

Figure 17.1 Diagram of a light-induced water splitting into hydrogen and oxygen by a tandem-absorber cell operating in (a) acidic and (b) alkaline solutions. The photoanode and photocathode are arranged in series with respect to the direction of the incident light. The light conversion catalysts are applied on the surface of the absorber, so that the catalysts are in contact with the electrolyte. The absorbers are embedded in a gas-impermeable ion-exchange membrane. The semiconductor electrodes are protected by a coating, which passivates the top and the sides of the photo-electrodes. The protective layers allow attachment of the catalysts on the outer surface. Source: Reproduced form (Hu et al. 2015) with permission from the publisher. © 2015 American Chemical Society.

and Pt nanoparticles on the TiO_2 nanotube array under AM 1.5 ($100\,mW\,cm^{-2}$) sunlight; this result was estimated to be about 20 times higher than that obtained by other authors, who used UV illumination.

Various solar-to-chemical reactions offer opportunity for solar energy accumulation and transmission. The most attractive process is the hydrogen and oxygen generation from water.

Figure 17.1 shows the basic principle of light-induced water splitting into hydrogen and oxygen by the semiconductor photo-electrodes arranged in series with respect to the direction of the incident light. The cell design includes the protective coatings with light conversion catalysts at the top. The absorbers are embedded in a gas-impermeable ion-exchange membrane. The schematics show the water-splitting reactions both in acid and alkaline solutions. In an acid medium, protons are reduced to hydrogen $H^+ \rightarrow H_2$ on the photocathode, while water is oxidized to oxygen $H_2O \rightarrow O_2$ on the photoanode. In alkaline solution, the photocathode produces hydrogen from water $H_2O \rightarrow H_2$, and the photoanode produces oxygen from hydroxide ions $OH^- \rightarrow O_2$. The photo-oxidation–reduction process can also take place in halide electrolytes, such as seawater, which contain on average 3.5% of NaCl. In such electrolytes, the halides are oxidized to halogens (e.g. $Cl^- \rightarrow Cl_2$) rather than water to oxygen.

Sunlight and seawater provide unlimited resources for sustainable energy generation and its storage in chemical form. Ocean saltwater can be used as the electrolyte and the hydrogen feedstock simultaneously; the resources are widely available for free. The solar energy is also for free. The solar radiation, which reaches the Earth, provides about 100 000 terawatts energy, which covers about 10 000 times the annual global energy consumption (Grätzel 2007). Breeze estimates that if 0.1% of the solar energy were converted into electricity with 10% efficiency, it would exceed the world energy needs (Breeze 2019). Solar energy provides the opportunity for decentralized supply of energy actually at any location around the globe.

The (photo)electrochemical hydrogen production can be performed with a very low carbon footprint. Nearly zero greenhouse emission is achieved when hydrogen is generated by splitting water using renewable energy sources such as solar, wind, hydro, or ocean. Renewable energy is irregular and sporadic in nature; thus, hydrogen generation is applicable for energy storage and load management. Technologies and demo plants have been demonstrated while connecting the electrolyzers to electricity and gas grids (Bertucioli et al. 2014). The technological processes integrating renewable electricity generation and the production of hydrogen have also been tested (Gahleitner 2012).

Electrochemical hydrogen generation means using electrons as absolutely clean agents, albeit, at slightly lower efficiency, but without pollutant emissions. The electrochemical process offers such advantages as high purity, flexibility, supply on-site, and on-demand as well as fast start-up and shutdown opportunities. Of particular interest is solar-to-fuel process, which occurs directly (wireless) on photo-electrodes.

A challenge is the development of sustainable electrodes for the electrolysis with such key performance indicators as high efficiency, cost reduction, and enhancement of lifetime. Materials used in solar devices have a larger share in the cost of the overall utilization process of solar energy. The creation and synthesis of new effective materials for solar energy applications, therefore, is very high on the agenda of materials scientists and engineers. Numerous photo- and electro-catalytic materials have been studied during the past decades: oxides, nitrides, III–V semiconducting compounds, sulfides, etc. Grätzel reviewed in 2001 the historical background and development prospects of photoelectrochemical cells (PECs) (Grätzel 2001). Sun et al. reviewed the major milestones for photocatalytic and photoelectrochemical water splitting (Sun et al. 2014). The effectiveness of solar energy conversion depends largely on the ability of PEC device to absorb the light. Silicon (both p and n types) as a photosensitive, chemically inert, low-cost, and robust material is very attractive for photoelectrochemical applications.

Fujishima and Honda proposed in 1972 the solar water splitting using titanium oxide (TiO_2, rutile) generation by photoelectrochemical splitting of water (Fujishima and Honda 1972). Furthermore, the photoelectrochemical splitting of water became one of the hottest multidisciplinary research fields (Dresselhaus and

Thomas 2001; Sun et al. 2014a,b). Fujishima's and Honda's article received about 23 700 citations, as of May 2022.

A monolithic photovoltaic-photoelectrochemical device for solar-to-hydrogen power conversion was proposed (Khaselev and Turner 1998). The design of the PEC was based on the p–n junction formed by GaAs and p-GaInP$_2$ semiconductors. The hydrogen production efficiency was estimated from the short-circuit current to be as high as 12.4% at illumination power of 500 mW cm^{-2}.

The first photocatalysts suitable for water splitting (e.g. (Fujishima and Honda 1972)) is operated at higher solar energy spectrum, that is, under UV illumination. This spectrum range accounts for only 4% of the incoming solar energy amount, which limits the overall process performance (Zou et al. 2001). Considerable efforts, therefore, have been directed toward developing the photocatalysts, which explore the visible spectrum of the light. This wavelength region is less active but more abundant when compared to the UV region. Visible light accounts for ~43% of the entire solar energy.

Zou et al. used the indium-tantalum-oxide ($In_{1-x}Ni_xTaO_4$, $x = 0 - 0.2$) electrodes with nickel oxide (NiO_y) nanoparticles as a dopant (Zou et al. 2001). They produced series of photocatalysts, which split the water into stoichiometric amounts of O_2 and H_2 under visible light. Such splitting is referred to as overall water splitting. The performance of 0.66% was achieved under visible irradiation in pure water.

Another material proposed for photo-electrodes was the solid solution of GaN and ZnO modified with RuO_2 nanoparticles (Maeda et al. 2005). One major advantage of this electrode is that overall water splitting (into stoichiometric amounts of H_2 and O_2) was achieved at a visible light region. The electrodes shoved the quantum yield up to 5.9% in pure water when illuminated with the visible light at 420–440 nm wavelength. The solid solution electrode, in contrast to CdS photocatalyst, showed stability during the water-splitting reactions; it was resistant against disintegration by the developed gaseous phase.

M. Grätzel's group proposed a highly active sophisticated photocathode for photoelectrochemical hydrogen production, which was based on the copper oxide layer (Cu_2O) deposited onto fluorine-doped tin oxide (FTO, NGS glass) substrate (Paracchino et al. 2011). The FTO substrate was thermally coated by a gold layer. The Cu_2O film 1.3 μm in thickness was electrodeposited from an alkaline (pH 12) sulfate solution with additives of 3 M lactic acid and 0.5 M K_2HPO_4 buffer. The Cu_2O surface texture was a pyramid-like one with distinctive grains of about 1 μm in size with predominant <111> orientation. Such texture promotes the light absorption due to light scattering and trapping; similar structures were achieved, for instance, when etching silicon in fluoride electrolytes, as discussed in Section 14.2. The copper oxide is electrochemically unstable; it can be reduced to copper $Cu_2O \rightarrow Cu$ during the electrolysis. To suppress this detrimental process, Cu_2O surface was passivated with a protective oxide layer, which prevents Cu_2O contact with the electrolyte. The protective layers of ZnO, Al_2O_3, and TiO_2 have been applied, which were deposited by the atomic layer deposition (ALD). The deposition temperature was 120 °C; the protective stability was further improved by subjecting the protective layers to mild heat treatment at 200 °C. The stability increased due to removal of the residual water from the ALD layers, as water has been used as an oxygen precursor in the ALD. Precursors of metals included diethyl zinc, trimethyl aluminum, and titanium tetraisopropoxide. To enhance hydrogen reduction, Pt nanoparticles were used as the catalysts. This has been done potentiostatically from a solution 1 mM H_2PtCl_6. The FTO and Au substrates were isolated from the Pt deposition by an epoxy resin. The photo-responsivity was measured using 450 W Xe lamp equipped with an UV filter. Photocurrents up to −7.6 mA cm^{-2} at 0 V vs. RHE were demonstrated. The authors believed that this was the highest photocurrent achieved on an oxide-based photoelectrode for water splitting under AM 1.5 (100 mW cm^{-2}) illumination. The power conversion efficiency at 0 V vs. RHE was measured as high as 9.3% under AM 1.5 illumination. The Faradaic efficiency was found to be close to 100% for the water reduction. The work discussed above marks a significant milestone in the development of photoelectrochemical water-splitting technology. At the same time, it should be noted that industrial-economical viability of the process is limited due to technical complexity and, in turn, high processing cost.

Black TiO_2 for solar-to-hydrogen power conversion with 24% efficiency in water–methanol electrolyte has been reported (Chen et al. 2011). It is known that TiO_2 absorbs only UV light. To narrow the TiO_2 band gap and enhance solar absorption, dopants, such as nitrogen, have been used. Chen et al. demonstrated conceptually different approach, which is based on disorder engineering of the TiO_2 surface on a nano-scale through hydrogenation. The disordered nanocrystals exhibited substantial photo-activities, such as photo-oxidation of organic compounds in water or hydrogen production (Chen et al. 2011). Efficiency of solar-to-hydrogen power conversion of 24% was achieved in water–methanol electrolyte.

Synthesis, properties, and applications of TiO_2 nanomaterials have been reviewed (Chen and Mao 2007). The authors surveyed syntheses of TiO_2 nanoparticles, nanorods, nanowires, and nanotubes. Preparation of meso-nano-porous TiO_2, as well as aerogels, opals, and photonic materials was also summarized. Wide scope of nanomaterials for renewable energy production with emphasis on the photocatalytic hydrogen production have been reviewed (Chen et al. 2012).

Das et al. proposed the sulfur-capped CdSe with cobalt benzene-1,2-dithiolate complex $[Co(bdt)_2]^-$ as a catalyst (Das et al. 2013). The electrode showed the quantum yield of 24% at visible light (520 nm) in a slightly acid water (pH 4.5). The one-sulfur or two-sulfur capping donors with attached carboxylate were tightly bound to CdSe surface. The donors stabilized dissociation of CdSe nanocrystals. The electrode showed high durability; it can withstand more than 300 000 cycles during 60 hours timespan.

Liao and coworkers studied the cobalt oxide nanoparticles (CoO NPs) as photocatalysts for solar-to-hydrogen production (Liao et al. 2014). The nano-catalysts have been synthesized from CoO micropowders using femtosecond laser ablation or mechanical ball milling. The electrode reached 5% performance at 100 mW cm^{-2} visible light illumination in neutral water. It has been shown that the nanoparticle catalyst can decompose pure water without any co-catalyst or sacrificial reagents. The authors reported that photoactivity of the photocatalyst arises from a shift in the band edge.

Silicon being a nontoxic, efficient, low cost, and robust material will remain dominant in the solar energy market for the next few decades (Schmalensee et al. 2015; Green 2016; Polman et al. 2016). Narrow band gap of silicon (~1.1 V) properly matches the solar spectrum. Mitigation of climate change, as a global task, could be achieved using technologies based on Earth-abundant materials. Silicon is the second most abundant in the Earth's crust; its content is 27%. These positive features make it possible to expand application of silicon to terawatt scale by 2050 (Schmalensee et al. 2015). Vast majority (over 90%) of solar cells are produced from silicon where total wafer cost represents more than 60% of the overall cell costs.

The energy of 1.229 V is needed for semiconductor/electrolyte interfaces in order to split electrochemically water into oxygen and hydrogen (Bard et al. 1985; Sun et al. 2014a,b). This is a fundamental thermodynamic value. In practice, a higher voltage is needed due to such limitations as the reaction overpotentials, the ohmic losses in the cell, as well as those due to transportation of reactants and products. It is assumed that the realistic voltage range is at least 1.5–1.6 V (Sun et al. 2014a,b). This translates to a requirement of minimum energy band gap of 1.6 V for a semiconductor in order to provide the energy of 1.229 V to the electrochemical cell, which split the water without external energy supply. Such evaluation was based on Wrighton's report of photoelectrochemical conversion of optical energy to electricity and fuels (Wrighton 1979). The band gap for crystalline silicon is 1.1 V, and, theoretically, it can provide the photovoltage of 0.8 V (Sun et al. 2014a,b). Thus, the energy is insufficient to split the water entirely into H_2 and O_2 without external energy assistance. However, the silicon photoelectrode is well suited for a half-reaction of the splitting: the energy of the Si conduction band is greater than the redox potential of the H^+/H electrode. The cathodic reaction can only be performed when another electrode – the electron donor – is used, on which oxygen is generated. The redox potential of this electrode should be more negative in comparison with the redox potential of the reaction of oxygen evolution.

Kinetics of (phot)electrochemical water splitting is determined mainly by the electron transfer: two electrons for the hydrogen reaction and four electrons for the oxygen reaction. These multi-electron processes are usually sluggish on semiconductor electrodes; thus, a heterogeneous catalyst is needed to enhance the electron transfer and improve process kinetics. Competing thermodynamic reactions and slow charge transfer on bare silicon are assumed to be main critical issues to be solved (Sun et al. 2014a,b). Another issue is poor light absorption. Surface engineering methods to improve the absorbance have been widely studied, as discussed in Section 14.2. The surface modifications on a nano-micro-level do not only increase the light absorbance by its internal trapping, it also modifies the density of surface states and improve the charge transfer kinetics, that is, the electrochemical reaction rate and efficiency.

Candea et al. first reported current–voltage measurements for heavily doped p-Si electrodes in NaCl solution (Candea et al. 1976). The voltage at open circuit depended logarithmically on the illumination power; this indicated the photoelectrochemical responsiveness of the silicon electrodes. The authors concluded that silicon was a poor hydrogen electrode, and this was attributed to its surface oxide. Thereafter, silicon has widely been investigated as a photoelectrode. Sun et al.

summarized in their review that silicon is one of the most studied electrodes in the semiconductor photoelectrochemistry (Sun et al. 2014a,b).

The main disadvantage of silicon is its poor intrinsic ability to absorb the light. More than 30% of incident light is reflected if silicon surface is not specifically textured or coated by antireflection coatings. Also, high processing complexity of silicon significantly contributes to the cost of PV devices.

Much effort has been made to increase the actual surface area of silicon. Absorbance of silicon was enhanced by increase in its roughness at micro- and nano-scale. The applied techniques include femtosecond laser engineering with subsequent sample etching, ALD, annealing in vacuum or coating by thermal oxide chloroalkylsilane layers, deposition of transparent conducting oxides using thermal ALD, deep reactive ion etching using oxygen inhibitor pulses, chemical etching using Au catalysts, inductively coupled plasma and cryogenic etching, as well as vapor–liquid–solid (VLS) reactions (Section 14.2). Most of these methods, however, are too costly and technically sophisticated to provide cost-competitive solutions for industrial-scale applications. Furthermore, etchings usually involve toxic chemicals, such as hydrofluoric acid, and rather expensive catalysts, such as gold or silver.

Of great interest is the nanoporous silicon, known as black silicon, which effectively absorbs the light in a wide range of wavelengths due to internal trapping by porous structures. An 18.2%-efficient black-silicon solar cell without an antireflection layer was reported (Oh et al. 2012a,b). The solar cell efficiency of 22.1% was achieved using black silicon covered by 20 nm Al_2O_3 film to passivate the silicon and reduce the effects of charge carrier recombination (Savin et al. 2015). Black silicon photocathode for hydrogen production has been reported (Oh et al. 2011) and patented (Oh and Branz 2014). It has been shown that nanostructured black silicon significantly accelerates hydrogen production with reduced need for surfactants.

Several techniques have been proposed to fabricate black silicon: laser texturization, plasma immersion ion implantation, metal-assisted etching, or cryogenic deep reactive ion etching (Section 14.2). These techniques, however, are rather costly. There is a great demand for the cost-effective and green technologies of black silicon mass production.

There is interest in development of PECs with effective light harvesting, which is based on silicon nano-micro architectures produced by means of molten salt electrochemistry (Fray and Juzeliūnas 2011, 2013, 2014a,b, 2015a,b,c). The proposed one-step methods of surface modifications are environment friendly, cost-effective, and have good prospects for commercialization. The field, however, is at the early stage of investigations; therefore, technical solutions are still pending.

17.2 Protection of Silicon Photoelectrodes in Solar-Fuel Generators

The solar-fuel generators, which capture the solar energy and convert it to chemical bonds, are composed of photocathodes and photoanodes. Two types of photoelectrochemical systems with semiconductor–liquid junctions are possible. The first type explores one semiconductor electrode in connection with a metal electrode. The second type uses two semiconductor electrodes, one is an n-type semiconductor and the second is a p-type semiconductor. Irradiation of an n-type electrode promotes photo-oxidation process, e.g. by reaction $H_2O \rightarrow O_2$, which results in a photoanodic current. The behavior of a p-type electrode is opposite: a photocathodic current flows due to promotion of a cathodic reaction, for instance, the hydrogen reduction reaction $H^+ \rightarrow H_2$. These phenomena relate to the electric field direction generated when a semiconductor comes into contact with an electrolyte, which contains a redox couple with a redox potential in the region of semiconductor's band gap. The space charge region of a p-type semiconductor moves the photoelectrons toward the surface and the holes in the opposite direction so that photo-reduction is enhanced. On the contrary, the field in an n-type semiconductor delivers holes to the surface where photo-oxidation is facilitated. These basic phenomena are widely discussed in handbooks and review articles, e.g. (Bard and Faulkner 2001; Nozik and Memming 1996; Hu et al. 2015).

Electrolytic properties are essential for the conversion efficiency of the solar-fuel generators. Efficient water splitting is considered when the electrolyte can support the photocurrents typically not less than 10 mA cm^{-2} under standard 1 Sun illumination (AM 1.5 or 1 kW m^{-2}). This value corresponds to efficiency of solar-to-hydrogen conversion not less than 12.3% (Coridan et al. 2015; Hu et al. 2015). Acidity gradients in the electrolyte lead to limitations in the transportations of protons and other ionic species; as a result, the process encounters the energy losses in the photo-electrochemical cell. To avoid such limitation, strongly acidic or alkaline electrolytes should be used. The media with $pH < 1$ or $pH > 13$ are capable of supporting the condition of the sufficiently high currents without significant pH gradients (Hernandez-Pagan et al. 2012; Hu et al. 2015). However, such media are highly corrosive, and a good number of semiconductors are unstable there. Silicon is covered by a native oxide film, which makes the material stable in acid and neutral solutions. However, the oxide dissolves in alkaline solutions leaving the substrate unprotected; silicon, therefore, actively dissolves in an alkaline medium. In the dark, silicon dissolves in alkaline electrolytes in a wide range of potentials so that shifting its potential cannot stop its dissolution. This implies a chemical process, which takes place along with an electrochemical process. The rate of chemical process does not depend on electrode potential, whereas the rate of the electrochemical process is determined by the potential.

Development of effective protection schemes and protective coatings is essential for creation of technologically viable PECs for water splitting. One approach of stabilization is using organic electrolytes. For instance, a solar cell was designed based on n-type Si in methanol (Gronet et al. 1983). The cell contained 0.2 M (1-hydroxyethyl)ferrocene and 0.5 mM (1-hydroxyethyl)ferricenium as electroactive species. The cell exhibited the efficiencies of 10.1% for optical energy conversion into electricity.

Another approach is surface functionalization by organic compounds. For instance, the HF-etched Si <111> was modified with olefins and organic diacyl peroxides (Linford et al. 1995). It was hypothesized that a self-assembled monolayer of Si-alkyls was formed. The surface of Si <111> can be almost completely functionalized by the methyl group. Bansal et al. reported a strategy to functionalize HF-etched Si, which involved halogenation and subsequent reaction with alkyl Grignard or alkyl lithium reagents (Bansal et al. 1996). The methyl group is stable against reductive degradation; thus, it can effectively protect the photo-cathodes. The aliphatic bonds are instable with respect to anodic oxidation; this makes the methyl-terminated silicon hardly applicable as photoanodes. Another example is the study by Bookbinder et al., who functionalized p-type silicon photocathodes with an *N,N′*-dialkyl-4,4′-bipyridinium redox reagent (Bookbinder et al. 1980). This reagent along with a Pt catalyst increased the conversion efficiency by 5% for photoelectrochemical H_2 generation with 632.8-nm light.

Furthermore, we will focus on the application of thin films as protective barriers for silicon stabilization in an aggressive medium. The films should be sufficiently conductive in order not to isolate the surface with respect to charge transfer reaction, but at the same time, the film should isolate the surface from direct contact with electrolyte. The protective films should also be resistant to electrochemical decomposition–disintegration. Other prerequisites are the optical transparence and the anti-reflectivity of the films. Catalytic acceleration of the charge transfer reaction is also highly desirable. Certainly, creation of the protective films, which would fulfill such complex factors, is not a trivial task for the researchers.

The protective films can be formed by the physical vapor deposition (PVD) techniques, such as magnetron sputtering (MS), pulsed laser deposition, or electron beam evaporation. Radio frequency (RF) sputtering is well suited to produce films of highly insulating oxides. The MS techniques are based on ejecting the atoms by sputtering from a target and deposition on a substrate, such as silicon wafer, from a gas phase. MS technique employs magnetrons that utilize strong electric and magnetic fields and confines plasma particles close to the surface of the sputter target. The method precisely controls the deposit composition; the deposited films have a composition close to that of the source material, that is, the composition of the target. The drawback, however, is that sputtered atoms and ions have a high energy and wide energy distribution range, which can damage the semiconductor surface structure.

Ultrathin passivating layers, whose thickness is on the order of nanometers, can be formed using the ALD. This method belongs to a family of chemical vapor deposition (CVD) techniques. ALD utilizes self-limiting surface reactions, which are introduced sequentially. In each alternate pulse, the precursor reacts with the substrate, and this reaction is self-limited; the reaction stops when the reactive sites on the substrate are completely used. The deposit is grown when repeating the pulses. A variety of materials can be deposited by ALD, such as oxides, sulfides, fluorides, as well as pure metals. Thickness of these films can be precisely controlled, as the film is formed layer-by-layer on an atomic scale. Another advantage of ALD is a possibility to deposit the films uniformly on porous materials. The precursor can reach the locations inside the pores with high-aspect ratio.

Hu et al. summarized comprehensively the achievements and new directions in the field of protection of photoelectrodes by thin-film materials (Hu et al. 2015). The data were compiled from numerous reports on the strategies of application of thin films to stabilize photoanodes and photocathodes. Of the variety of electrode materials, n-Si photoanodes and p-Si photocathodes have a distinctly dominating share.

17.2.1 Protection of Si Photoanodes

Yoshihiro et al. studied n-Si photoanodes coated with thin-film catalysts of precious metal, such as Au and Pd (Yoshihiro et al. 1975). A catalytic effect of the metal films was found, which induced photocurrents in aqueous solutions, as the authors demonstrated in borate buffered 0.5 M K_2SO_4 at pH 9.2. The photoelectrodes were stable in the alkaline solution; they did not react with alkaline solution, neither were they electrochemically dissolved. These results were of fundamental importance for the development of solar-energy electrochemical convertors.

Bard's group reported a series of works in the field of semiconductor electrodes. The work reported in 1977 deals with n-Si and p-Si single crystal electrodes evaporated with thin n-TiO_2 layer (Kohl et al. 1977). The n-type photoelectrodes were stabilized by TiO_2 layer, which was formed by a CVD. Studies have been performed in Na_2SO_4 electrolyte; no dissolution was observed during the photo-oxidation of water. The authors stated limitation of TiO_2 layer to be applied on a p-Si photocathode, which was due to limited band-bending requirements both at the p–n heterojunction and the oxide/solution interface.

Morisaki et al. reported the first application of n-Si as an efficient photoanode, which produced the oxygen evolution current as high as 50 mA cm^{-2} under 1 Sun (0.1 W cm^{2}) illumination (Morisaki et al. 1980). The measurements have been performed in an alkaline medium (pH 13), and the electrode was protected by Fe_2O_3 layer. The conversion efficiency was explained in terms of efficient hole injection from the silicon substrate across the Si-Fe_2O_3 interface. Thus, the iron oxide performed two important functions: as an anticorrosive agent and as a charge transfer promoter. The optical-to-electrical conversion was enhanced by 1.34–1.60% when Pd and RuO_2 were introduced into the n-Si/Fe_2O_3 heterojunction (Osaka et al. 1985). These experiments were performed in highly aggressive medium – 0.2 M KOH (pH 13.3).

Bockris group demonstrated high stability of n-Si electrode against photocorrosion, which was achieved by depositing a thin film of manganese oxide (MnO_x) on the n-Si substrate (Kainthla et al. 1986). The Si–MnO_x electrode was completely stable during oxygen evolution in alkaline solution (pH 13.7) under illumination of more than 650 hours. The electrodes have been studied by the scanning electron microscopy, X-ray photoelectron spectroscopy (XPS), which showed Mn was present in the +3 state. Other protective scheme for n-Si photoanode included the SiO_2 layer doped with Pt at 0.25 at.% (Contractor and Bockris 1984).

Li and Wang reported high currents of oxygen evolution, which were attained in 1 M KOH solution (pH 14) using the nickel-modified n-Si electrodes (Li and Wang 1987). Nickel film was sputtered on Si substrate; the film was converted into amorphous Ni(OH) during anodic polarization in the electrolyte (1 M KOH). The current density up to 160 mA cm^{-2} was measured at illumination power of 80 mW cm^{-2}. The effect was attributed to intensifying the separation of electron–hole pairs by n^+/p or n^+/n structure of Si/NiO(OH) heterojunction.

Other protective schemes of n-Si included application of such oxides as chromium Cr_2O_3 (Pozniak et al. 1989), indium-tin oxide (ITO) (Sun et al. 2014a, b), cobalt CoO_x (Yang et al. 2014), titanium and nickel–chromium $TiO_2/NiCrO_x$ (Shaner et al. 2015), and nickel–cobalt $NiCo_2O_4$ (Chen et al. 2015). Hu et al. summarized the possible candidates for photoanode protective coatings (Hu et al. 2015): TiO_2, ZrO_2, HfO_2, SnO_2, In_2O_3, FeO_x, MnO_x, NiO_x, CoO_x, WO_3, ZnO, Ta_2O_5, NbO_x, Al_2O_3, MgO, SiO_2, and BiO_x. These oxides are insoluble in sufficiently wide pH range; however, the authors did not consider their electronic transport properties. The thermodynamic stability of the oxides is not a sufficient property for effective protection of photoanodes. Electrode stability against disintegration during the anodic gas evolution and current passage is another important factor, which determines anticorrosion capabilities of the protective layers.

17.2.2 Protection of Si Photoanodes for Halide Reduction

An n-type or p–n junction silicon electrode is predominantly used for halide (Cl^-, Br^-, I^-) oxidation. These reactions more readily proceed when compared with the reaction of oxygen evolution from water. The water splitting in acidic or neutral electrolyte occurs according to the following reaction:

$$2H_2O \rightarrow O_2 + 4H^+ + 4e^- \tag{17.1}$$

This reaction is characterized by the standard electrode potential $E° = 1.229$ V relative to the standard hydrogen electrode at a temperature 25 °C, pH 1, and oxygen gas pressure 1 atm (1.013 25 bar) (Vetter 1967). The analogous reaction in an alkaline electrolyte is

$$4OH^- \rightarrow O_2 + 2H_2O + 4e^- \tag{17.2}$$

This reaction has the standard potential of $E° = 0.401$ V, which is defined for the effective concentration of hydroxide ions of 1 mol l^{-1}. The reactions of halide oxidation and their standard potentials are as follows (Vetter 1967):

$$2I^- \rightarrow I_2 + 2e^- \quad E° = 0.535\,V \tag{17.3}$$

$$2Br^- \rightarrow Br_2 + 2e^- \quad E° = 1.065\,V \tag{17.4}$$

$$2Cl^- \rightarrow Cl_2 + 2e^- \quad E° = 1.359\,V \tag{17.5}$$

Halide oxidation is performed in neutral or acid solutions; thus, the reaction (17.1) can be a competitive one for the halide processing. It is generally assumed that halide oxidation proceeds more readily when compared with the oxygen evolution

(Hu et al. 2015). It follows from the above reactions that, from thermodynamics point of view, iodide can be oxidized more easily. Note also that, chlorine is industrially produced by electrolysis of saturated NaCl brine.

Metal–organic compounds, noble metals, silicides, and conductive oxides have been used in order to protect the n-Si photoanodes for halide oxidation.

In the early 1980s, several innovative strategies were reported on n-Si protection for halide-splitting reactions. Bard and coworkers reported in 1982 the inhibition of photocorrosion of n-Si by a metal-free phthalocyanine (empirical formula: $C_{32}H_{18}N_8$) and copper phthalocyanine ($C_{32}H_{18}CuN_8$) (Leempoel et al. 1982). Thereafter, the group reported electrochemical and photoelectrochemical behavior of n-Si coated with platinum and iridium silicide layers as well as the electrodes modified with ruthenium oxide (RuO_2) (Fan et al. 1983). Photogeneration of I_2, Br_2, and Cl_2 with efficiency more than 5 as well as high electrode stability was reported. Density of the photocurrent was achieved as high as 14 mA cm^{-2} under 65 mW cm^{-2} simulated solar illumination. About the same time, Hodes's work on the application of ITO on n-Si appeared (Hodes et al. 1983). The authors studied the photoelectrodes comprising Si/SiO_x/ITO structure. While main focus was on the $[Fe(CN)_6]^{3-/4-}$ couple, chloride and iodide electrolytes have also been studied. Photo-oxidation of Cl^- from concentrated LiCl and NaCl electrolytes was performed in order to demonstrate chemical inertness of the produced electrodes. Gerischer's group reported n-Si electrode passivated with SnO_2 layer and $[Fe(CN)_6]^{3-/4-}$ and Cl_2/Cl^- as the redox systems (Decker et al. 1983). The authors demonstrated the charge transfer overvoltage effect as well as the limiting currents due to redox ion diffusion. They also discussed the feasibility of solar energy conversion using photoelectrolysis induced by heterojunction photoelectrodes. Bélanger et al. studied the stability of n-Si–SnO_2 system as well (Bélanger et al. 1986). They used various redox systems; a halide system, that was $I^-{}_3/I$, has also been studied. They found that SnO_2 stabilized n-Si surface against corrosion, but at the same time, limits the choice of redox systems: only the systems with good charge transfer kinetics can be used. Pt catalyst was used to enhance the photo-conversion process in $I^-/I^-{}_3$ electrolyte.

Simon and Wrighton turned to organic coatings and studied n-Si-polyacetylene electrode in 11 M LiCl solutions with $I^-{}_3/I^-$ couple (Simon and Wrighton 1984). The electrodes showed substantial improvement in stability compared to non-protected Si electrodes. Nakato et al. used a p^+–n junction silicon with deposited by e-beam evaporation Pt film (3–5 nm in thickness) on the p^+ surface (Nakato et al. 1985). Platinum silicide (PtSi and Pt_2Si) was formed then by heating at 320–350 °C. The photocurrent of c. 14 mA cm^{-2} was achieved in iodide/iodine solution; this current only slightly reduced during a half-year illumination.

Another organic anticorrosion film on n-Si was proposed, which was composed of a conducting polymer – poly(3,4-ethylenedioxythiophene):poly(styrenesulfonate), PEDOT:PSS (Mubeen et al. 2013). The electrodes were tested in highly acidic electrolytes: 8.4 M HBr ($pH < 0$) and HI (pH 2). A transparent film of the conducting polymer (20 nm in thickness) was produced by a spin-casting procedure. The film protected the photoelectrode from corrosion; it also served as Schottky contact or as a hole transport layer. The functionality depended on the choice and design of the semiconductor heterostructure. Importantly, the PEDOT:PSS film was found to have electrocatalytic activity; thus, it played two roles simultaneously – protection and activation of the photoelectrode. The proposed electrode is a low-cost one when compared with the electrodes, which are based on precious metal catalysts.

Tsubomura et al. reported photoelectrochemical properties of the p^+–n junction Si electrode coated with the oxides of titanium (TiO_2), tungsten (WO_3), and iron (Fe_2O_3) (Tsubomura et al. 1985). The oxides were deposited by electron-beam evaporation; the electrochemical studies were performed in iodide-iodine solutions (7.6 M HI and 0.03 M I_2). The PV performance was improved by Ru or RuO_2. Solar energy conversion of 9.3% was achieved at AM 1 sunlight illumination, and the photocurrent was stable for 350 hours.

When summarizing the protection means of n-Si photoanodes, one can say that various oxide films showed high protective performance. The amorphous TiO_2 layer grown by ALD showed the longest protection time frame of about 2200 hours at relevant current densities about 5 mA cm^{-2} in 1 M KOH (pH 13.6) (Shaner et al. 2015; Hu et al. 2015). High protection was achieved also with NiO_x, which showed in 1 M KOH the stability time of 1200 hours. An important focus is the simultaneous photoanode stabilization, which provides the catalytic activity for the photocharge transfer. Inactive metal oxides such as ZnO, ZrO_2, HfO_2, and BiO_x may be doped or supplied with defect states, which promote the hole transport (Hu et al. 2015). Combination of the catalytically active films with protective barrier layers provides another opportunity.

17.2.3 Protection of Si Photocathodes

P-type silicon is conductive with respect to electron transfer toward the electrolyte; this material, therefore, can be used as a photocathode. While silicon rapidly dissolves in alkaline solutions, it is stable in acid or neutral solutions. As a result of this feature, stabilization of p-Si photocathode received less attention when compared with the n-Si photoanodes. Typically the anodic dissolution of a metal is suppressed when electrode potential is shifted negatively; the method is known as a cathodic protection. In the case of silicon, however, its anodic dissolution in alkaline electrolyte cannot be completely arrested by a negative potential shift, as silicon oxidizes both by chemical and electrochemical mechanisms (Hu et al. 2015). Another important feature is spontaneous formation of SiO_2 film in acidic and neutral solutions, which protects silicon from corrosion. The oxide film is stable in acid and neutral solutions; however, it actively dissolves in alkaline medium. This film, however, is detrimental in terms of electron transfer; it reduces efficiency of solar-to-hydrogen conversion. It is desirable to replace this film by another one, which would protect the substrate and, at the same time, not inhibit substantially the electrochemical reaction of hydrogen evolution. The substitute should be electrochemically stable (non-reducible) in the cathodic region. Catalytic activity with respect to hydrogen evolution is another desirable function of the protective coating. Precious metal catalysts are widely used to facilitate the hydrogen evolution reaction. Dominey et al. reported in the early 1980s improvement of the photoelectrochemical hydrogen generation on platinized p-Si photocathodes (Dominey et al. 1982). Platinum was deposited either by photoelectrochemical plating or confining with *N,N′*-dialkyl-4,4′-bipyridine, $[(PQ^{2+/+})_n\ nPt(0)]_{surf}$. While the photoreaction output was pH dependent, the dependence did not follow the simple thermodynamic condition of 59 mV per pH.

Hu et al. surveyed systematically the materials used for protection of p-Si photocathodes (Hu et al. 2015). They refer to the following oxides: TiO_2, ZnO, SiO_2, Al_2O_3, and $SrTiO_3$. Other materials included: Mo_3S_4, Ti, Mo/MoS_2, CdS, and Ti/Ni.

Bard's group reported in 1977 application of evaporated TiO_2 layer for protection of both n-Si and p-Si single crystal electrodes (Kohl et al. 1977). This oxide received further considerable attention not only for Si but also for other semiconductors' protection such as p-Cu_2O, p-InP, p-CZTS (copper zinc tin sulfide), and p-InP.

Hafnium oxide (HfO_2) on p-Si has also recently been comprehensively studied (Staišiūnas et al. 2022; Kalinauskas et al. 2022). Amorphous ultrathin films of hafnium oxide in the range of 15–70 nm were deposited by the ALD on p-Si (100). The interface transformations were studied by the QCN measurements, which provided *in situ* information with nanogram resolution in real time. Measurements with Si and Au substrates indicated dynamics of the electrolyte intake into oxide film. The QCN did not indicate the oxide dissolution both in acidic and alkaline electrolytes. Perchlorate electrolytes have been applied, and Raman spectroscopy indicated that perchlorate does not enter appreciably the hafnia film. Mott–Schottky plots showed that dark Si surface adjacent to Si–HfO_2 interface is positively charged in acid solution and negatively charged in alkaline solution. The number of photoelectrons was determined to be much greater than the doping level of the p-Si. The cathodic photoactivity of p-Si–HfO_2 electrode was studied with respect to hydrogen reduction in acid and alkaline solutions. In acid solution, the film enhanced the reduction process when compared to that on the coating free electrode. The acceleration effect was explained in terms of the prevention of the highly insulating SiO_x formation. In alkaline electrolyte, hafnia film protected Si from corrosion; however, at the same time, the film reduced the hydrogen reduction efficiency.

A closer look has recently been taken at the classical p-Si–SiO_2 system, focusing on electrolytic transformations of the SiO_2 film as well as the interface (Kalinauskas et al. 2022). While silica is stable in acid and neutral electrolytes, its properties could significantly change due to electrolyte absorption. The authors reported the photoelectrochemical and nanogravimetric features that allow identification of electrolytic transformation of Si–SiO_2 interface. Photoresponsivity of the p-Si with 10 nm SiO_2 layer was compared with that of hydrogen-terminated Si–H surface. The QCN identified in neutral solution dynamics of the electrolyte absorption along with slight dissolution of the silica layer. No silica dissolution indications were observed in an acid electrolyte; its absorption was about one third of the dry oxide mass. The measurements have been performed in acid perchlorate, chloride, and sulfate electrolytes, and the photoresponses were similar in these electrolytes.

This demonstrated that the photoeffects are associated with the phenomena within the semiconductor-oxide phase and not at the oxide-electrolyte phase.

An important subject is the p-Si–Al_2O_3 photoelectrode. The films effectively passivated Si surface and increased PV conversion (Hoex et al. 2008; Schmidt et al. 2009; Dingemans et al. 2010; Repo et al. 2013; Savin et al. 2015). This material exhibited high passivation ability, especially for the p-type emitters, which is due to the low density of interface states and fixed negative charge present in the material (Repo et al. 2013). Effectiveness of such films has been demonstrated in the solar cells based on black silicon – a nanostructured material, which is capable of completely absorbing the light (Savin et al. 2015). The p-Si with 10 nm Al_2O_3 film was studied in acid perchlorate electrolyte (Kalinauskas et al. 2022). ALD technique was used to deposit ultrathin (10 nm) alumina films. The photopotential variation has been studied applying 505 nm wavelength illumination. The electrochemical systems were ideally polarizable; they did not include any specific redox couple to induce Faradaic process during the illumination. The electrode showed distinctively different photo-features from hydrogen-terminated silicon. When illuminated, an abrupt positive photopotential jump was observed, which was followed by slow potential relaxation to an initial value in the dark. When illumination was terminated, the potential dropped sharply down and relaxed to the initial value in the dark. The effects were explained in terms of charging of the alumna film by photoelectrons and their recombination. The obtained results enable us to properly understand photoresponsivity, passivity, and degradation of Si/oxide electrodes for applications in devices of solar energy conversion.

References

Bansal, A., Li, X., Lauermann, I. et al. (1996). Alkylation of Si surfaces using a two-step halogenation/Grignard route. *J. Am. Chem. Soc.* 118: 7225–7226. https://doi.org/10.1021/ja960348n.

Bard, A.J. and Faulkner, L.R. (2001). *Electrochemical Methods. Fundamentals and Applications*, 100–104. Wiley.

Bard, A.J., Parsons, R., and Jordan, J. (1985). *Standard Potentials in Aqueous Solution.* New York: International Union of Pure and Applied Chemistry.

Bélanger, D., Dodelet, J.P., and Lombos, B.A. (1986). Characteristics and stability of n-Si/SnO_2 and n-Si/SnO_2/Pt photoanodes. *J. Electrochem. Soc.* 133: 1113–1119. https://doi.org/10.1149/1.2108796.

Bertucioli, L., Chan, A., Hart, D., et al. (2014). Study on development of water electrolysis in the EU. Final report. E4tech Sàrl, Switzerland; elementenergy, Cambridge, UK.

Bookbinder, D.C., Bruce, J.A., Dominey, R.N., and Wrighton, M.S. (1980). Synthesis and characterization of a photosensitive interface for hydrogen generation: chemically modified p-type semiconducting silicon photoanodes. *Proc. Natl. Acad. Sci. U.S.A.* 77: 6280–6284. https://doi.org/10.1073/pnas.77.11.6280.

Breeze, P. (2019). Solar power, Chapter 13. In: *Power Generation technologies*, 3e, 293–321. Elsevier Ltd https://doi.org/10.1016/C2017-0-03267-6.

Candea, R.M., Kastner, M., Goodman, R., and Hickok, N. (1976). Photoelectrolysis of water: Si in salt water. *J. Appl. Phys.* 47: 2724. https://doi.org/10.1063/1.322995.

Chen, X. and Mao, S.S. (2007). Titanium dioxide nanomaterials: synthesis, properties, modifications, and applications. *Chem. Rev.* 107: 2891–2959. https://doi.org/10.1021/cr0500535.

Chen, X., Liu, L., Yu, P.Y., and Mao, S.S. (2011). Increasing solar absorption for photocatalysis with black hydrogenated titanium dioxide nanocrystals. *Science* 331: 746–750. https://doi.org/10.1126/science1200448.

Chen, X., Li, C., Grätzel, M. et al. (2012). Nanomaterials for renewable energy production and storage. *Chem. Soc. Rev.* 41: 7909–7937. https://doi.org/10.1039/c2cs35230c.

Chen, L., Yang, J., Klaus, S. et al. (2015). p-Type transparent conducting oxide/n-type semiconductor heterojunctions for efficient and stable solar water oxidation. *J. Am. Chem. Soc.* 137: 9595–9603. https://doi.org/10.1021/jacs.5b03536.

Contractor, A.Q. and Bockris, J.O.'.M. (1984). Investigation of a protective conducting silica film on n-silicon. *Electrochim. Acta* 29: 1427–1434. https://doi.org/10.1016/0013-4686(84)87022-X.

Coridan, R.H., Nielander, A.C., Francis, S.A. et al. (2015). Methods for comparing the performance of energy-conversion systems for use in solar fuels and solar electricity generation. *Energy Environ. Sci.* 8: 2886–2901. https://doi.org/10.1039/C5EE00777A.

Das, A., Han, Z., Haghighi, M.G., and Eisenberg, R. (2013). Photogeneration of hydrogen from water using CdSe nanocrystals demonstrating the importance of surface exchange. *Proc. Natl. Acad. Sci. U.S.A.* 110: 16716–16723. https://doi.org/10.1073/pnas.1316755110.

Decker, F., Fracastoro-Decker, M., Badawy, W. et al. (1983). The photocurrent-voltage characteristics of the heterojunction combination n-Si/SnO_2/redox – electrolyte. *J. Electrochem. Soc.* 130: 2173–2179. https://doi.org/10.1149/1.2119547.

Dingemans, G., Seguin, R., Engelhart, P. et al. (2010). Silicon surface passivation by ultrathin Al_2O_3 films synthesized by thermal and plasma atomic layer deposition. *Phys. Status Solidi RRL* 4: 10–12. https://doi.org/10.1002/pssr.200903334.

Dominey, R.N., Lewis, N.S., Bruce, J.A. et al. (1982). Improvement of photoelectrochemical hydrogen generation by surface modification of p-type silicon semiconductor photocathodes. *J. Am. Chem. Soc.* 104: 467–482. https://doi.org/10.1021/ja00366a016.

Dresselhaus, M.S. and Thomas, I.L. (2001). Alternative energy technologies. *Nature* 414: 332–337. https://doi.org/10.1038/35104599.

Fan, F.R.F., Keil, R.G., and Bard, A.J. (1983). Semiconductor electrodes. 48. Photooxidation of halides and water on n-silicon protected with silicode layers. *J. Am. Chem. Soc.* 105: 220–224. https://doi.org/10.1021/ja00340a013.

Fray, D.J. and Juzeliūnas, E. (2011). Method for texturing silicon surface to create black silicon for photovoltaic applications. International Patent WO 2011/161479.

Fray, D.J. and Juzeliūnas, E. (2013). European Patent EP 2586064 B1.

Fray, D.J. and Juzeliūnas, E. (2014a). US Patent US 8, 790, 953.

Fray, D.J. and Juzeliūnas, E. (2014b). UK Patent GB2494566.

Fray, D.J. and Juzeliūnas, E. (2015a). China Patent ZL201180030064.0.

Fray, D.J. and Juzeliūnas, E. (2015b). Japan Patent 5689530.

Fray, D.J. and Juzeliūnas, E. (2015c). Brazil Patent Br 11 2012 032955.3.

Fujishima, A. and Honda, K. (1972). Electrochemical photolysis ow water at a semiconductor electrode. *Nature* 238: 37–38. https://doi.org/10.1038/238037a0.

Gahleitner, G. (2012). Hydrogen from renewable electricity: an international review of power-to-gas pilot plants for stationary applications. *Int. J. Hydrogen Energ.* 38: 2039–2061. https://doi.org/10.1016/j.ijhydene.2012.12.010.

Grätzel, M. (2001). Photoeletrochemical cells. *Nature* 414: 338–344. https://doi.org/10.1038/35104607.

Grätzel, M. (2007). Photovoltaic and photoelectrochemical conversion of solar energy. *Philos. Trans. A Math. Phys. Eng. Sci.* 365: 993–1005. https://doi.org/10.1098/rsta.2006.1963.

Green, M. (2016). Commercial progress and challenges for photovoltaics. *Nat. Energy* 1: 15015. https://doi.org/10.1038/nenergy.2015.15.

Gronet, C.M., Lewis, N.S., Cogan, G., and Gibbons, J. (1983). N-type silicon photoelectrochemistry in methanol: design of a 10.1% efficient semiconductor/liquid junction solar cell. *Proc. Natl. Acad. Sci. U.S.A.* 80: 1152–1156. https://doi.org/10.1073/pnas.80.4.1152.

Hernandez-Pagan, E., Vargas-Barbosa, N., Wang, T. et al. (2012). Resistance and polarization losses in aqueous buffer-membrane electrolytes for water-splitting photoelectrochemical cells. *Energy Environ. Sci.* 5: 7582–7589. https://doi.org/10.1039/C2EE03422K.

Hodes, G., Thompson, L., DuBow, J., and Rajeshwar, K. (1983). Heterojunction silicon/indium tin oxide photoelectrodes: development of stable systems in aqueous electrolytes and their applicability to solar energy conversion and storage. *J. Am. Chem. Soc.* 105: 324–330. https://doi.org/10.1021/ja00341a004.

Hoex, B., Schmidt, J., Pohl, P. et al. (2008). Silicon surface passivation by atomic layer deposited Al_2O_3. *J. Appl. Phys.* 104: 044903–044901. https://doi.org/10.1063/1.2963707.

Hu, S., Lewis, N.S., Ager, J.W. et al. (2015). Thin-film materials for the protection of semiconducting photoelectrodes in solar-fuel generators. *J. Phys. Chem. C* 119: 24201–24228. https://doi.org/10.1021/acsjpcc.5b05976.

Inoue, T., Fujishima, A., Konishi, S., and Honda, K. (1979). Photoelectrocatalytic reduction of carbon dioxide in aqueous suspensions of semiconductor powders. *Nature* 277: 637–638. https://doi.org/10.1038/277637a0.

Kainthla, R.C., Zelenay, B., and Bockris, J.O.'.M. (1986). Protection of n-Si photoanode against photocorrosion in photoelectrochemical cell for water electrolysis. *J. Electrochem. Soc.* 133: 248–253. https://doi.org/10.1149/1.2108556.

Kalinauskas, P., Staišiūnas, L., Grigucevičienė, A. et al. (2022). Photoelectrochemical and nanogravimetric study of electrolytic transformation of silicon-oxide interface. *J. Electrochem. Soc.* 169: 036508. https://doi.org/10.1149/1945-7111/ac5938.

Khaselev, O. and Turner, J. (1998). A monolithic photovoltaic-photoelectrochemical device for hydrogen production via water splitting. *Science* 280: 425–427. https://doi.org/10.1126/science.280.5362.425.

Kohl, P.A., Frank, S.N., and Bard, A.J. (1977). Semiconductor electrodes: XI. Behavior of n- and p-type single crystal semiconductors covered with thin n-TiO_2 films. *J. Electrochem. Soc.* 124: 225–229.

Leempoel, P., Castro-Acuna, C., Fan, F.R.F., and Bard, A.J. (1982). Semiconductor electrodes. 43. The effect of light intensity and iodine doping on the stabilization of n-silicon by phthalocyanine films. *J. Phys. Chem.* 86: 1396–1400. https://doi.org/10.1021/j100397a036.

Li, G. and Wang, S. (1987). Photoelectrochemical characteristics of metal-modified epitaxial n-Si anodes: Part I. NiO(OH)-coated n^+/p^-Si and n^+/n^-Si electrodes for catalytic oxygen evolution. *J. Electroanal. Chem. Interfacial Electrochem.* 227: 213–221. https://doi.org/10.1016/0022-0728(87)80076-1.

Liao, L., Zhang, Q., Su, Z. et al. (2014). Efficient solar water-splitting using a nanocrystalline CoO photocatalyst. *Nat. Nanotechnol.* 9: 69–73. https://doi.org/10.1038/nnano.2013.272.

Linford, M.R., Fenter, P., Eisenberger, P.M., and Chidsey, C.E.D. (1995). Alkyl monolayers on silicon prepared from 1-alkenes and hydrogen-terminated silicon. *J. Am. Chem. Soc.* 117: 3145–3155. https://doi.org/10.1021/ja00116a019.

Maeda, K., Takata, T., Hara, M. et al. (2005). GaN:ZnO solid solution as a photocatalyst for visible-light driven water splitting. *J. Am. Chem. Soc.* 127 (23): 8286–8287. https://doi.org/10.1021/ja0518777.

Morisaki, H., Ono, H., Dohkoshi, H., and Yazawa, K. (1980). Iron-oxide coated n-Si as a heterostructure photoanode for the photoelectrolysis of water. *Jpn. J. Appl. Phys.* 19: L148–L150. https://doi.org/10.1143/JJAP.19.L148.

Mubeen, S., Lee, J., Singh, N. et al. (2013). Stabilizing inorganic photoelectrodes for efficient solar-to-chemical energy conversion. *Energy Environ. Sci.* 6: 1633–1639. https://doi.org/10.1039/C3EE40258D.

Nakato, Y., Hiramoto, M., Iwakabe, Y., and Tsubomura, H. (1985). ESCA and photoelectrochemical studies of p-n junction silicon electrodes protected by platinum deposition for use in solar energy conversion. *J. Electrochem. Soc.* 132: 330–334. https://doi.org/10.1149/1.2113832.

Nozik, A.J. and Memming, R. (1996). Physical chemistry of semiconductor – liquid interfaces. *J. Phys. Chem.* 100: 13061–13078. https://doi.org/10.1021/jp953720e.

Oh, J. and Branz, H.M. (2014). Anti-reflective nanoporous silicon for efficient hydrogen production. Patent US8729798 B2.

Oh, J., Deutsch, T.G., Yuan, H.-C., and Branz, H.M. (2011). Nanoporous black silicon photocathode for H_2 production by photoelectrochemical water splitting. *Energy Environ. Sci.* 4 (5): 1690–1694. https://doi.org/10.1039/C1EE01124C.

Oh, I., Kye, J., and Hwang, S. (2012a). Enhanced photoelectrochemical hydrogen production from silicon nanowire array photocathode. *Nano Lett.* 12 (1): 298–302. https://doi.org/10.1021/nl203564s.

Oh, J., Yuan, H.-C., and Branz, H.M. (2012b). An 18.2%-efficient black-silicon solar cell achieved through control of carrier recombination in nanostructures. *Nat. Nanotechnol.* 7: 743–748. https://doi.org/10.1038/nnano.2012.166.

Osaka, T., Hirota, N., Hayashi, T., and Eskildsen, S.S. (1985). Characteristics of photoelectrochemical cells with iron oxide/n-Si heterojunction photoanodes. *Electrochim. Acta* 30: 1209–1212. https://doi.org/10.1016/0013-4686(95)80015-8.

Paracchino, A., Laporte, V., Sivula, K. et al. (2011). Highly active oxide photocathode for photoelectrochemical water reduction. *Nat. Mater.* 10: 456–461. https://doi.org/10.1038/nmat3017.

Polman, A., Knight, M., Garnett, E.C. et al. (2016). Photovoltaic materials: present efficiencies and future challenges. *Science* 352 (6283): https://doi.org/10.1126/science.aad4424.

Pozniak, S.K., Makuta, I., and Kulak, A. (1989). Photoelectrochemical behavior of n-silicon photoanodes coated with chromium(III) oxide films and Cr_2O_3 containing composite layers. *Sol. Energy Matr.* 18: 357–364. https://doi.org/10.1016/0165-1633(89)90060-9.

Repo, P., Haarahiltunen, A., Sainiemi, L. et al. (2013). Effective passivation of black silicon surfaces by atomic layer deposition, IEEE. *J. Photovoltaics* 3: 90–94. https://doi.org/10.1109/JPHOTOV.2012.2210031.

Roy, S.C., Varghese, O.K., Paulose, M., and Grimes, C.A. (2010). Toward solar fuels: photocatalytic conversion of carbon dioxide to hydrocarbons. *ACS Nano* 4: 1259–1278. https://doi.org/10.1021/nn9015423.

Savin, H., Repo, P., Gastrow, G. et al. (2015). Black silicon solar cells with interdigitated back-contacts achieve 22.1% efficiency. *Nat. Nanoechnol.* 10: 624–629. https://doi.org/10.1038/nnano.2015.89.

Schmalensee, R., Bulovič, V., Armstrong, R. et al. (2015). *The Future of Solar Energy: An Interdisciplinary MIT Study*. Massachusetts Institute of Technology, MIT Energy Initiative http://mitei.mit.edu/futureofsolar.

Schmidt, J., Veith, B., and Brendel, R. (2009). Effective surface passivation of crystalline silicon using ultrathin Al_2O_3 fims and Al_2O_3/SiN_x stacks. *Phys. Status Solidi RRL* 3: 287–289. https://doi.org/10.1002/pssr.200903272.

Shaner, M.R., Hu, S., Sun, K., and Lewis, N.S. (2015). Stabilization of Si microwire arrays for solar-driven H_2O oxidation to $O_2(g)$ in 1.0 M KOH(Aq) using conformal coatings of amorphous TiO_2. *Energy Environ. Sci.* 8: 203–207. https://doi.org/10.1039/C4EE03012E.

Simon, R.A. and Wrighton, M.S. (1984). Stabilization of n-type silicon photoanodes against photoanodic decomposition with thin films of polyacetylene. *Appl. Phys. Lett.* 44: 930–932. https://doi.org/10.1063/1.94938.

Staišiūnas, L., Kalinauskas, P., Juzeliūnas, E. et al. (2022). Silicon passivation by ultrathin hafnium oxide layer for photoelectrochemical applications. *Front. Chem.* 10 (859023): https://doi.org/10.3389/fchem.2022.859023.

Sun, K., Shen, S., Liang, Y. et al. (2014a). Enabling silicon for solar-fuel production. *Chem. Rev.* 114: 8662–8719. https://doi.org/10.1021/cr300459q.

Sun, K., Shen, S., Cheung, J. et al. (2014b). Si photoanode protected by a metal modified ITO layer with ultrathin NiO_x for solar water oxidation. *Phys. Chem. Chem. Phys.* 16: 4612–4625. https://doi.org/10.1039/c4cp00033a.

Tsubomura, H., Nakato, Y., Hiramoto, M., and Yano, H. (1985). Metal oxide coated p–n junction silicon electrodes for photoelectrochemical solar energy conversion. *Can. J. Chem.* 63: 1759–1762. https://doi.org/10.1139/v85-295.

Varghese, O.K., Paulose, M., LaTempa, T.J., and Grimes, C.A. (2009). High-rate solar photocatalytic conversion of CO_2 and water vapor to hydrocarbon fuels. *Nano Lett.* 9: 731–737. https://doi.org/10.1021/nl803258p.

Vetter, K.J. (1967). *Electrochemical Kinetics: Theoretical Aspects*. New York: Academic Press Inc.

Wrighton, M.S. (1979). Photoelectrochemical conversion of optical energy to electricity and fuels. *Acc. Chem. Res.* 12: 303–310. https://doi.org/10.1021/ar50141a001.

Yang, J., Walczak, K., Anzenberg, A. et al. (2014). Efficient and sustained photoelectrochemical water oxidation by cobalt oxide/silicon photoanodes with nanotextured interfaces. *J. Am. Chem. Soc.* 136: 6191–6194. https://doi.org/10.1021/ja501513t.

Yoshihiro, N., Toshihiro, O., and Hiroshi, T. (1975). Photo-electrochemical behaviors of semiconductor electrodes coated with thin metal films. *Chem. Lett.* 4: 883–886. https://doi.org/10.1246/cl.1975.883.

Zou, Z., Ye, J., Sayama, K., and Arakawa, H. (2001). Direct splitting of water under visible light irradiation with an oxide semiconductor photocatalyst. *Nature* 414: 625–627. https://doi.org/10.1038/414625a.

18

Conclusions, Outlook, and Challenges

This book reviews the experimental and technological achievements in silicon electrochemistry, production and purification, highlighting subjects of technological significance and future perspectives.

Silicon lies at the heart of modern technology, and it will most probably remain dominant at least for the next few decades. Silicon being a nontoxic, efficient, low cost, and robust material will continue to dominate the solar-energy market. Mitigation of climate change, as a global task, could be achieved using technologies based on Earth-abundant materials. Silicon is the second most abundant in the Earth's crust.

World production of silicon over two decades increased from ~3.4 million metric tons (MT) in 2001 to 8 million MT in 2020. This amount was produced mainly by carbothermic silica reduction; this reaction translates to the direct emission of 11.8 MT of CO_2 per year. Non-direct CO_2 emissions are also significant, for instance, those related to heat generation. The carbothermic reduction is a very endothermic process with energy consumption of about 10.5 MWh per one ton of produced silicon. The carbothermic silicon extraction and purification requires large amount of energy and emits greenhouse gases. A fundamental challenge is to extract silicon electrochemically using electrons as absolutely clean reducing agents generated from sustainable renewable sources.

Silicon electrochemistry has the potential to significantly contribute to low-carbon economy; it offers an advancement in sustainable energy solutions, as well as environment friendly and secure technologies. Electrochemical methods of silicon surface engineering are attractive due to environmental friendliness, technical simplicity, and cost-effectiveness.

Silicon has poor intrinsic ability to absorb light – it reflects more than 30% of incident light if the surface is not specifically textured or coated by antireflection coatings. Much effort has been recently made to enhance silicon absorbance by forming various surface nano-micro-architectures. To this end, various techniques have been proposed, such as femtosecond laser engineering, atomic layer deposition (ALD), deposition of transparent conducting oxides, deep reactive ion etching, chemical etching using metal catalysts, inductively coupled plasma and cryogenic etching, as well as vapor–liquid–solid reactions. Most of these methods, however,

Silicon: Electrochemistry, Production, Purification and Applications, First Edition. Eimutis Juzeliūnas.

are too costly to provide competitive solutions for industrial-scale applications. Furthermore, etchings usually involve toxic chemicals, such as hydrofluoric acid, and rather expensive catalysts, such as gold or silver.

Of great interest is black silicon – the nano-micro-porous material, which effectively absorbs light in a wide range of wavelengths. This material is promising in applications where high surface-to-volume ratio is required. Electrochemically produced Si surface nano-architectures showed intrinsic quantum confinement effect. The solar cell efficiency of 22.1% was achieved using black silicon passivated with alumina film. This material is promising in solar-to-chemical conversions, such as photoelectrochemical hydrogen production from water or CO_2 reduction. Black silicon is also a promising material for solar thermo-generating applications. Flexible black silicon on insulator has been formed, which is a potential candidate for the use as an absorber of the solar-thermo generators, such as concentrating solar power parabolic trough. The nanostructured silicon significantly accelerates hydrogen production with a reduced need for surfactants. The fabrication techniques of black silicon include laser texturization, plasma immersion ion implantation, metal-assisted etching, or cryogenic deep reactive ion etching. Most of these techniques are costly. There is a great demand for the cost-effective and green technologies of black silicon mass production.

Important surface nano-micro-architectures including black silicon can be obtained by deoxidizing silica layer on silicon in molten salt electrolyte. The precursor layer can be formed by means of thermal oxidation, as well as sol–gel or physical vapor deposition.

The electrochemical silica–silicon conversion in a molten salt electrolyte is a dynamic process. The electrochemically driven conversion propagates into insulator oxide as a three-phase interline process. Most studies of silicon conversion in molten salts have predominantly been done using *ex situ* methods (e.g. scanning electron microscopy [SEM], scanning transmission electron microscopy [STEM], X-ray diffraction [XRD], or X-ray photoelectron spectroscopy [XPS]), which relate to different conditions from those during the processing in the electrochemical cell. It is a great challenge to obtain process information *in situ* and in real time. Synchrotron X-ray diffraction provides an opportunity whereby process information can be obtained. Structural properties of molten salts can also be studied by *in situ* Raman spectroscopy.

Silicon carbide (SiC, carborundum) can be synthesized electrochemically in high-temperature molten salts. The material is outstanding in numerous fields of applications: abrasive and cutting tools, automobile parts, power and hostile-environment electronics, catalytic supports, light detecting devices, photocatalytic hydrogen production, gas and chemical sensors, and biomedical applications. Electrochemical synthesis of Si–C systems is at an early stage of development. The Si–C–SiC composition was obtained when deoxidizing silica in molten calcium chloride at 900 °C using graphite anodes. The developed structure showed photo-responsiveness.

Silicides can also be formed by means of molten salt electrochemistry. Metal silicides gained attention as materials with excellent optical properties, high hydrogen storage capacity, and superconductivity at certain temperatures and pressure.

An environment friendly and secure approach provides electrochemical structuring of silicon surface in high-temperature molten salts. There is great interest in the development of photoelectrochemical electrodes with effective light harvesting by means of molten salt electrochemistry. The proposed one-step method of silicon surface modifications are being environment friendly, cost-effective, and have good prospects for commercialization. The technologies, however, are at the early stage of investigations; technical solutions are still pending.

Electrodeposition of thin Si layers can be used to replace multiple processes of Si wafer fabrication, such as production of metallurgical-grade silicon (MG-Si), upgrading to solar-grade silicon (SoG-Si), ingot casting, and slicing. Electrochemical deposition of doped silicon, as well as formation of p–n junction has recently been demonstrated. The approach has potential of reducing capital cost and energy consumption for the fabrication of solar cells when compared with the conventional manufacturing process. The investigations are on the initial technology readiness levels; basic principles and technological concepts have been formulated.

Molten fluorides with dissolved silica as a precursor are prospective electrolytes for electrodeposition of silicon thin layers. In spite of cost and energy management, Si deposition from fluoride electrolytes still remains at an early stage of practical development. The encountered problems related to the high aggressiveness of hydrofluoric acid, severe pitting as well as staining and clouding of silicon surface. Diffusion of silicon into substrate at high temperatures could also be an issue. Chloride molten salts have been recognized as promising electrolytes to electrodeposit silicon. Another promising composition is $SiCl_4$ precursor in molten KF–KCl electrolyte. High degree of $SiCl_4$ dissolution was achieved in this electrolyte.

Silicon deposition from room temperature ionic liquids remains problematic. While a wide electrochemical window (up to 6 V) makes Si deposition possible, the process is limited by low electric conductivity of the deposits at room temperature.

Significant efforts have been made to develop effective technologies for electrochemical silicon purification. A key driver was a cost-effective production of SoG-Si using MG-Si as a feedstock, which would outperform the energy efficiency of the classical purification methods, e.g. the Siemens process.

Electrochemical deoxidation of solid compounds in high-temperature molten salts has been discovered in the year 2000. The method was applied to a wide range of metal oxides. The process was named as the FFC Cambridge process. The technology has attracted global interest due to simplicity, energy efficiency, and rapidity of process. Over 30 metals or semimetals were extracted including silicon reduction from solid silicon dioxide. The electrochemical extraction provides a green alternative to conventional carbo-thermic silicon production, which is highly inefficient with regards to energy and harmful to the environment. Porous or powder materials can be obtained by the FFC method. Most of silica deoxidation studies were conducted in molten calcium chloride at temperatures of 850–900 °C. The solid-state deoxidation can also be used in space exploration; oxygen can be produced from the Moon or Mars regolith according to this route.

Purity of electrochemically-produced silicon is of great concern in photovoltaic (PV) applications. Researchers predominantly used graphite anodes, which can contaminate the electrolyte. Significant carbon impurity levels were found when deoxidizing silica using carbon anode. Also, the formation of silicon carbide on the cathode has been detected. Contaminations also can originate from other sources, such as stainless steel reactor. Quartz in some electrolytes is unstable as well; silica dissolution in the molten salt electrolytes containing calcium oxide was detected. The creation of substitutes of graphite anodes poses a challenge for future research.

Molten oxide electrolysis (MOE) provides an opportunity for the extraction of silicon from a feedstock in a liquid state. This route is technically simpler when compared to traditional extraction metallurgy. The primary MOE advantages include the reductions in energy consumption, the mitigation of greenhouse gas emission, and the opportunity to achieve high purity product while avoiding its additional purification. A challenging goal in MOE is the creation of stable electrode materials, which remain stable at high temperatures and operate without greenhouse gas emissions. The MOE cathodic reduction is associated with oxygen development on the anode provided it is made of inert material. This suggests an exciting application field that is oxygen production in space exploration. Lunar or Mars soils are composed of fine fractions of the regolith, that is, unconsolidated oxide minerals.

Electrochemical batteries are the key devices for low-carbon economy to reduce CO_2 emissions. Effective batteries are essential to store stationary energy produced by solar or wind. To meet the sustainability targets, batteries must exhibit high energy and power performance, outstanding lifetime, reliability, safety, and recyclability. Important requirements are the scalability to electricity grid level, as well as cost-effectiveness and sustainable battery production. The growth in global battery demand is anticipated, multiplied by a factor 14× from 2018 to 2030. The greatest aspect of this demand goes to the electric mobility sector. While lithium-ion based technology dominates the current battery market, there is great demand and a great challenge to create effective post-Li electrochemical energy storage systems. Metal-air or sodium-based batteries are under active development. Unlimited raw materials from the atmosphere and ocean are available for such batteries.

Silicon is an attractive material for anode production in lithium-ion batteries. The Li–Si binary system shows exceptionally high Li insertion capacity: one Si atom can accommodate 4.4 Li atoms forming the alloy $Li_{22}Si_5$, and the theoretical insertion capacity is as high as 4200 mA h g^{-1} (the analogous value for $Li_{15}Si_4$ is 3576 mA h g^{-1}). Such capacity outperforms significantly the capacity of the conventionally used graphite electrode (372 mA h g^{-1} for LiC_6). However, disintegration of the silicon electrode occurs due to 300–400% volume change during the lithiation–delithiation process. This detrimental process leads to battery capacity fading. Therefore, development of the restoration methods of electric properties (self-healing) is of great importance. Great efforts have been made in search of effective means to buffer the volume changes. To mention a few: graphene films, silicon-carbon nanocomposites, Si nanowires and nanotubes, solid core–shell structures, porous Si designs, controllable pores or patterned silicon films, etc.

PV electricity generation has and will continue the growth, which is close to exponential. PV accelerates the development of sustainable energy technologies with the scenarios to reach solar energy domination in the electricity market. Creation of efficient and low-cost PV devices plays a determining role in the political-social landscape of twenty-first century. There is a public awareness in PV technology as a provider of clean, sustainable, and secure energy from the most abundant source, which is for free. Analysts foresee that solar PV by 2050 would represent the second-largest power source behind wind power. The solar PV electricity will cover a quarter of the total electricity needs. The growth dynamics of PV capacity are expected from 480 GW in 2018 to 2840 GW projected by 2030 and 8519 GW projected by 2050.

Crystalline silicon has about 95% market share of PV solar production. Analysts forecast that thanks to technological development and a fall in the price of raw material, silicon will remain in a dominating position at least for the next few decades. Increased outputs of the Si-based panels also strengthen the positions of silicon: an average efficiency of multi-crystalline panels increased from ~13% in 2006 to ~17% in 2018, and this positive tendency is expected to continue in the future. Lowering the cost of Si-based PV devices remains a challenge for scientists and engineers.

The rapid expansion of PV resulted in an increasing amount of end-of-life solar panels, which poses a hazard for future PV development. It is a challenging goal to find solutions for PV modules recycling. There is a great need in conversion technologies of PV waste into feedstock for other processes, such as thermoelectrics. The technologies should match the principles of circular economy: elimination of waste and pollution by converting them into feedstock for other relevant processes.

Solar-to-chemical conversion offers important opportunity for solar energy accumulation and transmission, which mimics the process of photosynthesis in plants. The energy captured in chemical bonds can be converted to power in fuel cells with specifically high efficiency or in combustion engines. The (photo)electrochemical hydrogen production can be performed with nearly zero greenhouse emission when hydrogen is generated using renewable energy sources, such as solar, wind, hydro, or ocean.

Electrochemical hydrogen generation from water is very attractive due to unlimited electrolyte resources in ocean with salt concentration of about 3.5%. Electrochemical conversion means the usage of electrons as absolutely clean agents once they are generated from renewable sources. The electrochemical process offers such advantages as high purity, flexibility, supply on-site and on-demand, as well as fast start-up and shutdown opportunities. Of particular interest is solar-to-fuel process, which occurs directly on the photo-electrodes. Silicon, as a photosensitive, chemically stable, and inexpensive material, is attractive as a photocathode for hydrogen production from water.

A challenge is the development of sustainable electrodes of high efficiency, low-cost, and long-lifetime. Synthesis of new effective materials for photo-electrodes is very high on the agenda of materials scientists and engineers. Protection schemes by various oxides have been recently developed. Ultrathin protective layers, whose thickness is on the order of nanometers, were formed by the ALD.

Electrochemistry provides a range of opportunities for development of low- carbon economy. Most of the developments, however, are still at the conceptual level. Viable technological advances are a challenging goal for future research.

Index